ILLUSTRATED PHYSIOLOGY

ILLUSTRATED PHYSIOLOGY

B.R. Mackenna MB ChB PhD MRCP (Glas)
Senior Lecturer, Institute of Physiology,
University of Glasgow

R. Callander FFPh FMAA AIMBI

Formerly Director of Medical Illustration,
University of Glasgow

FIFTH EDITION

CHURCHILL LIVINGSTONE
EDINBURGH LONDON MELBOURNE AND NEW YORK 1990

CHURCHILL LIVINGSTONE
Medical Division of Longman Group HK Limited

Distributed in the United States of America by
Churchill Livingstone Inc., 1560 Broadway, New York,
N.Y. 10036, and by associated companies, branches
and representatives throughout the world.

First Edition 1963
   Italian translation 1966
Second Edition 1970
   Danish translation 1973
Third Edition 1975
   Japanese translation 1976
   Spanish translation 1981
Fourth Edition 1983
   Reprinted 1984, 1986
Fifth Edition 1990

ISBN 0-443-04095-8

British Library Cataloguing in Publication Data
Mackenna, B.R.
   Illustrated physiology. – 5th. ed.
   1. Man. Physiology. Illustrations
   I. Title II. Callander, Robin 1922 – III. McNaught, Ann
   B. (Ann Boyce). Illustrated physiology
   612

Library of Congress Cataloging in Publication Data
Mackenna, B.R.
   Illustrated physiology. – 5th ed./B.R. Mackenna, R. Callander.
      p. cm.
   Rev. ed. of: Illustrated physiology/by Ann B. McNaught and Robin
   Callander. 4th ed. 1983.
   1. Human physiology – Outlines, syllabi, etc. I. Callander,
   Robin. II. McNaught, Ann B. Illustrated physiology. III. Title.
   [DNLM: 1. Physiology – atlases. QT 17 M155i]
   QP41.M145 1990
   612–dc20

Produced by Longman Group (F.E.) Ltd
Printed in Hong Kong

PREFACE TO THE FIFTH EDITION

With the death of Dr Ann McNaught in December, 1986, medical education suffered a great loss. Her clear, concise lecturing was greatly appreciated by a generation of Glasgow medical students and, as co-author of Illustrated Physiology, her unique yet simple method of presentation contributed significantly to the education of many students throughout the world.

For a new author to be invited to revise the text of such a widely acclaimed book is an honour and at the same time a great responsibility, but his task has been considerably lightened by the help and guidance of the original illustrator.

Although the text of the book has been completely revised and updated its successful format has remained unchanged. Because of advances in our knowledge of function at the cellular and molecular levels some twenty pages dealing with these aspects of Physiology have been added. These include a description of cellular organelles, an account of the movement of substances through membranes and an explanation of the origin of action potentials and their role in carrying signals in nerves and in initiating contraction of muscle. Information on immunity has been expanded due to the recent highlighting of the immune system with the increase in organ transplantation and the appearance of the AIDS virus.

We hope that this edition of the book will continue to be useful to nurses and other health-care workers and that, as in the past, many medical and dental students will find the diagrams which summarize the salient points of each topic a valuable aid to learning and revision.

B.R. Mackenna
R. Callander
1990

PREFACE TO THE FIRST EDITION

This book grew originally from the need to provide visual aids for the large number of students in this department who come to the study of human physiology with no background of mammalian anatomy and often without any conventional training in either the biological or the physical and chemical sciences. Such students include postgraduates studying for the Diploma or Degree in Education, undergraduates working for Degrees in Science, laboratory technicians taking courses for the Ordinary and High National Certificates in Biology, and the increasing number of medical auxiliaries (physiotherapists, occupational therapists, radiographers, cardiographers, dietitians, almoners and social workers) many of whom are required to study to quite advanced levels at least regional parts of the subject.

Many medical, dental and pharmacy students, as well as nurses in training, have been good enough to indicate that they too find diagrams which summarize the salient points of each topic valuable aids to learning or revision. It is our hope that some of these groups may find our book helpful.

Each page is complete in itself and has been designed to oppose its neighbour. It is hoped that this will facilitate the choice of those pages thought suitable for any one course while making it easy to omit those which are too detailed for immediate consideration.

A book of this sort is largely derivative and it is impossible to acknowledge our wider debt. We wish to record our gratitude, however, to Professor R.C. Garry for his generous permission to borrow freely from the large collection of teaching diagrams built up over the years by himself and his staff; to Dr H.S.D. Garven and Dr G. Leaf for permission to use some of their own teaching material; and to Messrs Ciba Pharmaceutical Products Inc. from whose fine book of Medical Illustrations by Dr Frank H. Netter the diagram of the Cranial Nerves has been modified.

We are indebted to the following colleagues and friends who read parts of the original draft and offered helpful criticism:– Dr H.S.D. Garven, Dr J.S. Gillespie, Mr J.A. Gilmour, Dr M. Holmes, Dr B.R. Mackenna, Mr T. McClurg Anderson, Dr I.A. Boyd, Dr R.Y. Thomson, Dr J.B. deV. Weir.

We should like to express our gratitude to Mr Charles Macmillan and Mr James Parker of Messrs E. & S. Livingstone Ltd. for their unfailing courtesy and encouragement, and to Mrs Elizabeth Callander for help in preparing the index.

Ann B. McNaught
Robin Callander
1963

WHAT IS PHYSIOLOGY?

Physiology is the study of the function of living matter. Hence, there are many types of Physiology including Bacterial Physiology, Plant Physiology and **Human Physiology**.

To understand how human beings function it is necessary to appreciate that *all* living things are made of microscopic units of **protoplasm** called **cells**. There are some very simple living creatures which consist of just one cell, for example the **amoeba** which lives in pond water. These unicellular creatures exemplify the structure of all animal cells and show the phenomena which distinguish living from non-living things. Hence we start with a brief look at such creatures.

The human being is made up of 75 trillion cells which are arranged in various combinations and form various degrees of organized structure. Cells with similar properties aggregate to form **tissues** (e.g. muscular tissue, nervous tissue). Different tissues combine to form **organs** (e.g. kidneys, brain, heart). Organs are linked together to form **organ systems** (e.g. the heart and blood vessels form the cardiovascular system).

56% of the adult human body is fluid. Most is **inside** the cells (**intracellular fluid**). However about one third is **outside** the cells (**extracellular fluid**) and consists of the **plasma** of the blood which circulates in the cardiovascular system, plus the fluid which surrounds the cells (**intercellular** or **interstitial fluid**). Cells receive their nutrients from the interstitial fluid and as the nutrients are used up more must be brought to this surrounding fluid. Likewise the cells pass waste products to their bathing fluid and this waste must not be allowed to accumulate or it will poison the cells. In addition, the concentration of salts in the interstitial fluid must be kept constant for the cells to function normally.

The extracellular fluid was given the special name **The Internal Environment** of the body or the **Milieu Intérieur** in the 19th century by the French physiologist Claude Bernard. Physiologists use the term **homeostasis** to mean maintenance of constant conditions in the internal environment.

Thus the main function of most of the tissues and organ systems of the human body is to maintain the constancy of the internal environment so that its cells can function normally. However, to do so the systems must be controlled and regulated. The nervous and hormonal systems are specialized for this regulatory function.

How cells function, how the tissues and organ systems maintain homeostasis and how the systems are regulated is basically what human physiology is about and that is what is illustrated in the following pages.

CONTENTS

WHAT IS PHYSIOLOGY? vii

1. **Introduction: Cells, Tissues and Systems**
   The amoeba   2
   The phenomena of life   3
   The paramecium   4
   The cell   5
   Fine structure of cells   6
   Organelles   7, 8
   Cell division (mitosis)   9
   Differentiation of animal cells   10
   Epithelia   11, 12
   Connective tissues   13-15
   Muscular tissues   16
   Nervous tissues   17-19
   Cell division (meiosis)   20
   Development of the individual   21
   The body systems   22

2. **Nutrition and Metabolism – the Sources, Release and Uses of Energy**
   Basic constituents of protoplasm   24
   Carbohydrates   25
   Lipids   26
   Proteins   27
   Nucleic acids and mixed organic
       molecules   28
   Source of energy: photosynthesis   29
   Carbon 'cycle' in nature   30
   Nitrogen 'cycle' in nature   31
   Nutrition   32
   Energy-giving foods   33
   Body-building foods   34
   Vitamins   35, 36
   Digestion   37
   Protein metabolism   38
   Carbohydrate metabolism   39
   Fat metabolism   40

   Energy from food   41
   Formation of ATP   42
   Heat balance   43
   Maintenance of body
       temperature   44, 45
   Growth   46, 47
   Energy requirements   48, 49
   Balanced diet   50

3. **Cell Membrane Functions**
   Transport through membranes   52, 53
   Ions and charges   54
   Equilibrium potential   55
   Resting potential   56
   Electrotonic potentials   57
   The action potential   58
   Propagation of the nerve impulse   59
   Communication between cells   60
   Second messengers   61

4. **Digestive System**
   Digestive system   64
   Progress of food along alimentary
       canal   65
   Digestion in the mouth   66
   Control of salivary secretion   67
   Oesophagus   68
   Swallowing   69
   Stomach   70
   Gastric juice   71
   Movements of the stomach   72
   Vomiting   73
   Pancreas   74
   Pancreatic juice   75
   Liver and gall bladder   76
   Expulsion of bile   77
   Small intestine   78
   The basic pattern of the gut wall   79
   Intestinal secretions   80

Movements of the small intestine	81
Absorption in small intestine	82
Food absorption by cells of small intestine	83
Transport of absorbed foodstuffs	84
Large intestine	85
Movements of the large intestine	86
Innervation of the gut wall	87
Nervous control of gut movements	88

5. Transport System – the Heart, Blood Vessels and Body Fluids: Haemopoietic System

Cardiovascular system	90
General course of the circulation	91
Heart	92-94
Cardiac cycle	95
Heart sounds	96
Origin and conduction of the heartbeat	97
Electrocardiogram	98
Nervous regulation of action of heart	99
Cardiac reflexes	100, 101
Cardiac output	102
Blood vessels	103
Blood pressure	104
Measurement of arterial blood pressure	105
Elastic arteries	106
Nervous regulation of arterioles	107
Reflex and chemical regulation of arteriolar tone	108, 109
Capillaries	110
Veins: venous return	111
Blood flow	112
Pulmonary circulation	113
Distribution of water and electrolytes in body fluids	114
Water balance	115
Blood	116
Haemostasis and blood coagulation	117
Factors required for normal haemopoiesis	118
Haemopoiesis	119
Blood groups	120, 121
Rhesus factor	122
Lymphatic system	123
Spleen	124
Thymus	125
Immune system	126
Cerebrospinal fluid	127

6. Respiratory System

Respiratory system	130
Air conducting passages	131
Lungs: respiratory surfaces	132
Thorax	133
Mechanism of breathing	134
Artificial respiration	135
Volumes and capacities of lungs	136
Composition of respired air	137
Movement of respiratory gases	138
Dissociation of oxygen from haemoglobin	139
Uptake and release of carbon dioxide	140
Carriage and transfer of oxygen and carbon dioxide	141, 142
Nervous control of respiratory movements	143
Chemical regulation of respiration	144
Voluntary and reflex factors in the regulation of respiration	145

7. Excretory System

Excretory system	148
Kidney	149
Formation of urine	150-152
Water reabsorption – proximal tubule	153

The 'clearance' of inulin in the
   nephron 154
Urea 'clearance' 155
PAH 'clearance' 156
Maintenance of acid-base
   balance 157, 158
Defence of body fluid tonicity 159
Defence of body fluid volume 160
Urinary bladder and ureters 161
Storage and expulsion of urine 162
Urine 163

8. **Endocrine System**
Endocrine system 166
Thyroid 167
Underactivity of thyroid 168
Overactivity of thyroid 169
Parathyroids 170
Underactivity of parathyroids 171
Overactivity of parathyroids 172
Adrenal cortex 173
Underactivity of adrenal cortex 174
Overactivity of adrenal cortex 175
Adrenal medulla 176
Adrenaline 177
Development of pituitary 178
Anterior pituitary 179
Underactivity of anterior pituitary 180
Overactivity of pituitary somatotroph
   cells 181
Overactivity of pituitary corticotroph
   cells 182
Panhypopituitarism 183
Posterior pituitary 184
Oxytocin 185
Antidiuretic hormone 186
Underactivity of posterior pituitary 187
Aldosterone and antidiuretic hormone
   (ADH) in the maintenance of blood
   volume 188
Pancreas: islets of Langerhans 189

9. **Reproductive System**
Male reproductive system 192
Testis 193
Male accessory sex organs 194
Control of events in the testis 195
Female reproductive system 196
Adult pelvic sex organs in ordinary
   female cycle 197
Ovary in ordinary adult cycle 198
Ovary in pregnancy 199
Control of events in the ovary 200
Ovarian hormones 201
Uterus and uterine tubes 202, 203
Uterine tubes in cycle ending in
   pregnancy 204
Uterus 205, 206
Placenta 207
Uterus 208
Fetal circulation 209
Uterus 210-212
Mammary glands 213-215
Menopause 216
Pituitary, ovarian and endometrial
   cycles 217

10. **Central Nervous System –
Locomotor System**
Nervous sytem 220
Development of the nervous system 221
Cerebrum 222
Horizontal section through brain 223
Vertical section through brain 224
Coronal section through brain 225
Cranial nerves 226
Spinal cord 227
Synapse 228
Reflex action 229

Stretch reflexes 230
Spinal reflexes 231
'Edifice' of the CNS 232
Reflex action 233
Arrangement of neurons 234
Sense organs 235
Smell 236
Taste 237
Pathways and centres for taste 238
Eye 239
Protection of the eye 240
Muscles of eye 241
Control of eye movements 242
Iris, lens and ciliary body 243
Action of lens 244
Fundus oculi 245
Retina 246
Mechanism of vision 247, 248
Visual pathways to the brain 249
Stereoscopic vision 250
Light reflex 251
Ear 252
Cochlea 253
Mechanism of hearing 254
Auditory pathways to brain 255
Special proprioceptors 256
Organ of equilibrium: mechanism of action 257
Vestibular pathways to brain 258
General proprioceptors 259
Proprioceptor pathways to brain 260
Cutaneous sensation 261
Pain 262
Sensory pathways from skin of face 263
Pain and temperature pathways from trunk and limbs 264

Touch and pressure pathways from trunk and limbs 265
Sensory cortex 266
Motor cortex 267
Motor pathways to head and neck 268
Motor pathways to extremities 269
Motor unit 270
Final common pathway 271
Multineural motor pathways 272
Pathways controlling motor activity 273
Cerebellum 274-276
Control of muscle movement 277
Skeletal system 278
Skeletal muscles 279, 280
Muscular movements 281
Reciprocal innervation 282
Skeletal muscle and the mechanism of contraction 283
Skeletal muscle – molecular basis of contraction 284
Autonomic nervous system 286, 287
Autonomic reflex 288
Chemical transmission at nerve endings 289

Index 290

CHAPTER 1

INTRODUCTION: CELLS, TISSUES AND SYSTEMS

THE AMOEBA

All living things are made of **protoplasm**. Protoplasm exists in microscopic units called **cells**. The **amoeba** (which lives in pond water) consists of just one cell but exemplifies the basic structure of all animal cells and shows the phenomena which distinguish living from non-living things.

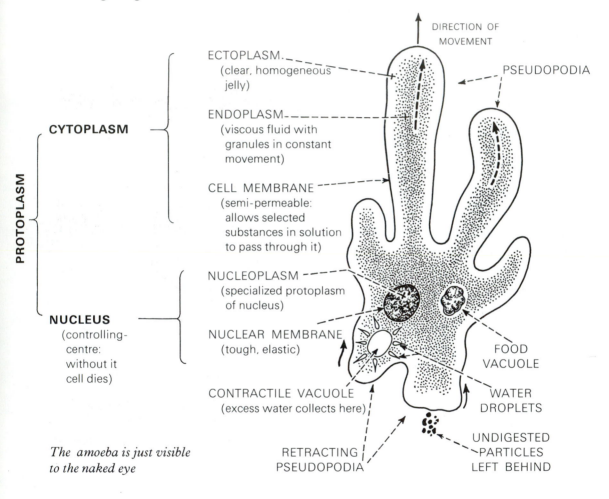

PROTOPLASM

CYTOPLASM

ECTOPLASM.
(clear, homogeneous jelly)

ENDOPLASM
(viscous fluid with granules in constant movement)

CELL MEMBRANE
(semi-permeable: allows selected substances in solution to pass through it)

NUCLEUS
(controlling-centre: without it cell dies)

NUCLEOPLASM
(specialized protoplasm of nucleus)

NUCLEAR MEMBRANE
(tough, elastic)

CONTRACTILE VACUOLE
(excess water collects here)

RETRACTING PSEUDOPODIA

DIRECTION OF MOVEMENT

PSEUDOPODIA

FOOD VACUOLE

WATER DROPLETS

UNDIGESTED PARTICLES LEFT BEHIND

The amoeba is just visible to the naked eye

2

THE PHENOMENA WHICH CHARACTERIZE ALL LIVING THINGS ARE SHOWN BY THE AMOEBA

1 ORGANIZATION
Autoregulation —
inherent ability to control
all life processes.

2 IRRITABILITY
Ability to respond to
stimuli (from changes
in the environment).

3 CONTRACTILITY
Ability to move.

4 NUTRITION
Ability to ingest,
digest, absorb and
assimilate food.

**5 METABOLISM AND
GROWTH**
Ability to liberate potential
energy of food and to convert
it into mechanical work
(*e.g. movement*) and to
rebuild simple absorbed
units into the complex
protoplasm of the living cell.

6 RESPIRATION
Ability to take in oxygen
for oxidation of food
with release of energy;
and to eliminate the
resulting carbon dioxide.

7 EXCRETION
Ability to eliminate
waste products of
metabolism.

8 REPRODUCTION
Ability to reproduce the
species.

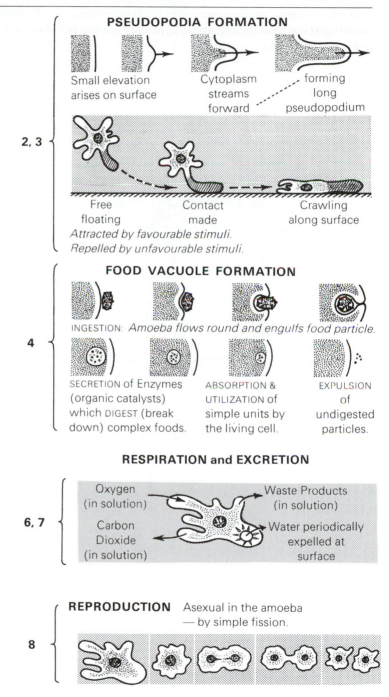

PSEUDOPODIA FORMATION

Small elevation arises on surface — Cytoplasm streams forward — forming long pseudopodium

2, 3

Free floating — Contact made — Crawling along surface

Attracted by favourable stimuli.
Repelled by unfavourable stimuli.

FOOD VACUOLE FORMATION

4

INGESTION: *Amoeba flows round and engulfs food particle.*

SECRETION of Enzymes (organic catalysts) which DIGEST (break down) complex foods.

ABSORPTION & UTILIZATION of simple units by the living cell.

EXPULSION of undigested particles.

RESPIRATION and EXCRETION

6, 7

Oxygen (in solution) — Waste Products (in solution)

Carbon Dioxide (in solution) — Water periodically expelled at surface

REPRODUCTION Asexual in the amoeba — by simple fission.

8

THE PARAMECIUM

The paramecium (another one-celled fresh water creature) shows:–
Modification and localization of structure for specialization and localization of certain functions

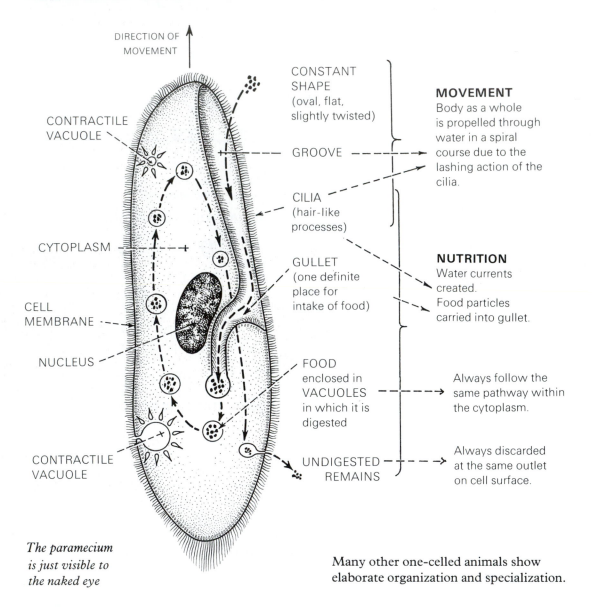

DIRECTION OF MOVEMENT

CONTRACTILE VACUOLE

CYTOPLASM

CELL MEMBRANE

NUCLEUS

CONTRACTILE VACUOLE

CONSTANT SHAPE (oval, flat, slightly twisted)

GROOVE

CILIA (hair-like processes)

GULLET (one definite place for intake of food)

FOOD enclosed in VACUOLES in which it is digested

UNDIGESTED REMAINS

MOVEMENT
Body as a whole is propelled through water in a spiral course due to the lashing action of the cilia.

NUTRITION
Water currents created.
Food particles carried into gullet.

Always follow the same pathway within the cytoplasm.

Always discarded at the same outlet on cell surface.

The paramecium is just visible to the naked eye

Many other one-celled animals show elaborate organization and specialization.

4

THE CELL

The cell is the **structural** and **functional** unit of the many-celled animal. Higher animals, including man, are made up of millions of living cells which vary widely in structure and function but have certain features in common.

The cell is divided into a **nucleus** (a spherical or oval organelle often near the centre) and **cytoplasm**, the region outside the nucleus. Cell **organelles** (little organs) are in the cytoplasm suspended in a fluid, the **cytosol**.

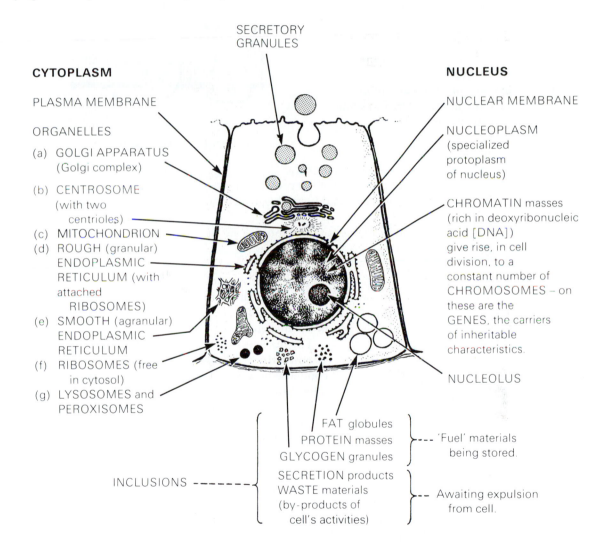

SECRETORY GRANULES

CYTOPLASM

PLASMA MEMBRANE

ORGANELLES

(a) GOLGI APPARATUS
    (Golgi complex)

(b) CENTROSOME
    (with two
        centrioles)

(c) MITOCHONDRION

(d) ROUGH (granular)
    ENDOPLASMIC
    RETICULUM (with
    attached
    RIBOSOMES)

(e) SMOOTH (agranular)
    ENDOPLASMIC
    RETICULUM

(f) RIBOSOMES (free
    in cytosol)

(g) LYSOSOMES and
    PEROXISOMES

NUCLEUS

NUCLEAR MEMBRANE

NUCLEOPLASM
(specialized
protoplasm
of nucleus)

CHROMATIN masses
(rich in deoxyribonucleic
acid [DNA])
give rise, in cell
division, to a
constant number of
CHROMOSOMES – on
these are the
GENES, the carriers
of inheritable
characteristics.

NUCLEOLUS

INCLUSIONS

FAT globules
PROTEIN masses
GLYCOGEN granules

'Fuel' materials
being stored.

SECRETION products
WASTE materials
(by-products of
cell's activities)

Awaiting expulsion
from cell.

FINE STRUCTURE OF CELLS

PLASMA MEMBRANE

The wall of the cell is the **plasma membrane** which controls the rate and type of ions and molecules passing into and out of the cell. It consists of two layers of **phospholipid** interspersed with **protein** molecules.

Note the clothes-pin shape of the phospholipid molecules. The head is the phosphate portion–relatively soluble in water (polar, hydrophilic). The tails are the lipid, relatively insoluble (non-polar, hydrophobic) and they meet in the interior of the membrane. Note that some proteins are in the inner surface, some in the outer, and some extend through the membrane.

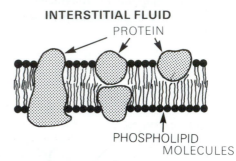

The proteins can function as:

  (a) structure of membrane,

  (b) pumps transporting ions across the membrane,

  (c) channels for ions. Changes in the shape of the protein can open or close the channel.

  (d) receptors, binding nerve transmitters and hormones which can initiate changes inside the cell,

  (e) enzymes catalysing chemical reactions at the membrane surface.

NUCLEUS

A **nuclear envelope** (or membrane), which is really two membranes separated by a space, surrounds the nucleus. At numerous points these membranes are joined forming the rims of circular openings, the **nuclear pores**, through which molecular messages can pass in and out of the nucleus.

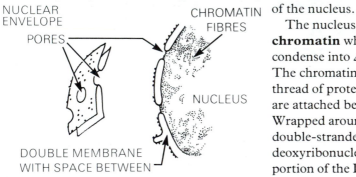

The nucleus is packed with fibres of **chromatin** which, when the cell divides, condense into 46 rod-shaped **chromosomes**. The chromatin fibres consist of a central thread of proteins (mainly **histone**) to which are attached beads of other histone molecules. Wrapped around the histone are double-stranded, helically coiled molecules of deoxyribonucleic acid (DNA). Each small portion of the DNA molecule is called a **gene**.

Each gene provides the code for the assembly of one protein. The exact structure of the proteins thus assembled determines the inherited characteristics of an individual.

DNA molecules are too large to pass out of the nucleus. Hence a smaller nucleic acid, ribonucleic acid (RNA) is assembled by part of the DNA molecule. This process is called **transcription**. This type of RNA is called messenger RNA (mRNA) because it carries the code for protein assembly into the cytoplasm where the protein molecules are assembled from amino acids. This later process is called **translation**.

6

NUCLEOLUS

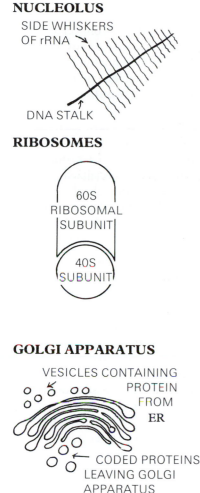

SIDE WHISKERS
OF rRNA

DNA STALK

The nucleolus is a spherical body within the nucleus. It has no membrane and is **packed** with 'fern-like' structures each consisting of a stalk of DNA and side whiskers of ribosome ribonucleic acid (rRNA). The rRNA molecules are combined with protein to form ribosomes which pass into the cytoplasm.

RIBOSOMES

60S
RIBOSOMAL
SUBUNIT

40S
SUBUNIT

Each ribosome consists of over 70 protein molecules and rRNA molecules, and is divided into two subunits called 40S and 60S on the basis of their sedimentation rates in a centrifuge. Some ribosomes are bound to a structure called **endoplasmic reticulum** (ER) and some are free in the cytosol. Ribosomes synthesize proteins from amino acids using information carried by rRNA molecules from genes in the nucleus. Proteins synthesized by endoplasmic reticulum ribosomes pass into the ER lumen then to the **Golgi apparatus** where they are processed as described below. Proteins manufactured by free ribosomes perform their functions in the cytosol.

GOLGI APPARATUS

VESICLES CONTAINING
PROTEIN
FROM
ER

CODED PROTEINS
LEAVING GOLGI
APPARATUS

This apparatus consists of a collection of membrane-enclosed sacs like 4–6 stacked saucers. Proteins from the endoplasmic reticulum have their structure altered here. This alteration is a kind of code which determines whether the protein will be (a) passed into **lysosomes** (see below), (b) stored in secretory granules or (c) inserted into the plasma membrane.

LYSOSOMES

MEMBRANE

Lysosomes are large membrane-bound organelles of various sizes which act as intracellular scavengers. They contain digestive enzymes which digest e.g. bacteria, which have been engulfed by the cell, and cellular debris such as damaged organelles.

PEROXISOMES

Peroxisomes are similar in structure to the lysosomes. They contain (a) enzymes which combine oxygen and hydrogen to form hydrogen peroxide (H_2O_2) and (b) an enzyme which converts H_2O_2 to water.

ORGANELLES

ENDOPLASMIC RETICULUM

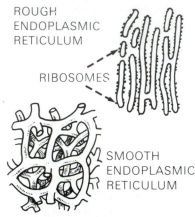

ROUGH ENDOPLASMIC RETICULUM

RIBOSOMES

SMOOTH ENDOPLASMIC RETICULUM

The endoplasmic reticulum is a network of interconnected tubular and flattened sac-like channels. The space between their walls is continuous with the space of the nuclear membrane and can transport substances from one part of the cell to another. One form of ER, **rough** or **granular** endoplasmic reticulum, has ribosomes attached to its outer surface and the other form, **smooth** or **agranular**, has no ribosomes. The spaces between both types are connected. Ribosomes on rough ER synthesize proteins while smooth ER is involved in carbohydrate metabolism. Specialized types of ER are present in various cells, e.g. in skeletal muscle cells smooth ER is involved in binding calcium ions which are liberated to initiate contraction of muscle cells.

MITOCHONDRIA

Mitochondria are sausage or oval shaped organelles with a smooth outer membrane and an inner membrane which is folded to form shelves or **cristae** which extend into the internal space or **matrix**. The inner membrane is the power plant of the cell. Enzymes in the matrix function

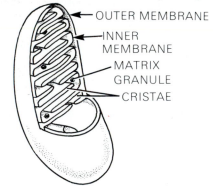

OUTER MEMBRANE

INNER MEMBRANE

MATRIX GRANULE

CRISTAE

in association with oxidative enzymes on the cristae to convert the products of fat, protein and carbohydrate metabolism to carbon dioxide and water via the citric acid cycle. Energy is thus liberated and used to synthesize a high energy substance **adenosine triphosphate** (ATP). ATP is transported out of the mitochondria and diffuses throughout the cell to release its energy wherever it is required.

CENTROSOME

(CROSS-SECTION)

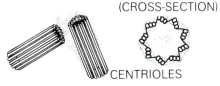

CENTRIOLES

The centrosome consists of two rod-like structures called **centrioles** arranged at right angles to one another. It is concerned with the synthesis of microtubules, e.g. the spindle and aster microtubules present during cell division.

SECRETORY VESICLES

All secretory substances are formed by the endoplasmic reticulum – Golgi apparatus system. They are then released from the Golgi apparatus into the cytoplasm inside storage vesicles called **secretory vesicles** or **secretory granules**.

In addition to the above organelles the cytoplasm may contain any of a variety of rod-like filaments, microfilaments and microtubular structures, depending on the function of the cell.

CELL DIVISION (MITOSIS)

All cells arise from the division of pre-existing cells. In **mitosis** there is an exact *qualitative* division of the **nucleus** and a less exact *quantitative* division of the **cytoplasm**.

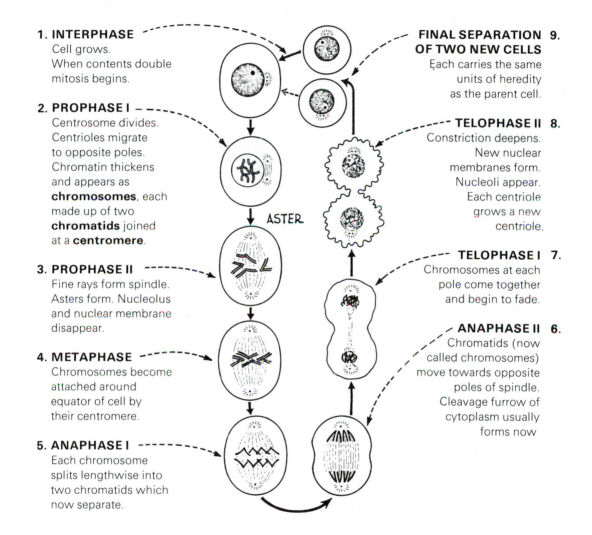

1. INTERPHASE
Cell grows.
When contents double mitosis begins.

2. PROPHASE I
Centrosome divides.
Centrioles migrate
to opposite poles.
Chromatin thickens
and appears as
chromosomes, each
made up of two
chromatids joined
at a **centromere**.

3. PROPHASE II
Fine rays form spindle.
Asters form. Nucleolus
and nuclear membrane
disappear.

4. METAPHASE
Chromosomes become
attached around
equator of cell by
their centromere.

5. ANAPHASE I
Each chromosome
splits lengthwise into
two chromatids which
now separate.

ASTER

**FINAL SEPARATION 9.
OF TWO NEW CELLS**
Each carries the same
units of heredity
as the parent cell.

TELOPHASE II 8.
Constriction deepens.
New nuclear
membranes form.
Nucleoli appear.
Each centriole
grows a new
centriole.

TELOPHASE I 7.
Chromosomes at each
pole come together
and begin to fade.

ANAPHASE II 6.
Chromatids (now
called chromosomes)
move towards opposite
poles of spindle.
Cleavage furrow of
cytoplasm usually
forms now

The longitudinal halving of chromosomes and genes ensures that each new cell receives the same hereditary factors as the original cell.
The number of chromosomes is constant for any one species.
The cells of the human body (somatic cells) carry 23 pairs – i.e. 46 chromosomes.
For clarity only 4 chromosomes (2 pairs) are shown in these diagrams.

DIFFERENTIATION OF ANIMAL CELLS

Specialization distinguishes multicellular creatures from more primitive forms of life.

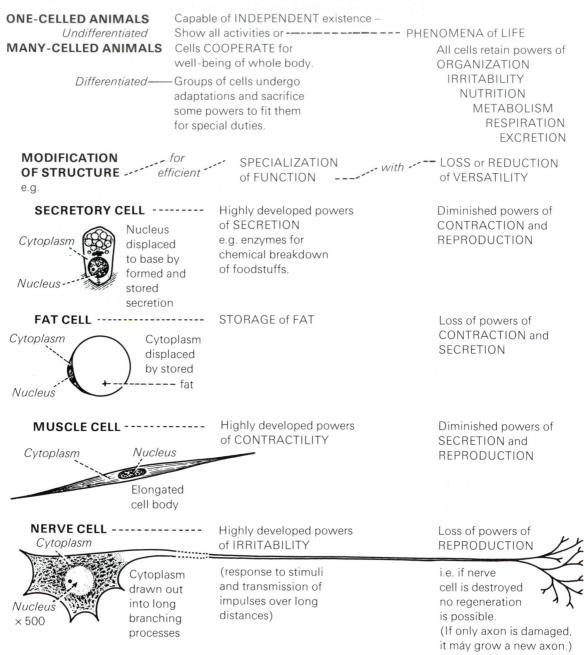

ONE-CELLED ANIMALS — Capable of INDEPENDENT existence –
Undifferentiated — Show all activities or ------------------ PHENOMENA of LIFE

MANY-CELLED ANIMALS — Cells COOPERATE for well-being of whole body.

Differentiated—— Groups of cells undergo adaptations and sacrifice some powers to fit them for special duties.

All cells retain powers of
ORGANIZATION
IRRITABILITY
NUTRITION
METABOLISM
RESPIRATION
EXCRETION

MODIFICATION OF STRUCTURE ---- *for efficient* ---- SPECIALIZATION of FUNCTION ---- *with* ---- LOSS or REDUCTION of VERSATILITY
e.g.

SECRETORY CELL --------

Cytoplasm

Nucleus displaced to base by formed and stored secretion

Nucleus

Highly developed powers of SECRETION
e.g. enzymes for chemical breakdown of foodstuffs.

Diminished powers of CONTRACTION and REPRODUCTION

FAT CELL ----------------

Cytoplasm

Cytoplasm displaced by stored
------ fat

Nucleus

STORAGE of FAT

Loss of powers of CONTRACTION and SECRETION

MUSCLE CELL ------------

Cytoplasm

Nucleus

Elongated cell body

Highly developed powers of CONTRACTILITY

Diminished powers of SECRETION and REPRODUCTION

NERVE CELL ------------

Cytoplasm

Cytoplasm drawn out into long branching processes

Nucleus
× 500

Highly developed powers of IRRITABILITY

(response to stimuli and transmission of impulses over long distances)

Loss of powers of REPRODUCTION

i.e. if nerve cell is destroyed no regeneration is possible.
(If only axon is damaged, it may grow a new axon.)

10

Different cell types are not mixed haphazardly in the body. Cells which are alike are arranged together to form **tissues**. There are *four* main types of tissue:– 1. EPITHELIA or LINING, 2. CONNECTIVE or SUPPORTING, 3. MUSCULAR, 4. NERVOUS.

EPITHELIA

STRUCTURAL MODIFICATIONS Sheets of cells with minimum intercellular substance	*SITE*	*SPECIALIZED FUNCTIONS* Line all internal and external surfaces of body
A. **SIMPLE** Single layer		
(a) SQUAMOUS	Lining of blood vessels, heart and cavities of body.	Reduces friction between surfaces.
(b) CUBICAL		
Simple	Small ducts, e.g. of salivary glands.	Protects underlying tissues: non-secretory.
SECRETORY Serous	Many glands, e.g. serous secreting part of mixed salivary gland.	Forms watery secretion containing e.g. digestive enzymes.
Mucous	Mucous glands, e.g. in mixed salivary gland.	Forms viscous lubricant, protective mucus.
(c) COLUMNAR		
	Large ducts of kidney.	Protects and lines.
Simple	Surface lining of stomach.	Secretes carpet of mucus to protect stomach from its own acid and digestive enzymes.
Mucous / Goblet Cells	E.g. in lining of intestine.	Form a viscous, lubricating, protective mucus.
Striated or Brush Bordered	Small intestine.	Absorbs foodstuffs.
Ciliated	Uterine tube.	Forms currents – wafts ovum towards uterus.

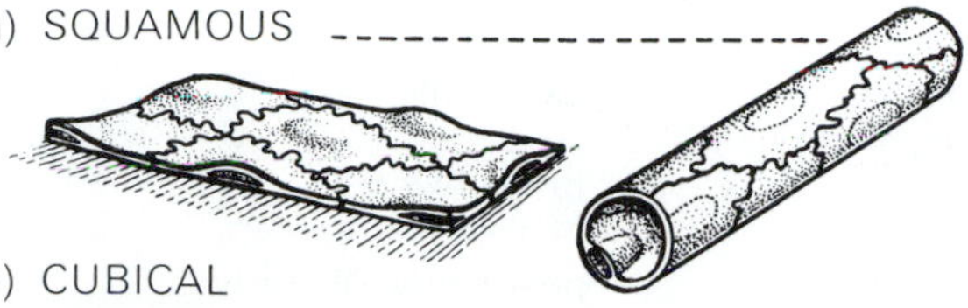

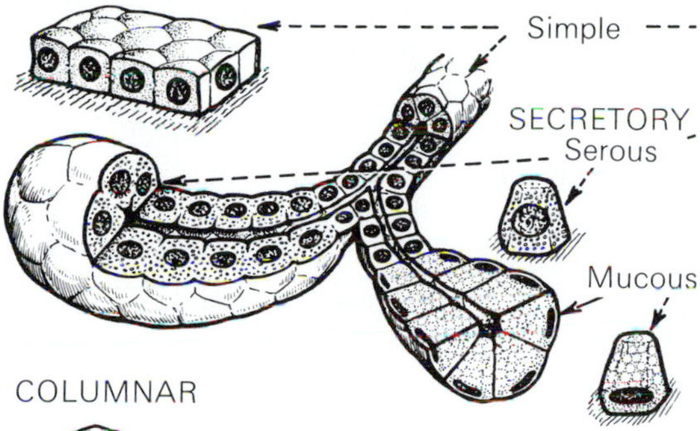

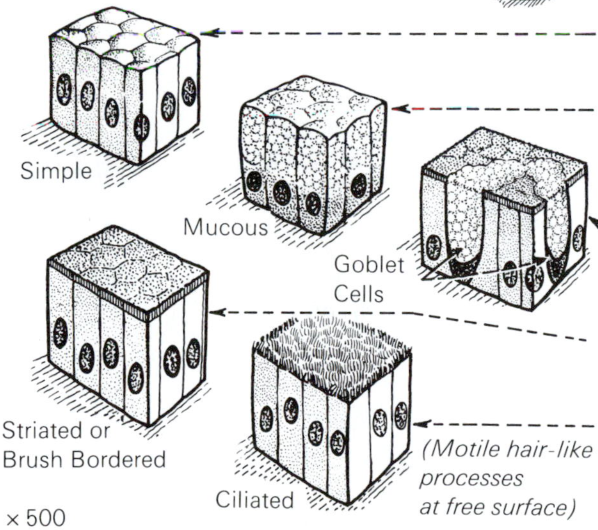

(Motile hair-like processes at free surface)

× 500

EPITHELIA

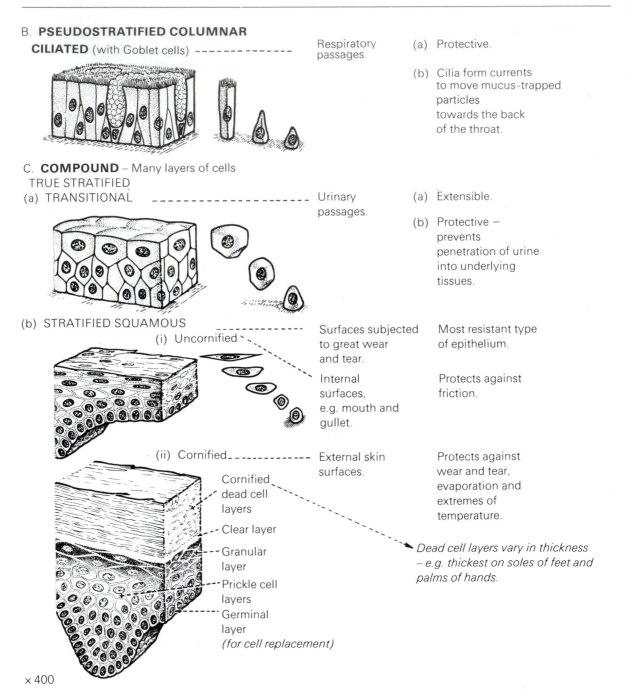

B. **PSEUDOSTRATIFIED COLUMNAR**
CILIATED (with Goblet cells) – – – – – – – – – – – – – Respiratory passages.

(a) Protective.

(b) Cilia form currents to move mucus-trapped particles towards the back of the throat.

C. **COMPOUND** – Many layers of cells
TRUE STRATIFIED
(a) TRANSITIONAL – Urinary passages.

(a) Extensible.

(b) Protective – prevents penetration of urine into underlying tissues.

(b) STRATIFIED SQUAMOUS

(i) Uncornified – – – – – – – – – – Surfaces subjected to great wear and tear.

Most resistant type of epithelium.

Internal surfaces, e.g. mouth and gullet.

Protects against friction.

(ii) Cornified – – – – – – – – – – External skin surfaces.

Protects against wear and tear, evaporation and extremes of temperature.

Cornified dead cell layers

Clear layer

Granular layer

Prickle cell layers

Germinal layer
(for cell replacement)

Dead cell layers vary in thickness – e.g. thickest on soles of feet and palms of hands.

× 400

CONNECTIVE TISSUES

STRUCTURAL MODIFICATIONS
Cells plus large amount of intercellular matrix and extracellular elements.

SPECIALIZED FUNCTIONS
Form framework, connecting, supporting and packing tissues of the body.

1. CELLS (floating free) in a FLUID MATRIX

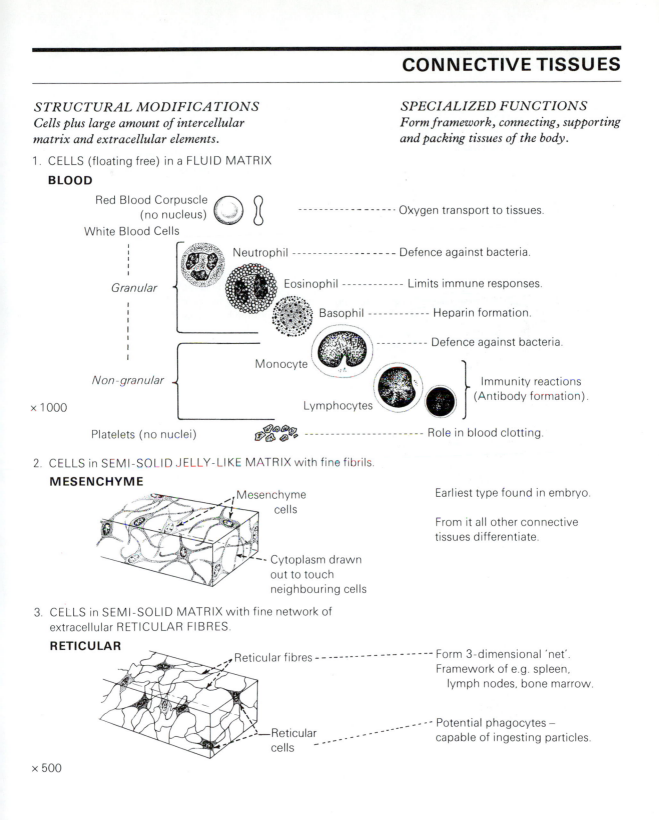

BLOOD

Red Blood Corpuscle (no nucleus) ---------------- Oxygen transport to tissues.

White Blood Cells

Granular

Neutrophil ------------------ Defence against bacteria.

Eosinophil ------------ Limits immune responses.

Basophil ------------ Heparin formation.

---------- Defence against bacteria.

Monocyte

Non-granular

Lymphocytes

Immunity reactions (Antibody formation).

× 1000

Platelets (no nuclei) ---------------------- Role in blood clotting.

2. CELLS in SEMI-SOLID JELLY-LIKE MATRIX with fine fibrils.

MESENCHYME

Mesenchyme cells

Cytoplasm drawn out to touch neighbouring cells

Earliest type found in embryo.

From it all other connective tissues differentiate.

3. CELLS in SEMI-SOLID MATRIX with fine network of extracellular RETICULAR FIBRES.

RETICULAR

Reticular fibres --------------------- Form 3-dimensional 'net'. Framework of e.g. spleen, lymph nodes, bone marrow.

Reticular cells

-------- Potential phagocytes – capable of ingesting particles.

× 500

CONNECTIVE TISSUES

4. CELLS in SEMI-SOLID MATRIX with thicker collagenous and elastic fibres.

(a) **LOOSE FIBROUS**

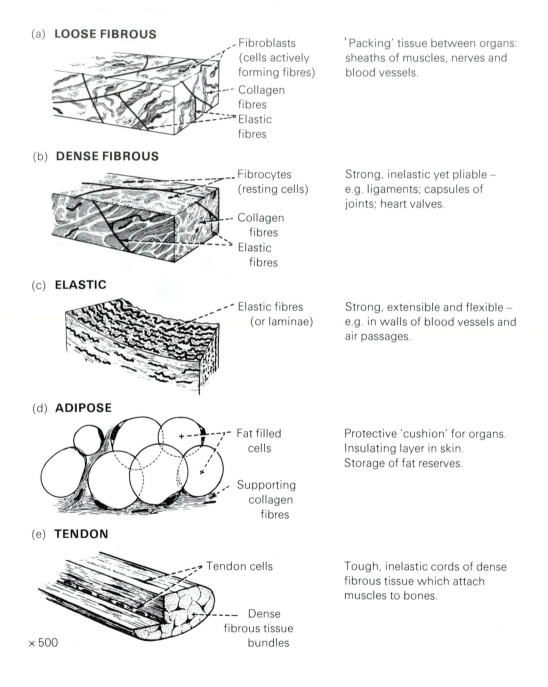

Fibroblasts (cells actively forming fibres)

Collagen fibres

Elastic fibres

'Packing' tissue between organs: sheaths of muscles, nerves and blood vessels.

(b) **DENSE FIBROUS**

Fibrocytes (resting cells)

Collagen fibres

Elastic fibres

Strong, inelastic yet pliable – e.g. ligaments; capsules of joints; heart valves.

(c) **ELASTIC**

Elastic fibres (or laminae)

Strong, extensible and flexible – e.g. in walls of blood vessels and air passages.

(d) **ADIPOSE**

Fat filled cells

Supporting collagen fibres

Protective 'cushion' for organs. Insulating layer in skin. Storage of fat reserves.

(e) **TENDON**

Tendon cells

Dense fibrous tissue bundles

× 500

Tough, inelastic cords of dense fibrous tissue which attach muscles to bones.

5. CELLS in SOLID ELASTIC MATRIX with fibres.

(a) **HYALINE CARTILAGE**

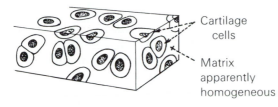

Cartilage cells

Matrix apparently homogeneous

Firm yet resilient –
e.g. in air passages; ends of bones at joints.
Transition stage in bone formation.

(b) **WHITE FIBRO-CARTILAGE**

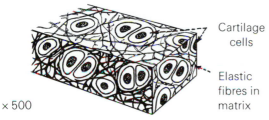

Cartilage cells

Collagen fibres in matrix

Tough. Resistant to stretching. Acts as shock absorber between vertebrae.

(c) **ELASTIC or YELLOW FIBRO-CARTILAGE**

Cartilage cells

Elastic fibres in matrix

More flexible, resilient –
e.g in larynx and ear.

× 500

6. CELLS in SOLID RIGID MATRIX impregnated with Calcium and Magnesium Salts.

BONE

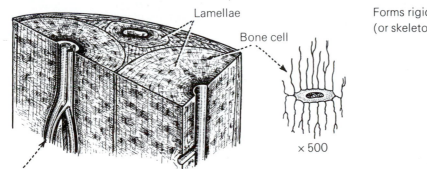

Lamellae

Bone cell

Forms rigid framework (or skeleton) of body

× 500

Haversian canal containing blood vessels.

× 100

MUSCULAR TISSUES

All have **elongated** cells with special development of **contractility**.

1. SMOOTH, UNSTRIPED, VISCERAL or INVOLUNTARY muscle

Nucleus

Least specialized. Shows slow rhythmical contraction and relaxation.
Not under voluntary control.
Found in walls of viscera and blood vessels.

2. CARDIAC or HEART muscle

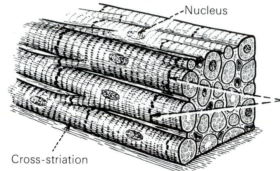

Nucleus

Cross-striation

More highly specialized.
Rapid rhythmical contraction (and relaxation) spreads through whole muscle mass.
Not under voluntary control.
Found only in heart wall.

Cells adhere, end to end, at intercalated discs to form long 'fibres' which branch and connect with adjacent 'fibres'.

There are GAP junctions (page 60) between the 'fibres'.

3. STRIATED, STRIPED, SKELETAL or VOLUNTARY muscle

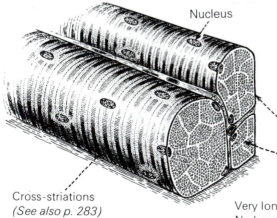

Nucleus

Cross-striations
(See also p. 283)

×500

Most highly specialized.
Very rapid, powerful contractions of individual fibres.
Under voluntary control.
Found in e.g. muscles of trunk, limbs, head.

Thick covering membrane (sarcolemma)

Many myofibrils embedded in sarcoplasm

Very long, large multi-nucleated units. No branching.

16

NERVOUS TISSUES

Nervous tissue is divided into:-

(a) **Neurons** or **Nerve cells**
  specialized in
  IRRITABILITY
  CONDUCTION
  INTEGRATION

{ MOTOR – pass messages from brain and spinal cord to effector organs (muscles and glands).
ASSOCIATION – relay messages between neurons.
SENSORY – receive and pass messages from environment to brain and spinal cord.

(b) **Accessory** or Supporting cells
  Not RECEPTIVE
  Not CONDUCTING

{ NEUROGLIA in Central Nervous System (Brain and Spinal Cord).
SHEATH (Schwann) cells on peripheral nerve fibres, i.e. outside CNS.
SATELLITE cells in ganglia of peripheral nervous systems.

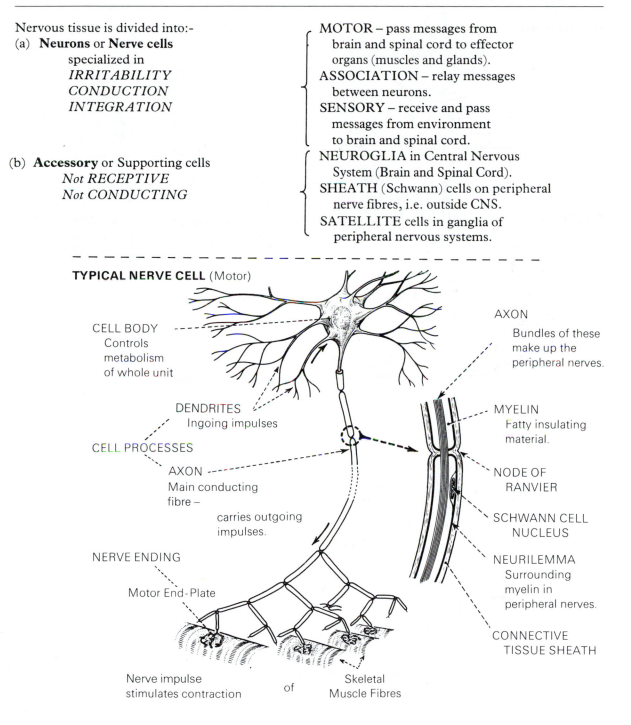

TYPICAL NERVE CELL (Motor)

AXON
  Bundles of these make up the peripheral nerves.

CELL BODY
Controls metabolism of whole unit

DENDRITES
Ingoing impulses

CELL PROCESSES

AXON
Main conducting fibre –
  carries outgoing impulses.

MYELIN
Fatty insulating material.

NODE OF RANVIER

SCHWANN CELL NUCLEUS

NEURILEMMA
Surrounding myelin in peripheral nerves.

NERVE ENDING

Motor End-Plate

CONNECTIVE TISSUE SHEATH

Nerve impulse stimulates contraction   of   Skeletal Muscle Fibres

NERVOUS TISSUES

MULTIPOLAR (many cell processes)
NEURONS

MODIFICATIONS with SITE

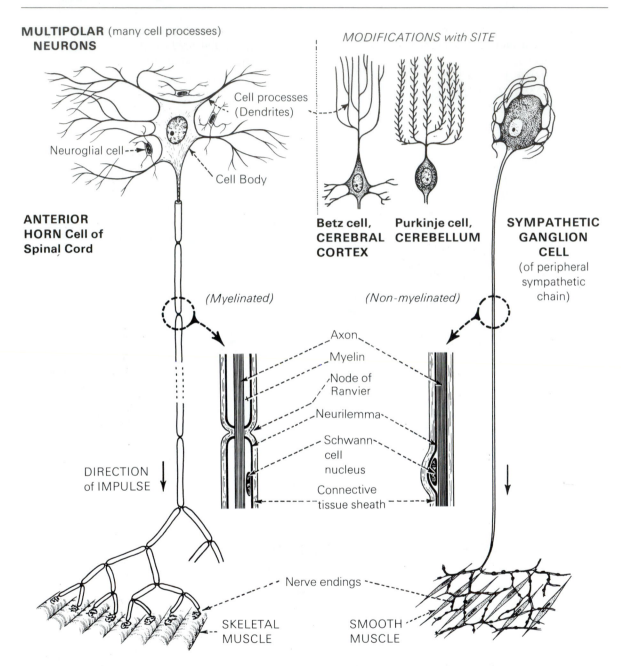

Cell processes
(Dendrites)

Neuroglial cell

Cell Body

**ANTERIOR
HORN Cell of
Spinal Cord**

**Betz cell,
CEREBRAL
CORTEX**

**Purkinje cell,
CEREBELLUM**

**SYMPATHETIC
GANGLION
CELL**
(of peripheral
sympathetic
chain)

(Myelinated)

(Non-myelinated)

Axon

Myelin

Node of
Ranvier

Neurilemma

Schwann
cell
nucleus

Connective
tissue sheath

DIRECTION
of IMPULSE

Nerve endings

SKELETAL
MUSCLE

SMOOTH
MUSCLE

Most multipolar neurons are **motor** (efferent) or **association** in function.

18

UNIPOLAR NEURON (one main process leaves cell body and divides in T-shaped manner)

BIPOLAR NEURON (two main processes)

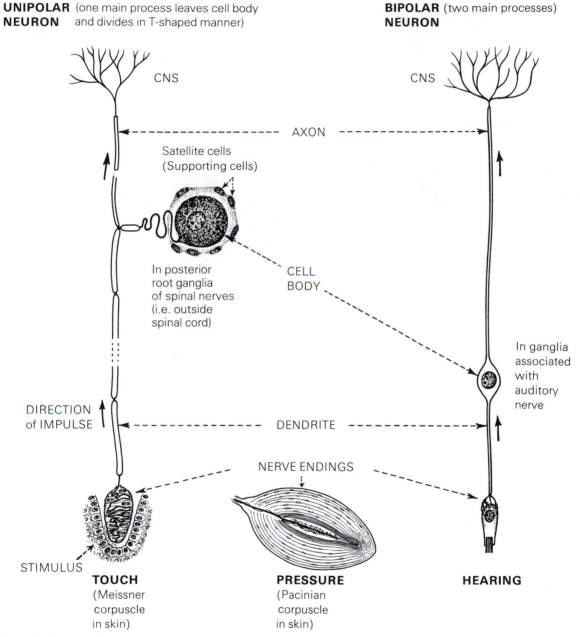

CNS

CNS

AXON

Satellite cells
(Supporting cells)

In posterior
root ganglia
of spinal nerves
(i.e. outside
spinal cord)

CELL
BODY

In ganglia
associated
with
auditory
nerve

DIRECTION
of IMPULSE

DENDRITE

NERVE ENDINGS

STIMULUS

TOUCH
(Meissner
corpuscle
in skin)

PRESSURE
(Pacinian
corpuscle
in skin)

HEARING

All unipolar and bipolar neurons are **sensory (afferent)** in function i.e. carry information TO the Central Nervous System (CNS).

19

CELL DIVISION (MEIOSIS)

New individuals develop after fusion of 2 specialized cells – the **gametes**. During their formation the **ovum** (female) and the **spermatozoon** (male) undergo two special cell divisions to reduce the chromosome content of each to the haploid number of 23. In fusion, the mingling of male and female chromosomes restores the normal diploid number – 46.

MATURATION of the MALE GAMETES (only one pair of chromosomes illustrated)

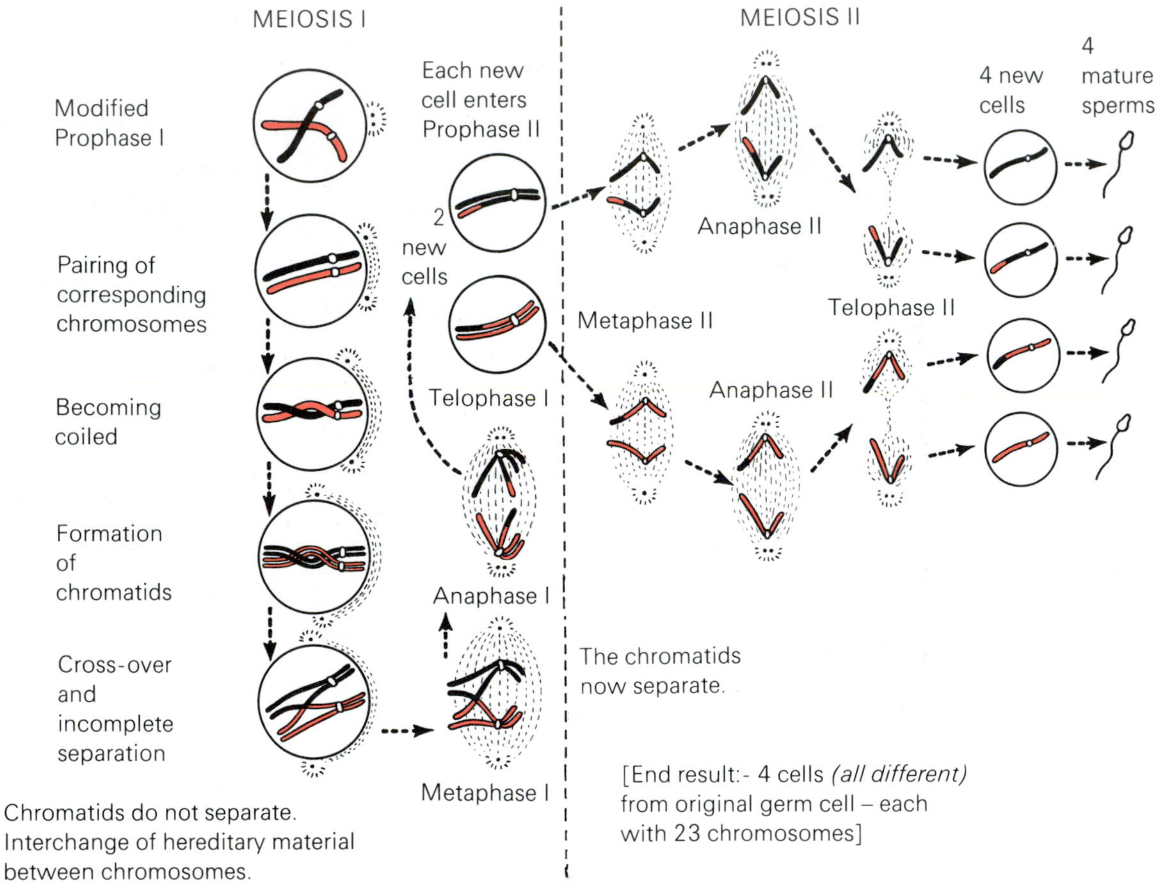

MEIOSIS I

MEIOSIS II

Modified
Prophase I

Pairing of
corresponding
chromosomes

Becoming
coiled

Formation
of
chromatids

Cross-over
and
incomplete
separation

Each new
cell enters
Prophase II

2
new
cells

Telophase I

Anaphase I

Metaphase I

Chromatids do not separate.
Interchange of hereditary material
between chromosomes.

4 new
cells

4
mature
sperms

Anaphase II

Metaphase II

Telophase II

Anaphase II

The chromatids
now separate.

[End result:- 4 cells *(all different)*
from original germ cell – each
with 23 chromosomes]

In the **female**, three of the 'cells' are small polar bodies which are discarded and disintegrate.

One single mature **ovum** receives most of the cytoplasm.

DEVELOPMENT OF THE INDIVIDUAL

All tissues of the human body are derived from the single cell – the **fertilized ovum**.

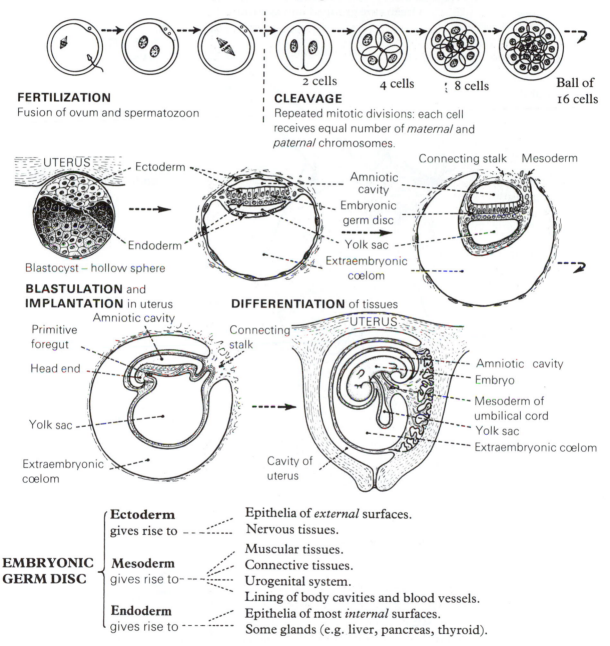

FERTILIZATION
Fusion of ovum and spermatozoon

CLEAVAGE
Repeated mitotic divisions: each cell receives equal number of *maternal* and *paternal* chromosomes.

2 cells

4 cells

8 cells

Ball of 16 cells

UTERUS

Ectoderm

Endoderm

Blastocyst – hollow sphere

BLASTULATION and
IMPLANTATION in uterus

Amniotic cavity

Embryonic germ disc

Yolk sac

Extraembryonic cœlom

Connecting stalk   Mesoderm

DIFFERENTIATION of tissues

Primitive foregut

Head end

Yolk sac

Extraembryonic cœlom

Amniotic cavity

Connecting stalk

UTERUS

Amniotic cavity

Embryo

Mesoderm of umbilical cord

Yolk sac

Extraembryonic cœlom

Cavity of uterus

EMBRYONIC GERM DISC

Ectoderm
gives rise to
Epithelia of *external* surfaces.
Nervous tissues.

Mesoderm
gives rise to
Muscular tissues.
Connective tissues.
Urogenital system.
Lining of body cavities and blood vessels.

Endoderm
gives rise to
Epithelia of most *internal* surfaces.
Some glands (e.g. liver, pancreas, thyroid).

21

THE BODY SYSTEMS

The tissues are arranged to form **organs**.
Organs are grouped into **systems**.

The **ESSENTIAL LIFE PROCESSES** - - - - - - - - are delegated to - - - - - - - - - **SEPARATE SYSTEMS**

1. **Irritability and control**
   (response to stimuli: and integration)

2. **Metabolism and growth**
   (energy release for work and growth)

3. **Respiration**
   (oxygen intake for release of energy; CO_2 loss)

4. **Transport**
   of materials (e.g. waste, food, respiratory gases)

5. **Nutrition**
   (source of energy ingested, digested and absorbed)

6. **Excretion**
   (waste products of metabolism eliminated)

7. **Reproduction**
   (propagation of species)

8. **Contractility**
   (movement)

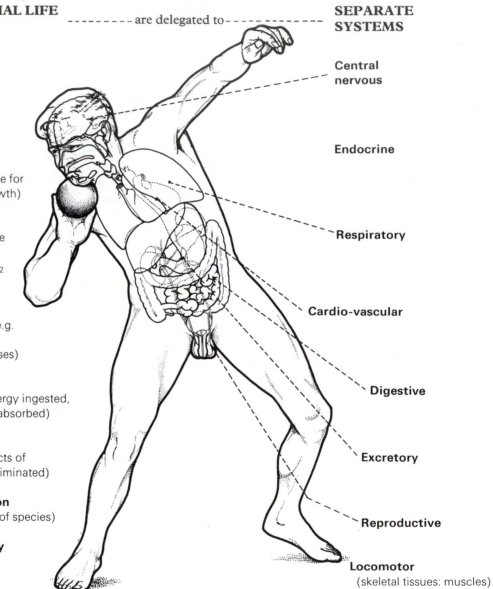

Central nervous

Endocrine

Respiratory

Cardio-vascular

Digestive

Excretory

Reproductive

Locomotor
(skeletal tissues: muscles)

The systems do not work independently. The body works as a whole. Health and well-being depend on the coordinated effort of every part.

CHAPTER 2

NUTRITION AND METABOLISM: THE SOURCES, RELEASE AND USES OF ENERGY

BASIC CONSTITUENTS OF PROTOPLASM

Protoplasm is made up of certain
ELEMENTS —— present mainly in —— **CHEMICAL COMBINATION**

	e.g. in a 70 kg man average amounts (grams)			

Chemical Symbol

		e.g. in a 70 kg man average amounts (grams)		C, H, O and N combine to form PROTEINS (main BUILDING constituents of all protoplasm)	These make up most of the Body Weight
H	– HYDROGEN	6580	A large amount of the H and O is present as WATER (H_2O)		
O	– OXYGEN	43 550		C, H & O combine chemically to form CARBOHYDRATES and LIPIDS (chief sources of ENERGY in living protoplasm)	
C	– **CARBON**	12 590			
N	– NITROGEN	1815			

Ca – CALCIUM	1700	Important constituents of blood and of hard tissues – e.g. bones and teeth.		These eight make up much of the remaining Body Weight
P – PHOSPHORUS	680			
Cl – CHLORINE	115	Important constituents of body fluids.		
Na – SODIUM	70			
K – POTASSIUM	70	Important constituent of all cells.		
S – SULPHUR	100			
Mg – MAGNESIUM	42	Important for activity of Brain, Nerves and Muscles.		
Fe – IRON	7	Important component of haemoglobin in red blood cells		

Trace elements make up the last few grams or so.
These include: – Manganese, copper, iodine, zinc, cobalt, molybdenum, nickel, aluminium, chromium, titanium, silicon, rubidium, lithium, arsenic, fluorine, bromine, selenium, boron, barium and strontium.
NOTE: Many of these inorganic substances in the protoplasm are in chemical combination. Apart from water, the chief constituents are present as compounds of **carbon**, i.e. they are **organic substances**.
NB: Cl element is chlor**ine**; Salt, e.g. NaCl is chlor**ide**; Ion is chlor**ide** ion (Cl^-).

Carbohydrates consist of atoms of C, H and O.

MONOSACCHARIDES (sugars) are the simplest. Most common in the diet are the **hexoses,** which have six carbon atoms, e.g. **glucose, fructose** and **galactose.** Four or five carbon atoms lie in a flat plane, linked with an oxygen atom. Remaining atoms form side groups projecting above or below the ring.

Glucose is the major fuel required by cells to provide energy.

GLUCOSE

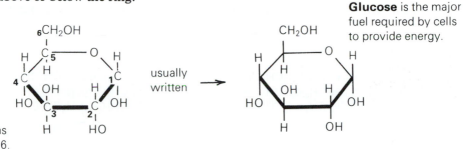

usually written →

NB: The carbon atoms are numbered 1-6.

Very large organic molecules can be made by linking together smaller molecular subunits forming chains known as **polymers.**

Larger carbohydrate molecules can be formed by linking monosaccharides together.

DISACCHARIDES Two monosaccharide molecules linked together, e.g.
galactose + glucose = lactose (milk sugar),
glucose + fructose = sucrose (table sugar).

LACTOSE

[NB: Same constituent elements as glucose but difference in orientation of H and O groups on carbon 4.]

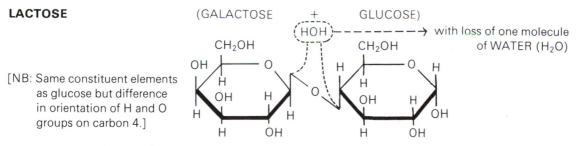

(GALACTOSE   +   GLUCOSE)
HOH - - - - - - - → with loss of one molecule of WATER (H_2O)

POLYSACCHARIDES Long chains of **glucose** units can form **glycogen, starch** and **cellulose.**

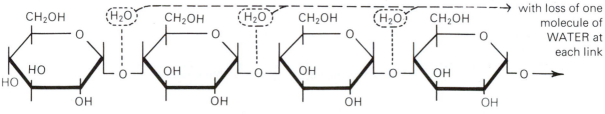

with loss of one molecule of WATER at each link

Cellulose is a straight chain without branches; important in the structure of plants.
Glycogen is a chain of glucose units with frequent branches along the molecule; form in which **animals** store glucose.
Starch is less branched; form in which **plants** store glucose.

25

LIPIDS

LIPIDS can be subdivided into three classes: (1) **Triglycerides** (Triacylglycerols or Neutral Fats), (2) **Phospholipids** and (3) **Steroids**. They all contain mainly H and C and are insoluble in water.

    TRIGLYCERIDES – the most common form of fat in the body. They consist of one molecule of **glycerol** linked with 3 molecules of **fatty acid**. The three fatty acids may be the same or may be different. A fatty acid consists of a chain of carbon atoms with a carboxyl group at one end. In triglycerides the carboxyl group is linked to a hydroxyl group of glycerol.

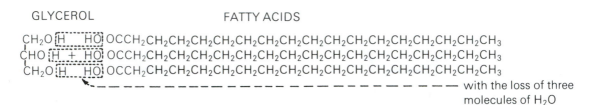

GLYCEROL   FATTY ACIDS

$CH_2O\ [H\quad HO]\ OCCH_2CH_2CH_2CH_2CH_2CH_2CH_2CH_2CH_2CH_2CH_2CH_2CH_2CH_2CH_2CH_2CH_3$
$CHO\ [H\ +\ HO]\ OCCH_2CH_2CH_2CH_2CH_2CH_2CH_2CH_2CH_2CH_2CH_2CH_2CH_2CH_2CH_2CH_2CH_3$
$CH_2O\ [H\quad HO]\ OCCH_2CH_2CH_2CH_2CH_2CH_2CH_2CH_2CH_2CH_2CH_2CH_2CH_2CH_2CH_2CH_2CH_3$

with the loss of three molecules of H_2O

    When all the carbons in a fatty acid chain are linked by single bonds, the fatty acid is said to be **saturated**. If the fatty acid chain contains double bonds it is **unsaturated**. If the molecule contains more than one double bond, the fatty acid is **polyunsaturated**. Animal fats contain saturated fatty acids and vegetable fats contain polyunsaturated fatty acids.

    PHOSPHOLIPIDS   In these lipids glycerol is linked to two fatty acids and the third hydroxyl group of the glycerol molecule is linked to a phosphate group which is in turn linked to a nitrogen containing molecule.

$CH_2O\ [H\quad HO]\ OCCH_2CH_2 - - - -\ \text{fatty acid} - - - -\ CH_3$

$C_2HO\ [H\ +\ HO\ ]\ OCCH_2CH_2 - - -\ \text{fatty acid} - - -\ CH_3$

$$CH_2O - \overset{O}{\underset{O^{\ominus}}{\overset{\|}{P}}} - O\text{-}CH_2\text{-}CH_2\text{-}\overset{\oplus}{N}\text{-}(CH_3)_3$$

NB: Both the phosphate and nitrogen groups are electrically charged.

    STEROIDS   Four interconnected rings of carbon atoms form the basic structure of all steroids. The steroid family includes cholesterol, bile acids, some hormones (e.g. oestrogen and testosterone) and some vitamins.

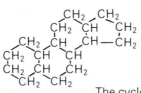

The cyclopentanoperhydrophenanthrene nucleus.

PROTEIN is the chief organic material of all protoplasm – plant or animal. It is composed of atoms of C, H, O, N and sometimes other atoms, especially S. The protein molecule is assembled by linking subunits called **amino acids**, i.e. proteins are **polymers** of amino acids.

All the 20 or so amino acids found in nature have an amino and a carboxyl group. Each has a different **amino acid side chain**.

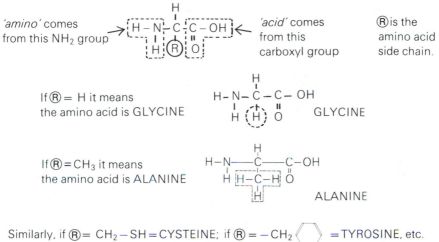

'amino' comes from this NH_2 group

'acid' comes from this carboxyl group

Ⓡ is the amino acid side chain.

If Ⓡ = H it means the amino acid is GLYCINE — GLYCINE

If Ⓡ = CH_3 it means the amino acid is ALANINE — ALANINE

Similarly, if Ⓡ = $CH_2 - SH$ = CYSTEINE; if Ⓡ = $- CH_2$ ⬡ = TYROSINE, etc.

In a protein molecule the amino acids are linked by combination of the amino group from one amino acid with the carboxyl group of the next amino acid. This junction is a **peptide bond** or **link**.

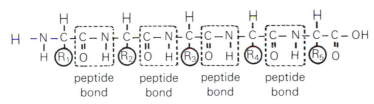

peptide bond   peptide bond   peptide bond   peptide bond

Thus a **polypeptide chain** is formed. If there are fewer than 50 amino acids in the chain the molecule is a **peptide**; if there are more than 50 the polypeptide is a **protein**.

Since each Ⓡ in the above diagram can be any one of the 20 amino acids found in nature thousands of different kinds of protein can be formed. Thus differences between proteins depend on which amino acids are present and on their number and arrangement.

Although proteins are made up of a chain of units they have also a three-dimensional shape which is determined by electrical attraction between charged groups on the chain. There may be more than one chain in the molecule, e.g. the hormone insulin has two polypeptide chains. The three-dimensional shape of proteins is functionally important.

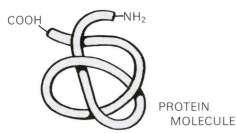

COOH   NH₂

PROTEIN MOLECULE

27

NUCLEIC ACIDS AND MIXED ORGANIC MOLECULES

NUCLEIC ACIDS are responsible for the storage of genetic information and passage of this information from cell to cell and from one generation to the next. There are two types: **Deoxyribonucleic Acid** (**DNA**) and **Ribonucleic Acid** (**RNA**). Both types consist of linked chains of subunits called **nucleotides**. Each nucleotide has a phosphate group, a sugar and a ring of carbon and nitrogen atoms called a **Base**.

DNA

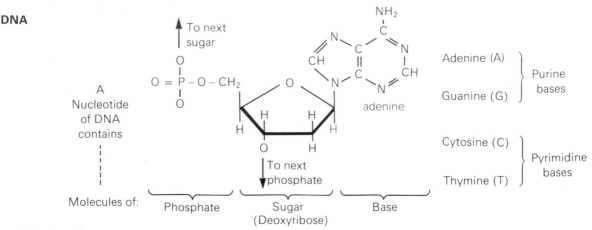

The phosphate group of one nucleotide is linked to the sugar of the adjacent nucleotide to form a chain. The DNA molecule consists of two such chains joined at their bases and coiled in a double helix. A can only pair with T, and G can only pair with C. DNA stores genetic information in code form, made from the sequence of its bases – like a pattern.

RNA – a single chain of nucleotides similar to DNA in which the sugar is **Ribose** instead of deoxyribose, and the pyrimidine base thymine in DNA is replaced by the base *uracil* (U) in RNA.

NB: Ribose has OH where deoxyribose has H.

RNA is smaller than DNA and can pass through the nuclear membrane into the cytosol.

RNA molecules decode genetic information into instructions for ribosomes to link amino acids in the specific sequence to make a protein.

MIXED ORGANIC MOLECULES

Glycoproteins (protein plus monosaccharide). Most protein secreted by cells is combined in this way.

Lipoproteins (lipid molecules coated with layer of protein). Involved in the transport of blood lipids.

Glycolipids (lipid plus monosaccharide). Found in plasma membrane. Involved in recognition of cells by defence cells and viruses, etc.

Nucleoproteins (DNA plus protein). Occurs in ribosomes.

SOURCE OF ENERGY: PHOTOSYNTHESIS

The essential life processes or the phenomena which characterize life depend on the use of **energy**. The **SUN** is the **source** of energy for **all** living things.

Only green **plants** can **trap** and **store** the sun's energy and build simple **carbohydrates** from carbon dioxide and water. This process is called **photosynthesis**. These simple carbohydrates are converted by additional metabolic processes of the plant into lipids, proteins, nucleic acids, etc. Thus **PLANTS** are the primary source of the energy-rich body building compounds required by **protoplasm.**

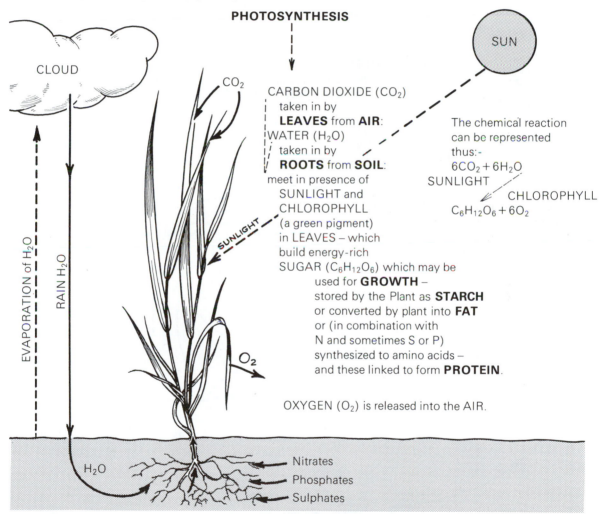

PHOTOSYNTHESIS

CLOUD

CO_2

CARBON DIOXIDE (CO_2)
taken in by
LEAVES from **AIR**:
WATER (H_2O)
taken in by
ROOTS from **SOIL**:
meet in presence of
SUNLIGHT and
CHLOROPHYLL
(a green pigment)
in LEAVES – which
build energy-rich
SUGAR ($C_6H_{12}O_6$) which may be
used for **GROWTH** –
stored by the Plant as **STARCH**
or converted by plant into **FAT**
or (in combination with
N and sometimes S or P)
synthesized to amino acids –
and these linked to form **PROTEIN**.

OXYGEN (O_2) is released into the AIR.

SUN

The chemical reaction
can be represented
thus:-
$6CO_2 + 6H_2O$
SUNLIGHT
CHLOROPHYLL
$C_6H_{12}O_6 + 6O_2$

SUNLIGHT

EVAPORATION of H_2O

RAIN H_2O

O_2

H_2O

Nitrates
Phosphates
Sulphates

When plants or their products are eaten this stored energy becomes available to animals and man.

CARBON 'CYCLE' IN NATURE

Animal bodies are unable to build proteins, carbohydrates or fats directly from **atoms**. They must be built up for them by **plants**.
Carbon is the basic building unit of all these compounds.

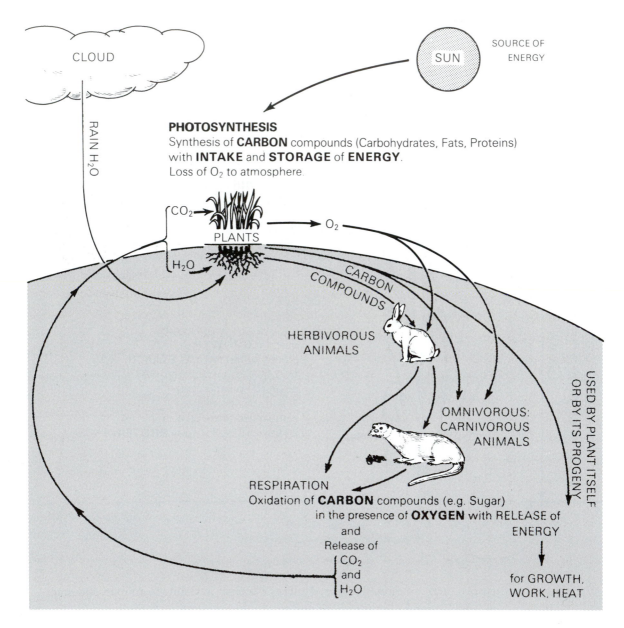

CLOUD

SUN

SOURCE OF ENERGY

RAIN H_2O

PHOTOSYNTHESIS
Synthesis of **CARBON** compounds (Carbohydrates, Fats, Proteins) with **INTAKE** and **STORAGE** of **ENERGY**.
Loss of O_2 to atmosphere.

CO_2 →

PLANTS

O_2

H_2O

CARBON COMPOUNDS

HERBIVOROUS ANIMALS

OMNIVOROUS: CARNIVOROUS ANIMALS

USED BY PLANT ITSELF OR BY ITS PROGENY

RESPIRATION
Oxidation of **CARBON** compounds (e.g. Sugar) in the presence of **OXYGEN** with RELEASE of
and
Release of
CO_2
and
H_2O

ENERGY

for GROWTH, WORK, HEAT

Although **animals** are surrounded by **nitrogen** in the air they cannot use that nitrogen to build nitrogen-containing compounds. They can only use the nitrogen trapped by plants to make these compounds.

NITROGEN in the **Atmosphere**

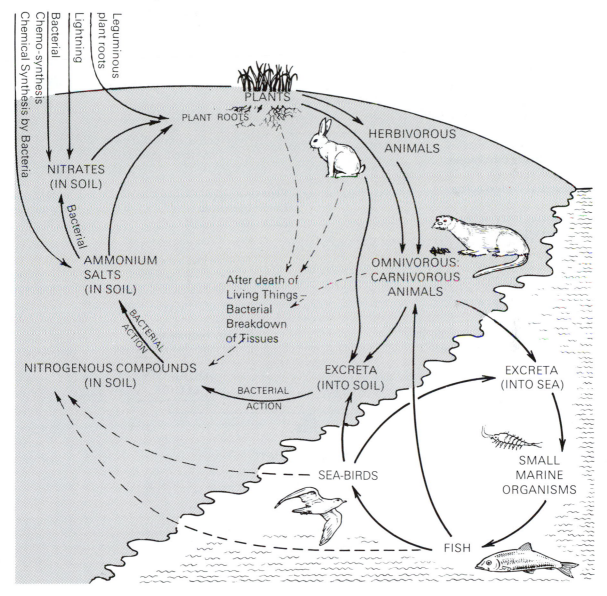

31

NUTRITION

Man eats as **FOOD** the substances which **plants** (and, through them, **animals**) have made. These are broken down in man's body into simpler chemical units which provide:

BUILDING and PROTECTIVE MATERIALS

Man requires more or less the same elements as plants. (Some, such as the minerals iodine, sodium, iron, calcium, he assimilates in **inorganic form**.)

Most must be built up for him by plants:-

Organically combined
Carbon, Nitrogen, and Sulphur, etc.
Essential amino acids

Essential fatty acids
Certain **vitamins**

These are
used to
BUILD, MAINTAIN or REPAIR
PROTOPLASM

Body-building requirements
of the individual determine

QUALITY
of DIET

ENERGY

Stored originally
by plants

RELEASED in man's cells
by
OXIDATION

When food is 'burned' it gives up
its stored energy

Proteins yield 17 kJ (4 kcal)	units of
Carbohydrates yield 17 kJ (4 kcal)	Energy
Fats yield 38 kJ (9 kcal)	per gram

Most of this appears as HEAT
and is used for
KEEPING BODY WARM;
some is used for WORK of CELLS

Energy requirements
of the individual determine

QUANTITY
of DIET

For a
BALANCED DIET
TOTAL INTAKE
of
Essential Constituents and Energy Units
must balance...
...amounts *stored* plus amounts *lost* from body plus amounts *used* as Work or Heat.

1 kilocaloire (kcal) = 4.2 kilojoules (kJ)
1000 kilojoules = 1 megajoule (MJ)

All the main foodstuffs yield **energy** – the energy originally trapped by plants.
Carbohydrates and **fats** are the chief energy-giving foods. **Proteins** can give energy but are mainly used for building and repairing protoplasm.

CARBOHYDRATES are the
   PRIMARY SOURCE of ENERGY –
More easily and quickly digested
and utilized than fats.

PLANT SOURCES:
   SUGAR is found in
   leaves, fruits and **roots** of **plants**

– *and in*
foods made from them by man
e.g. jam, treacle,
sweets, syrup

i.e. especially those products
made by plants for development
of next generation.
   Sugar can be stored in plants as
   – – – – – CARBOHYDRATE or converted to FAT and stored as such

STARCH is found in
grain, seeds and **roots** of **plants**
e.g.   e.g.
wheat   potatoes

FATS are a
   SECONDARY SOURCE of ENERGY –
Ideal energy storage material. Weight
for weight, give twice as many energy
units as carbohydrates.

OILS in
seeds   **nuts**
e.g. olives,   e.g. peanuts,
cotton seeds   coconuts

and in
foods made from them by man
flour e.g. bread, cakes.
cereals e.g. cornflakes.   cooking fats   peanut butter

ANIMAL SOURCES
SUGAR (glucose) is found in **tissues** and **blood** of **animals** and in
                    foods made from them by man
            e.g chop   e.g. black pudding

and in products made by animals for themselves or the next generation
Milk sugar (lactose)   **milk fat**

Honey (fructose)   milk – – – – – – – – → butter, cheese

Sugar can be stored in animals as
   – – – – CARBOHYDRATE or converted to FAT and stored (together with fat
ANIMAL STARCH (glycogen)   built from Dietary Fatty Acids and
is found in **muscles**   Glycerol) in **fat depots** of body.
      e.g. steak   **liver**   e.g. suet, lard, mutton-fat, fish oil

            and in **foods** made from them by man
      e.g. sausages   e.g. margarine

BODY-BUILDING FOODS

PROTEIN is the chief body-building food. Because it is the chief constitutent of protoplasm, tissues of plants and animals are the richest sources.

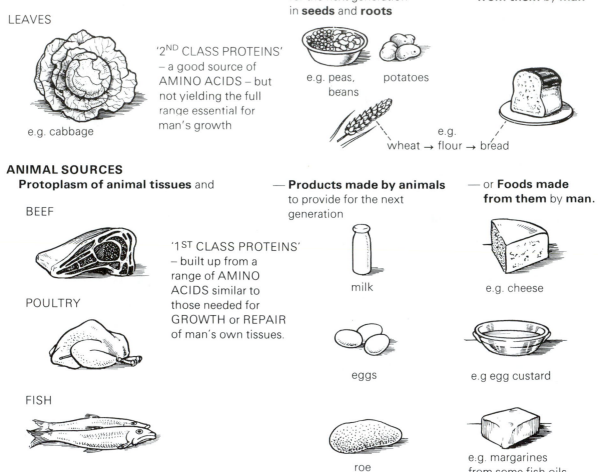

PLANT SOURCES
Protoplasm of plant tissues and

LEAVES

'2ND CLASS PROTEINS' – a good source of AMINO ACIDS – but not yielding the full range essential for man's growth

e.g. cabbage

— **Stores made by plants** for the next generation in **seeds** and **roots**

e.g. peas, beans   potatoes

wheat → flour → bread

— or **Foods made from them** by **man**

e.g.

ANIMAL SOURCES
Protoplasm of animal tissues and

BEEF

'1ST CLASS PROTEINS' – built up from a range of AMINO ACIDS similar to those needed for GROWTH or REPAIR of man's own tissues.

POULTRY

FISH

— **Products made by animals** to provide for the next generation

milk

eggs

roe

— or **Foods made from them** by **man.**

e.g. cheese

e.g egg custard

e.g. margarines from some fish oils

These foods between them also contain other important body-building elements:– e.g. **calcium** and **phosphorus** for making **bones** and **teeth** hard; **iron** for building **haemoglobin** – the **oxygen carrying** substance in **red blood cells**; **iodine** for building the **thyroid hormone**.

Even when provided with body-building and energy-giving foods, human tissues cannot build or repair their protoplasm or release their energy in the absence of organic subtances called *vitamins*. These substances cannot be manufactured in the cells of the body. They were thought to be 'amines essential to life', hence named *vitamine*. This name was changed to *vitamin* when it was found that they were not all amines. They are essential to complete the metabolism of fats, proteins and carbohydrates. *Vitamin* deficient diets can cause specific metabolic defects:

FAT-SOLUBLE VITAMINS	**A** ANTIXEROPH-THALMIC	**D** ANTI-RACHITIC	**E** ANTISTERILITY (α tocopherol)	**K** ANTI-HAEMORRHAGIC
	(Both A_1 and A_2) Protects **skin** and **mucous membranes** (especially front of eye and lining of digestive and respiratory tracts) Essential for regeneration of **visual purple**.	(also D_2 and D_3) Important in Ca and P Metabolism. Essential for deposition Ca and P in bones and teeth. Promotes absorption of Ca from intestine.	Perhaps important for normal reproduction in adult. Prevents oxidation of unsaturated fats.	Essential for production of Prothrombin and Factors VII, IX and X in liver- important for normal blood clotting.
PLANT SOURCES	Present as precursor **carotene**.	Vegetables, fruits and cereals contain negligible amounts.	Green leaves (e.g. lettuce) peas. Richest source-germ of various Cereals e.g. wheat germ oil.	Spinach, kale,. cabbage, cauliflower, cereals, tomatoes, carrots, potatoes.
Leaves –	Green, yellow vegetables (esp. spinach and kale).			
Seeds & Fruits –	Yellow maize, peas, beans.			
Roots –	Carrots.			
ANIMAL SOURCES	Animal liver breaks down **carotene** ($C_{40}H_{56} + 2H_2O \rightarrow 2C_{20}H_{30}O$)	Formed when ultra-violet rays fall on Sterols in man's skin Stored in liver of	Small amounts in meat and dairy produce	Synthesized by bacteria in man's intestine then absorbed in presence of bile.
Tissues –	Stored in **liver** of animals and **fish**	**fish** and animals		
Products for Progeny –	Secreted in milk egg yolk	cod liver oil Milk Eggs		
Foods made from them –	butter cream	Butter Cream Cheese } Vitamin content of dairy produce varies with plant in animal's diet.		
DIETARY DEFICIENCY of any or all vitamins retards **growth** and leads to **severe disease** and eventually **death**.	**XEROPHTHALMIA** Corneal epithelium thickened, dry and infected Night blindness. Dry skin. Epithelia have low resistance to infection.	**RICKETS** in children **OSTEOMALACIA** in **ADULTS** Bones become soft and deformed.	Little evidence of deficiency syndrome in human beings. (In some female animals developing embryos die, and, in male, testes may show degenerative changes.) May cause abnormalities of mitochondria, lysosomes and plasma membrane.	Antibiotic drugs cause deficient absorption from intestine and lead to upset in mechanism for blood clotting.

VITAMINS II

WATER-SOLUBLE VITAMINS

B Complex

Together form essential parts of tissue **enzyme systems** – concerned in metabolism and release of energy by the cells

Includes
THIAMINE (B$_1$)
RIBOFLAVIN (B$_2$)
NIACIN (nicotinic acid)
Each forms part of a coenzyme system – concerned with release of energy from foodstuffs.

PYRIDOXINE (B$_6$) – Coenzyme in synthesis of certain amino acids.

PANTOTHENIC ACID – Constituent of coenzyme A.

BIOTIN – Catalyses CO_2 incorporation into organic molecules.

CHOLINE – Probably necessary for fat utilization.

FOLIC ACID – Necessary for DNA synthesis, normal red blood cell formation.

VITAMIN B$_{12}$ (cyanocobalamin) – Necessary for DNA synthesis Antipernicious Anaemia factor. Involved in maturation of red cells

C ASCORBIC ACID

Activates an enzyme involved in forming a constituent of collagen. Essential for formation and maintenance of **intercellular cement** and **connective tissue**. Especially necessary for healthy blood vessels, wound healing, bone growth.

Green vegetables, e.g. Parsley. Citrus fruits, e.g. oranges, lemons, grapefruits, limes Tomatoes, rosehips, blackcurrants, red & green peppers. Peas. Turnips. Potatoes.

Stored in body – high concentration in adrenal glands. Found in meat, liver. Secreted in milk.

PLANT SOURCES

Common natural sources for most members of B Complex:-
Pulses, e.g. green peas (B$_1$, B$_2$).
Seeds and outer coats of grain, e.g. rice, wheat (B$_1$, B$_2$, Niacin).
Nuts, e.g. peanuts (B$_1$).
Yeast and yeast extracts (B$_1$, B$_2$, Niacin).

ANIMAL SOURCES

B$_2$ and Niacin are synthesized by bacteria in the human intestine. Found in liver (B$_1$, B$_2$, niacin, B$_{12}$, etc.), bacon and lean meat, meat extract, milk (B$_2$, Niacin).
Eggs (B$_2$)
Cheese (B$_2$)

DIETARY DEFICIENCY

Deficiency in Pyridoxine, Pantothenic acid, Biotin, Choline not observed in man

B$_1$ BERI-BERI

Gross disturbance of function of **nervous tissue** with great **muscular paralysis** and **weakness** and disturbance in **sensation. Oedema.** Gastro-intestinal upsets. **Heart failure**

B$_{12}$ PERNICIOUS ANAEMIA

Members of B Complex are frequently absent together, giving grave disturbances of chemical reactions in all tissues.

B$_2$ RIBOFLAVINOSIS

Roughening of skin. Cornea becomes cloudy. Cracks and fissures around lips & tongue.

NIACIN PELLAGRA

Roughening and reddening of skin. Tongue red and sore. In severe cases – gastro-intestinal upsets and mental derangement. **FOLIC ACID -** macrocytic anaemia

SCURVY

Intercellular cement breaks down. Capillary walls leak → haemorrhages into tissues – e.g. gums swell and bleed easily. Wounds heal slowly.

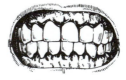

The organic substances of man's food are chemically *similar* to those which his body will form from them. In detail they *differ*.

Conversion of FOOD SUBSTANCE – to – BODY SUBSTANCE

Chemical **BREAKDOWN**
to simpler organic units
in **DIGESTIVE TRACT** under the action of

ENZYMES

PEPSIN

TRYPSIN
CHYMOTRYPSIN

e.g.
PROTEIN: EGG -- breakdown --> **AMINO ACIDS** -- -> Absorption to blood stream.
contains protein –   Synthesis by Body's Cells to
ALBUMIN   e.g. **SERUM ALBUMIN**
                                                              in liver
              Both **Proteins** built up
          from about 20 Amino Acids

AMYLASE ⎰ Saliva in Mouth
        ⎱ Pancreatic Juice
          (*Intestinal Juice*)

LACTASE, SUCRASE
α-DEXTRINASE (iso-maltase)
GLUCOAMYLASE

CARBOHYDRATE: **POTATO** -- breakdown --> **GLUCOSE** -- -- -> Absorption: Synthesis by
contains   (MONOSACCHARIDE)   muscle and liver cells to
STARCH   **GLYCOGEN**

              Both **Polysaccharides** built
              up from Glucose units

LIPASE

LIPID: **BUTTER** -- Partial breakdown --> **GLYCERIDES** -- -> Absorption to lymph or blood
              or   stream: Synthesis to
              Total breakdown --> **FATTY ACIDS**   **BODY FAT**
                                  and **GLYCEROL**

              Both **Triglycerides** built up
              from Fatty Acids and Glycerol

DIGESTION is brought about by **SPECIFIC ENZYMES** themselves made of **Protein** – each acts as a **CATALYST** for speeding up one particular chemical breakdown without effect on any others.

PROTEIN METABOLISM

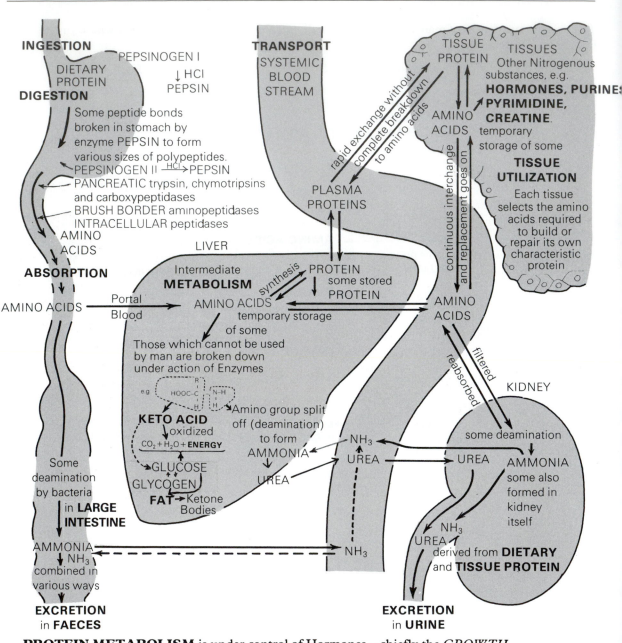

INGESTION

DIETARY PROTEIN

PEPSINOGEN I

↓ HCl

PEPSIN

DIGESTION

Some peptide bonds broken in stomach by enzyme PEPSIN to form various sizes of polypeptides.

PEPSINOGEN II —HCl→ PEPSIN

PANCREATIC trypsin, chymotripsins and carboxypeptidases

BRUSH BORDER aminopeptidases

INTRACELLULAR peptidases

AMINO ACIDS

ABSORPTION

AMINO ACIDS

Portal Blood

TRANSPORT

SYSTEMIC BLOOD STREAM

rapid exchange without complete breakdown to amino acids

PLASMA PROTEINS

LIVER

Intermediate **METABOLISM**

synthesis

PROTEIN some stored PROTEIN

AMINO ACIDS temporary storage of some

Those which cannot be used by man are broken down under action of Enzymes

e.g. $HOOC-C \begin{smallmatrix} R \\ | \\ | \\ H \end{smallmatrix} \begin{smallmatrix} N-H \\ | \\ H \end{smallmatrix}$

KETO ACID ↓oxidized

$CO_2 + H_2O + ENERGY$

→GLUCOSE

GLYCOGEN

FAT→Ketone Bodies

Amino group split off (deamination) to form AMMONIA ↓ UREA

TISSUE PROTEIN

TISSUES Other Nitrogenous substances, e.g.

HORMONES, PURINES PYRIMIDINE, CREATINE. temporary storage of some

AMINO ACIDS

TISSUE UTILIZATION

Each tissue selects the amino acids required to build or repair its own characteristic protein

continuous interchange and replacement goes on

AMINO ACIDS

filtered

reabsorbed

KIDNEY

some deamination

NH_3

UREA

UREA

AMMONIA some also formed in kidney itself

NH_3 UREA↗ derived from **DIETARY** and **TISSUE PROTEIN**

Some deamination by bacteria

in **LARGE INTESTINE**

AMMONIA ↓ NH_3 combined in various ways

EXCRETION in **FAECES**

NH_3

EXCRETION in **URINE**

PROTEIN METABOLISM is under control of Hormones – chiefly the *GROWTH HORMONE (SOMATOTROPHIN)* of the **ANTERIOR PITUITARY**

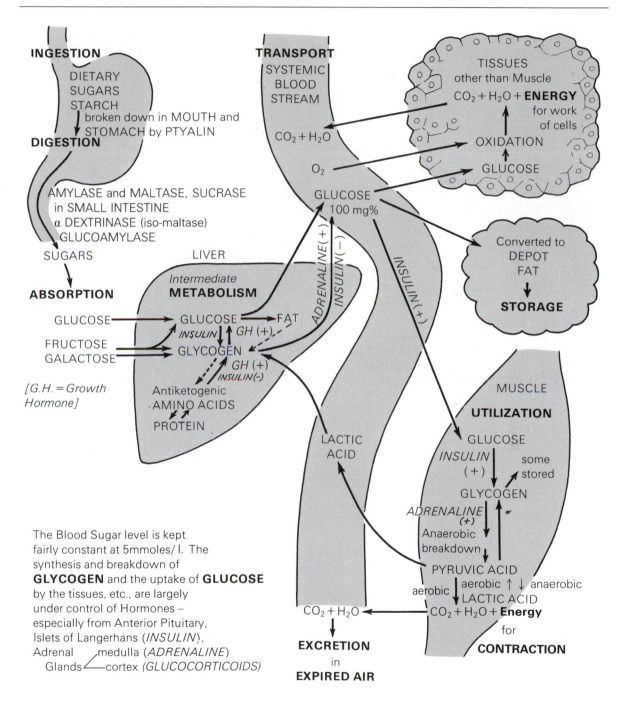

INGESTION

DIETARY
SUGARS
STARCH
broken down in MOUTH and
STOMACH by PTYALIN

DIGESTION

AMYLASE and MALTASE, SUCRASE
in SMALL INTESTINE
α DEXTRINASE (iso-maltase)
GLUCOAMYLASE

SUGARS

ABSORPTION

GLUCOSE

FRUCTOSE
GALACTOSE

[G.H. = Growth Hormone]

LIVER

Intermediate
METABOLISM

GLUCOSE → FAT
INSULIN GH (+)
GLYCOGEN
GH (+)
INSULIN (−)

Antiketogenic
AMINO ACIDS

PROTEIN

TRANSPORT

SYSTEMIC
BLOOD
STREAM

$CO_2 + H_2O$

O_2

GLUCOSE
100 mg%

ADRENALINE (+)
INSULIN (−)

INSULIN (+)

LACTIC
ACID

TISSUES
other than Muscle
$CO_2 + H_2O +$ **ENERGY**
for work
of cells

OXIDATION

GLUCOSE

Converted to
DEPOT
FAT

STORAGE

MUSCLE

UTILIZATION

GLUCOSE

INSULIN
(+)
some
stored

GLYCOGEN

ADRENALINE
(+)
Anaerobic
breakdown

PYRUVIC ACID
aerobic ↑ ↓ anaerobic
aerobic LACTIC ACID
$CO_2 + H_2O +$ **Energy**
for
CONTRACTION

$CO_2 + H_2O$ ←

EXCRETION
in
EXPIRED AIR

The Blood Sugar level is kept
fairly constant at 5mmoles/l. The
synthesis and breakdown of
GLYCOGEN and the uptake of **GLUCOSE**
by the tissues, etc., are largely
under control of Hormones –
especially from Anterior Pituitary,
Islets of Langerhans (INSULIN),
Adrenal   medulla (ADRENALINE)
    Glands   cortex (GLUCOCORTICOIDS)

FAT METABOLISM

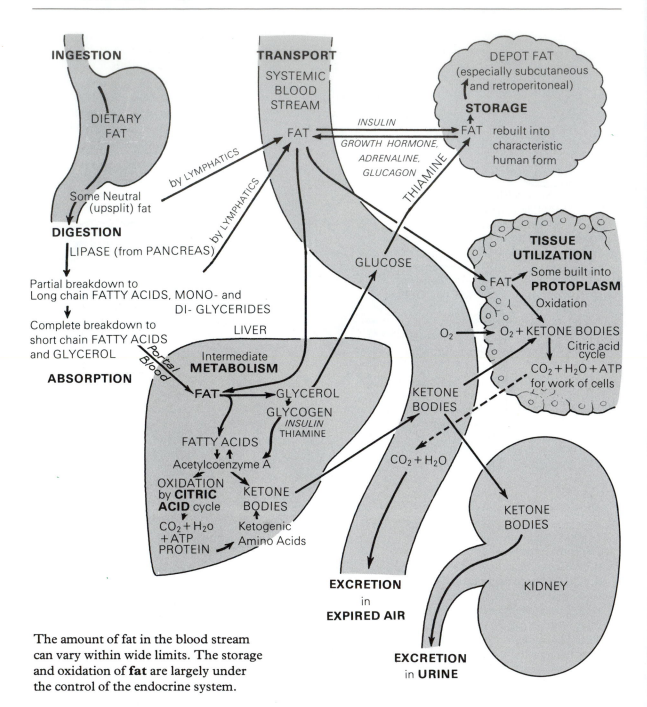

INGESTION

DIETARY FAT

Some Neutral (upsplit) fat

DIGESTION

LIPASE (from PANCREAS)

Partial breakdown to Long chain FATTY ACIDS, MONO- and DI- GLYCERIDES

Complete breakdown to short chain FATTY ACIDS and GLYCEROL

ABSORPTION

Portal Blood

LIVER

Intermediate **METABOLISM**

FAT → GLYCEROL

GLYCOGEN

INSULIN THIAMINE

FATTY ACIDS

Acetylcoenzyme A

OXIDATION by **CITRIC ACID** cycle

$CO_2 + H_2O$ +ATP PROTEIN → KETONE BODIES Ketogenic Amino Acids

by LYMPHATICS

by LYMPHATICS

TRANSPORT

SYSTEMIC BLOOD STREAM

FAT

INSULIN

GROWTH HORMONE, ADRENALINE, GLUCAGON

THIAMINE

GLUCOSE

KETONE BODIES

$CO_2 + H_2O$

EXCRETION in **EXPIRED AIR**

DEPOT FAT (especially subcutaneous and retroperitoneal)

STORAGE

FAT rebuilt into characteristic human form

TISSUE UTILIZATION

FAT → Some built into **PROTOPLASM** Oxidation

O_2 → O_2 + KETONE BODIES Citric acid cycle

$CO_2 + H_2O$ +ATP for work of cells

KETONE BODIES

KIDNEY

EXCRETION in **URINE**

The amount of fat in the blood stream can vary within wide limits. The storage and oxidation of **fat** are largely under the control of the endocrine system.

40

Organic molecules have chemical energy locked in their structure. This energy can be transferred to adenosine triphosphate (ATP) when food molecules are broken down. From ATP the energy can be transferred to operate energy-requiring cell functions, e.g. muscle contraction, the active transport of molecules across membranes, etc.

Proteins, carbohydrates and **fats** can all provide energy for cells through ATP synthesis. In addition the products (**intermediates**) of *each* of these types of molecule can, to a large extent, provide the raw materials necessary to synthesize members of other classes. NB: the *two-way* arrows:

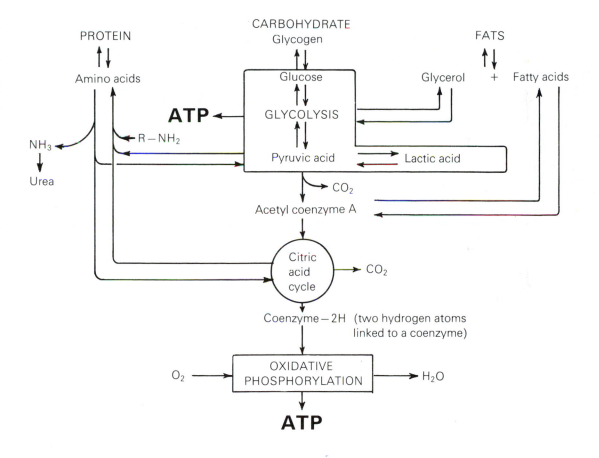

The main mechanism for producing ATP is **oxidative phosphorylation**. This occurs when **oxygen** is available. ATP can also be produced by **GLYCOLYSIS**, a process in which carbohydrate is broken down to **lactic acid** under **anaerobic** (without oxygen) conditions.

FORMATION OF ATP

ENZYME SYSTEMS exist within cells which can convert (in a series of steps) Fats, Proteins and Carbohydrates into intermediate compounds suitable for entering the **'ENERGY-PRODUCING' CITRIC ACID CYCLE** (Krebs cycle).

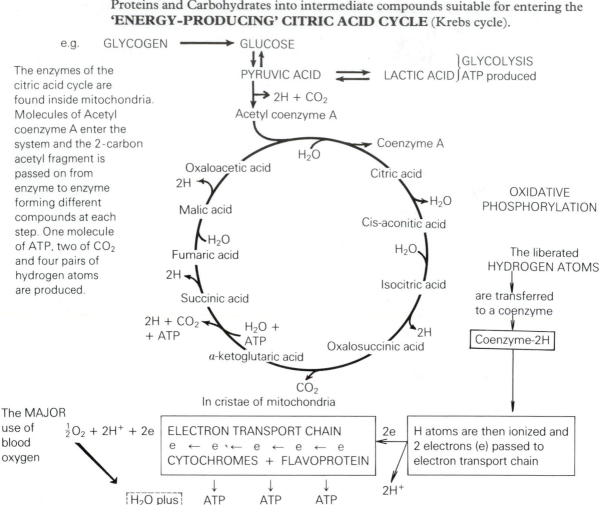

e.g.   GLYCOGEN ⟶ GLUCOSE

The enzymes of the citric acid cycle are found inside mitochondria. Molecules of Acetyl coenzyme A enter the system and the 2-carbon acetyl fragment is passed on from enzyme to enzyme forming different compounds at each step. One molecule of ATP, two of CO_2 and four pairs of hydrogen atoms are produced.

PYRUVIC ACID ⇌ LACTIC ACID | GLYCOLYSIS ATP produced

→ 2H + CO_2

Acetyl coenzyme A

→ Coenzyme A

H_2O

Oxaloacetic acid

2H

Malic acid

H_2O

Fumaric acid

2H

Succinic acid

2H + CO_2 + ATP

H_2O + ATP

α-ketoglutaric acid

Citric acid

H_2O

Cis-aconitic acid

H_2O

Isocitric acid

2H

Oxalosuccinic acid

CO_2

In cristae of mitochondria

OXIDATIVE PHOSPHORYLATION

The liberated HYDROGEN ATOMS

are transferred to a coenzyme

Coenzyme-2H

The MAJOR use of blood oxygen

$\frac{1}{2}O_2 + 2H^+ + 2e$

ELECTRON TRANSPORT CHAIN		2e	H atoms are then ionized and 2 electrons (e) passed to electron transport chain
e ← e ← e ← e ← e			
CYTOCHROMES + FLAVOPROTEIN			

H₂O plus   ATP   ATP   ATP

2H⁺

The **ENERGY** produced in the electron transport chain is used to link inorganic phosphate to ADP (adenosine diphosphate) to form the energy-rich compound ATP. The energy 'trapped' in ATP is used as required.

For example:

(a) for MEMBRANE TRANSPORT. Sodium, potassium etc. require the expenditure of energy to transport them across cell membranes.

(b) for SYNTHESIS of CHEMICAL COMPOUNDS. Many thousands of ATP molecules must release their energy to form one protein molecule.

(c) for MECHANICAL WORK. Contraction of a muscle fibre requires expenditure of tremendous quantities of ATP.

The energy stored in food is thus released by cells to make their own energy-rich phosphorus compound — **ADENOSINE TRIPHOSPHATE** (ATP).

Energy is released in cells by **oxidation**. It is used for work and to keep the body warm. The body temperature is kept relatively constant (with a slight fluctuation throughout the 24 hours) in spite of wide variations in environmental temperature and heat production.

HEAT PRODUCTION ————————→ *must balance* ————————→ **HEAT LOSS**

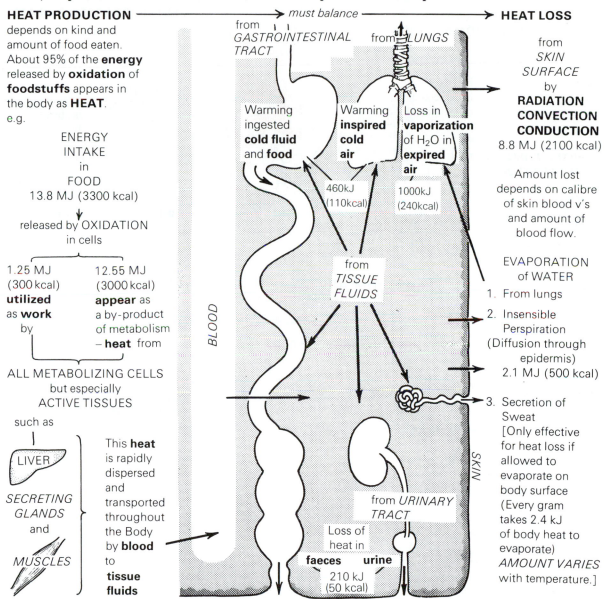

HEAT PRODUCTION

depends on kind and amount of food eaten. About 95% of the **energy** released by **oxidation** of **foodstuffs** appears in the body as **HEAT**. e.g.

ENERGY
INTAKE
in
FOOD
13.8 MJ (3300 kcal)

↓

released by OXIDATION
in cells

| 1.25 MJ (300 kcal) **utilized** as **work** by | 12.55 MJ (3000 kcal) **appear** as a by-product of metabolism – **heat** from |

ALL METABOLIZING CELLS
but especially
ACTIVE TISSUES

such as

LIVER

SECRETING GLANDS and *MUSCLES*

This **heat** is rapidly dispersed and transported throughout the Body by **blood** to **tissue fluids**

BLOOD

from
GASTROINTESTINAL TRACT

Warming ingested **cold fluid** and **food**

from *LUNGS*

Warming **inspired cold air**

Loss in **vaporization** of H_2O in **expired air**

460kJ (110kcal)   1000kJ (240kcal)

from
TISSUE FLUIDS

SKIN

from *URINARY TRACT*

Loss of heat in **faeces**   **urine**
210 kJ (50 kcal)

HEAT LOSS

from
SKIN SURFACE
by
RADIATION CONVECTION CONDUCTION
8.8 MJ (2100 kcal)

Amount lost depends on calibre of skin blood v's and amount of blood flow.

EVAPORATION of WATER

1. From lungs

2. Insensible Perspiration (Diffusion through epidermis) 2.1 MJ (500 kcal)

3. Secretion of Sweat [Only effective for heat loss if allowed to evaporate on body surface (Every gram takes 2.4 kJ of body heat to evaporate) *AMOUNT VARIES* with temperature.]

[1 kilocalorie = 4.1855 kJ (kilojoules)]

Normal oral temperature = 35.8–37.7°C

MAINTENANCE OF BODY TEMPERATURE

Any tendency for the **body temperature** to *rise*
as by

1. **INCREASED HEAT PRODUCTION** —— is balanced by —— **INCREASED HEAT LOSS**

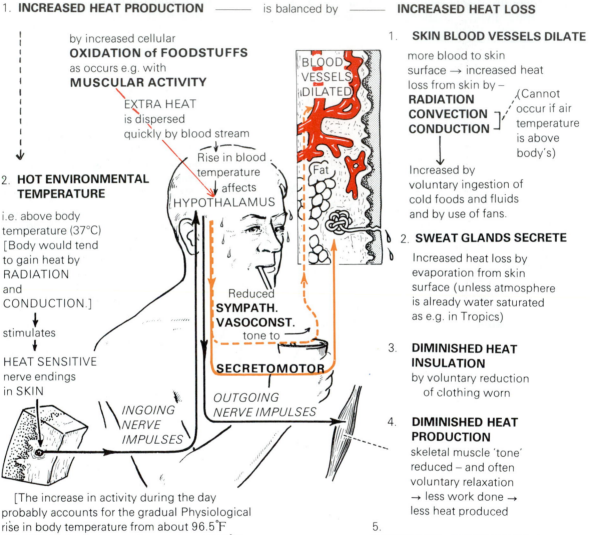

by increased cellular
OXIDATION of FOODSTUFFS
as occurs e.g. with
MUSCULAR ACTIVITY

EXTRA HEAT
is dispersed
quickly by blood stream

Rise in blood
temperature
↓ affects
HYPOTHALAMUS

2. **HOT ENVIRONMENTAL
TEMPERATURE**

i.e. above body
temperature (37°C)
[Body would tend
to gain heat by
RADIATION
and
CONDUCTION.]

↓

stimulates

↓

HEAT SENSITIVE
nerve endings
in SKIN

Reduced
**SYMPATH.
VASOCONST.**
tone to

SECRETOMOTOR

*INGOING
NERVE
IMPULSES*

*OUTGOING
NERVE IMPULSES*

[The increase in activity during the day
probably accounts for the gradual Physiological
rise in body temperature from about 96.5°F
(35.8°C) in the early morning to about 99.2°F
(37.3°C) in the late afternoon.]

Unless exercise is very strenuous or environment
is very hot and humid these measures ⟶

BLOOD
VESSELS
DILATED

Fat

1. **SKIN BLOOD VESSELS DILATE**

more blood to skin
surface → increased heat
loss from skin by –
**RADIATION
CONVECTION
CONDUCTION**

(Cannot
occur if air
temperature
is above
body's)

Increased by
voluntary ingestion of
cold foods and fluids
and by use of fans.

2. **SWEAT GLANDS SECRETE**

Increased heat loss by
evaporation from skin
surface (unless atmosphere
is already water saturated
as e.g. in Tropics)

3. **DIMINISHED HEAT
INSULATION**
by voluntary reduction
of clothing worn

4. **DIMINISHED HEAT
PRODUCTION**
skeletal muscle 'tone'
reduced – and often
voluntary relaxation
→ less work done →
less heat produced

5.
REDUCTION of 'ENERGY INTAKE'
by voluntary restriction of
protein in diet

RESTORE BODY TEMPERATURE
to **normal**

MAINTENANCE OF BODY TEMPERATURE

Any tendency for the **body temperature** to *fall*
as by

1. DIMINISHED HEAT PRODUCTION —— is balanced by —— **DIMINISHED HEAT LOSS**

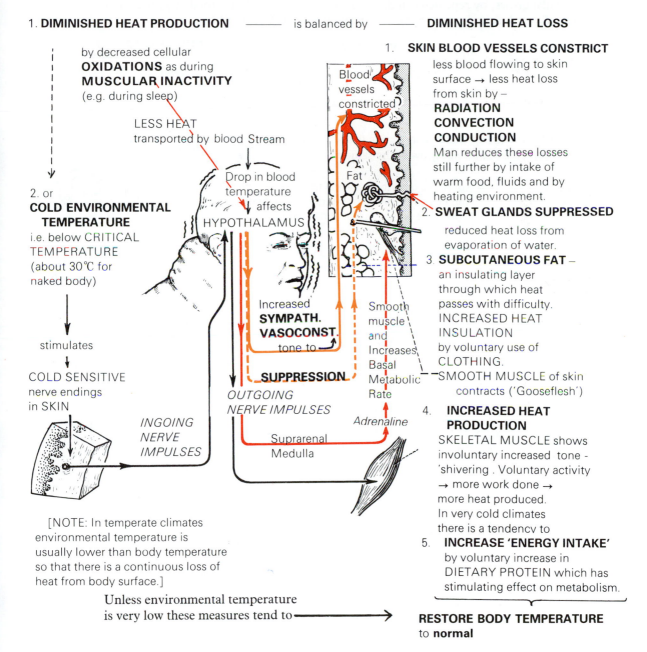

by decreased cellular
OXIDATIONS as during
MUSCULAR INACTIVITY
(e.g. during sleep)

LESS HEAT
transported by blood Stream

Drop in blood
temperature
affects
HYPOTHALAMUS

**2. or
COLD ENVIRONMENTAL
TEMPERATURE**
i.e. below CRITICAL
TEMPERATURE
(about 30°C for
naked body)

stimulates

COLD SENSITIVE
nerve endings
in SKIN

*INGOING
NERVE
IMPULSES*

Increased
**SYMPATH.
VASOCONST.**
tone to

SUPPRESSION

*OUTGOING
NERVE IMPULSES*

Suprarenal
Medulla

Adrenaline

Smooth
muscle
and
Increases
Basal
Metabolic
Rate

Blood
vessels
constricted

Fat

1. SKIN BLOOD VESSELS CONSTRICT
less blood flowing to skin
surface → less heat loss
from skin by –
**RADIATION
CONVECTION
CONDUCTION**
Man reduces these losses
still further by intake of
warm food, fluids and by
heating environment.

2. SWEAT GLANDS SUPPRESSED
reduced heat loss from
evaporation of water.

3. SUBCUTANEOUS FAT –
an insulating layer
through which heat
passes with difficulty.
**INCREASED HEAT
INSULATION**
by voluntary use of
CLOTHING.
SMOOTH MUSCLE of skin
contracts ('Gooseflesh')

**4. INCREASED HEAT
PRODUCTION**
SKELETAL MUSCLE shows
involuntary increased tone -
'shivering . Voluntary activity
→ more work done →
more heat produced.
In very cold climates
there is a tendency to

5. INCREASE 'ENERGY INTAKE'
by voluntary increase in
DIETARY PROTEIN which has
stimulating effect on metabolism.

[NOTE: In temperate climates
environmental temperature is
usually lower than body temperature
so that there is a continuous loss of
heat from body surface.]

Unless environmental temperature
is very low these measures tend to ———→

RESTORE BODY TEMPERATURE
to **normal**

GROWTH

Each individal grows, by repeated cell divisions, from a single cell to a total of 75 trillion or more cells. Growth is most rapid before birth and during 1ST year of life.

The proportion of ENERGY INTAKE in food used to build and maintain tissue	INFANCY		CHILDHOOD		
	At Birth	3 months	1 year	2 years	9-11 years
——40%	40%	20%	20%	4-10%
Average WEIGHT——3 kg	5 kg	10.4 kg	12.4 kg	27.1 kg
Average HEIGHT——50 cm	58 cm	73 cm	84 cm	129 cm

A baby is born with epithelia, connective tissues, muscles, nerves and organs all present and formed – but all tissues do not grow at the same rate.

Differential ———— e.g. Rapid growth of skeletal tissue during childhood.
growth and   Nervous tissue develops rapidly in first 2 years.
functional   Most rapid growth is first at the head
development   then legs begin to lengthen.
of **tissues**
            Chiefly **PROTEIN** being laid down
lead to   or retained

change in
body
proportions

Lymphoid Tissue Growth Spurt 10 YEARS

2 YEARS

1 YEAR

3 MONTHS

At BIRTH

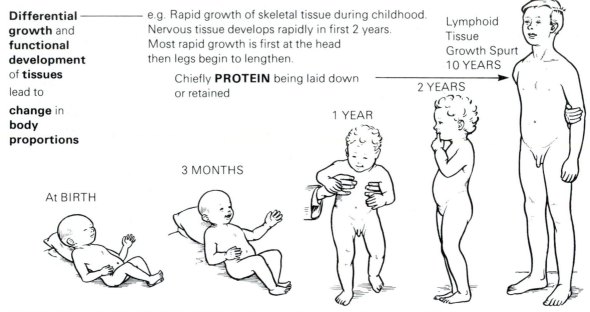

FIRST (Neutral) GROWTH PHASE (Infancy to Puberty). No marked difference between
  sexes – Regulated by *Growth-Stimulating Hormone* of Anterior Pituitary – stimulates
  Growth of all Tissues and Organs.
FACTORS INFLUENCING NORMAL GROWTH:
  Heredity: To large extent rate of growth/sequence of events is predetermined.
        Inherited factors control pattern and limitations of growth.
  Environment: Nutrition: For optimal growth the body requires an adequate and
          balanced diet.
  Endocrine glands: All play part: mental as well as physical development is influenced.

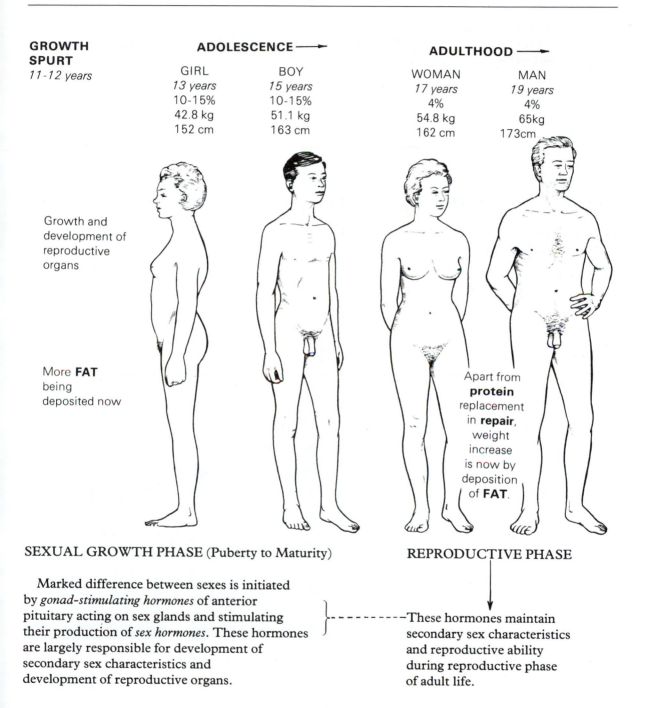

GROWTH SPURT
11-12 years

ADOLESCENCE ⟶

GIRL	BOY
13 years	*15 years*
10-15%	10-15%
42.8 kg	51.1 kg
152 cm	163 cm

ADULTHOOD ⟶

WOMAN	MAN
17 years	*19 years*
4%	4%
54.8 kg	65kg
162 cm	173cm

Growth and development of reproductive organs

More **FAT** being deposited now

Apart from **protein** replacement in **repair**, weight increase is now by deposition of **FAT**.

SEXUAL GROWTH PHASE (Puberty to Maturity)

REPRODUCTIVE PHASE

Marked difference between sexes is initiated by *gonad-stimulating hormones* of anterior pituitary acting on sex glands and stimulating their production of *sex hormones*. These hormones are largely responsible for development of secondary sex characteristics and development of reproductive organs.

These hormones maintain secondary sex characteristics and reproductive ability during reproductive phase of adult life.

ENERGY REQUIREMENTS — MALE

DAILY FOOD INTAKE
must supply **total energy requirements**
for **1. ACTIVITY**

SPECIAL – individual requirements vary with *type of work or play* and, from minute to minute, on the **intensity** of **work** and frequency and length of **rest pauses**.

EVERYDAY ACTIVITIES – such as **sitting, standing, walking, etc.**

2. SPECIFIC DYNAMIC ACTION of FOOD (S.D.A.)
The mere taking of food stimulates metabolism of cells so that heat production increases (30% by protein, 4-5% by carbohydrate and fat).
Must allow 10% above basal requirements on average mixed diet.

3. BASAL METABOLISM
(measured when body is at rest)
Energy expenditure of cells doing 'vegetative' processes of living – e.g. tasks involved in **respiration, circulation, digestion, excretion, secretion, synthesis** of special substances, **keeping body temperature** at 37°C, **growth** and **repair**.

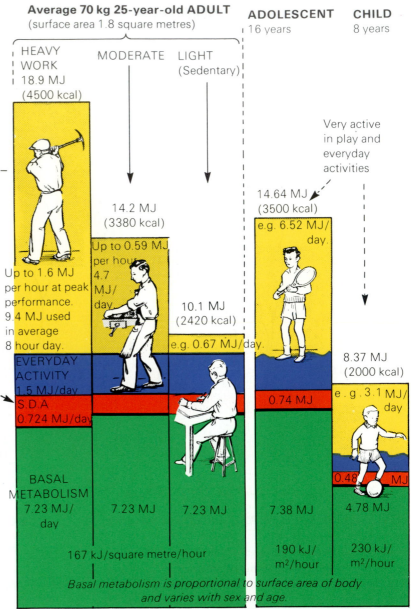

Average 70 kg 25-year-old ADULT
(surface area 1.8 square metres)

ADOLESCENT
16 years

CHILD
8 years

HEAVY WORK
18.9 MJ
(4500 kcal)

MODERATE   LIGHT
(Sedentary)

Very active in play and everyday activities

14.2 MJ
(3380 kcal)

14.64 MJ
(3500 kcal)

Up to 0.59 MJ per hour
4.7 MJ/day

e.g. 6.52 MJ/day.

Up to 1.6 MJ per hour at peak performance. 9.4 MJ used in average 8 hour day.

10.1 MJ
(2420 kcal)

e.g. 0.67 MJ/day.

8.37 MJ
(2000 kcal)

EVERYDAY ACTIVITY
1.5 MJ/day

e.g. 3.1 MJ/day

S.D.A
0.724 MJ/day

0.74 MJ

BASAL METABOLISM
7.23 MJ/day

7.23 MJ

7.23 MJ

7.38 MJ

4.78 MJ

0.48 MJ

167 kJ/square metre/hour

190 kJ/m²/hour

230 kJ/m²/hour

Basal metabolism is proportional to surface area of body and varies with sex and age.

All figures are approximate and intended only as a general guide.

Work or Play

Everyday activities

Specific Dynamic Action

Basal Metabolism

48

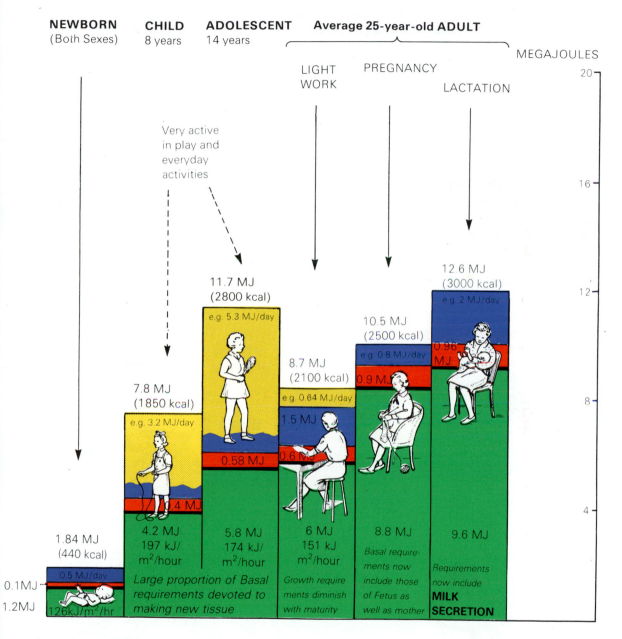

Proportion needed for growth diminishes in both sexes with age.

Because a person's Basal Metabolic Rate is not constant, to find total daily energy expenditure, some researchers now measure Resting Metabolic Rate (in bed) = $293 \times$ (body weight in kg)$^{0.75}$ + energy output at work + non-occupational work output.

BALANCED DIET

The individual's daily energy requirements are best obtained by eating well-balanced meals which contain the essential **body-building** and **energy-giving** foods plus **vitamins** and **minerals**.

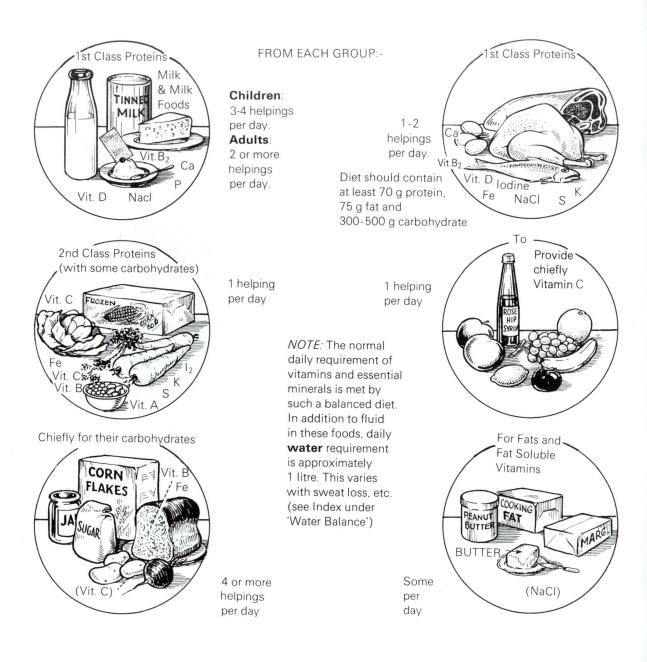

FROM EACH GROUP:-

1st Class Proteins
Milk & Milk Foods
Vit.B₂ — Ca — P
Vit. D — NaCl

Children:
3-4 helpings per day.
Adults:
2 or more helpings per day.

1st Class Proteins
1-2 helpings per day.
Ca — Vit.B₂ — Vit. D — Iodine — Fe — NaCl — S — K

Diet should contain at least 70 g protein, 75 g fat and 300-500 g carbohydrate

2nd Class Proteins
(with some carbohydrates)
Vit. C — FROZEN CORN COB — Fe — Vit. C — Vit. B — I₂ — K — S — Vit. A

1 helping per day

1 helping per day

To Provide chiefly Vitamin C
ROSE HIP SYRUP

NOTE: The normal daily requirement of vitamins and essential minerals is met by such a balanced diet. In addition to fluid in these foods, daily **water** requirement is approximately 1 litre. This varies with sweat loss, etc. (see Index under 'Water Balance')

Chiefly for their carbohydrates
CORN FLAKES — Vit. B — Fe — JAM — SUGAR — (Vit. C)

4 or more helpings per day

Some per day

For Fats and Fat Soluble Vitamins
PEANUT BUTTER — COOKING FAT — MARG. — BUTTER
(NaCl)

CHAPTER 3

CELL MEMBRANE FUNCTIONS

TRANSPORT THROUGH MEMBRANES I – DIFFUSION

The PLASMA MEMBRANE of a cell is semipermeable and consists of a double layer of phospholipids with protein molecules embedded in it (see p. 6). Transport through membranes takes several forms.

SIMPLE DIFFUSION

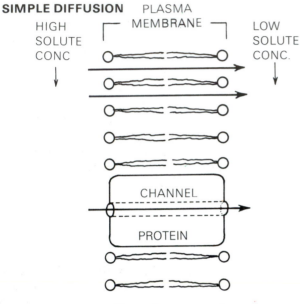

Channels are opened or closed by a change in shape of the protein.

Small uncharged molecules can diffuse between the phospholipid molecules of the membrane by random thermal motion. This requires a **concentration** gradient. Ions can move down both a concentration and an electrical gradient, i.e. an **electrochemical** gradient. Because of their charge, ions *cannot* move between the phospholipid molecules. However some membrane proteins span the whole membrane and can form in their structure water-filled **channels** or **pores** which allow the passage of ions across the membrane, e.g. Na^+, K^+, Cl^-, Ca^{2+}.

Movement through some channels is altered by the membrane potential (see p. 56), i.e. **voltage-gated** channels. Others are opened or closed by the binding of a hormone or neurotransmitter (**ligand-gated** channels). Many channels remain open permanently.

FACILITATED DIFFUSION

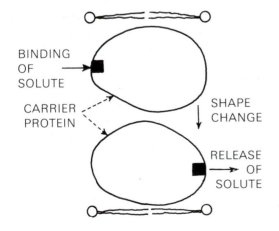

Other membrane proteins are called **carrier** proteins. The solute to be transported binds to the carrier which then changes its shape and by so doing moves the solute to the other side of the membrane. Glucose is an important substance transported by this mechanism. What form the change in shape takes is unknown. The diagram is not meant to represent the shape change – only the result of the change.

TRANSPORT THROUGH MEMBRANES II – ACTIVE TRANSPORT

Carrier proteins are involved also in transporting molecules 'uphill' *against* an electrochemical gradient from a region of *low* concentration to a region of *high* concentration. Such a mechanism requires the energy of ATP and is called ACTIVE TRANSPORT.

PRIMARY ACTIVE TRANSPORT

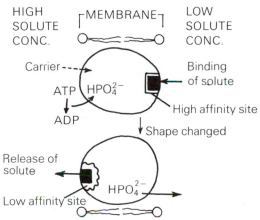

The process is called **PRIMARY ACTIVE TRANSPORT** if ATP is used directly in the mechanism. ATP produces a **high affinity** binding site on the carrier protein on the **low solute** concentration side of the membrane. The transported solute thus binds *tightly* to the carrier – the carrier then changes its shape – the binding site becomes a **low affinity site** on the opposite side of the membrane where the concentration of solute is *high* – the solute molecule can therefore dissociate into the high concentration. These active transport mechanisms are often referred to as PUMPS, e.g. Na, K-pump; Ca-pump; H-pump.

SECONDARY ACTIVE TRANSPORT

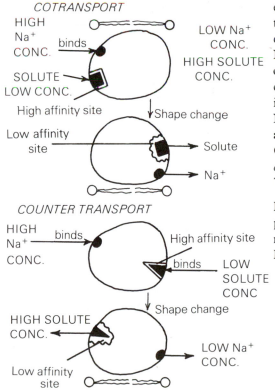

SECONDARY ACTIVE TRANSPORT uses the energy of a concentration gradient (often Na^+) to energize the carrier protein. This is similar to the use of a waterfall to energize a water wheel to perform work. The binding of Na^+ produces a **high affinity** solute binding site on the outside of the membrane where this time solute concentration is **low**. Na^+ and solute are transported to the inside of the membrane by a change in shape of the carrier. Na^+ dissociates – the solute binding site then becomes a **low affinity** site and solute dissociates. This is called COTRANSPORT. A solute can be transported in the *opposite* direction to Na^+. This is called COUNTER TRANSPORT.

ENDOCYTOSIS is the transport of large molecules or particles in a membrane bound vesicle through the membrane from the *outside* to the *inside* of a cell. **EXOCYTOSIS** is similar but in the opposite direction.

IONS AND CHARGES

IONS have electrical charges which may be **positive** or **negative**.
E.g. Na$^+$ is a **postive** ion called a CATION,
     Cl$^-$ is a **negative** ion called an ANION.
   Like the poles of a magnet, *unlike* charges are *drawn towards* each other. *Like* charges *repel* each other.

   If the numbers of **positive** and **negative** charges inside a cell were equal the inside of the cell would be electrically neutral.

INTRACELLULAR FLUID

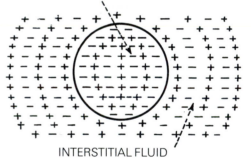

   However, essentially all cells of the body have an *excess* of **negative** charges *inside* the cell and an *excess* of **positive** charges *outside*. Thus a POTENTIAL DIFFERENCE exists between the inside and the outside of the cell. Inside is **negative** to the outside.

INTERSTITIAL FLUID

   The excess **negative** charge *inside* the cell and the excess **positive** charges *outside* the cell are attracted to each other. Hence the excess ions collect in a thin layer on the inside and outside surfaces of the plasma membrane. The bulk of the interstitial and intracellular fluid is electrically neutral. The total number of positive and negative charges that account for the potential difference is a minute fraction of the K$^+$ and Na$^+$ present and therefore cannot be detected chemically. The potential difference across the membrane is measured in millivolts (mV).
   Nerve and muscle cells are 'excitable', i.e. their membrane potential can change rapidly in response to stimulation, and they employ these potential changes to transmit signals along their membranes.

To understand how the **resting membrane potential** of an excitable cell is established it is necessary to consider first what potentials would be produced if the membrane were (a) permeable only to K^+ and (b) permeable only to Na^+.

Inside an excitable cell there is a *high* concentration of K^+, *outside* a *low* concentration. If the membrane was permeable only to K^+ some K^+ would diffuse out of the cell down its concentration gradient carrying **positive** charge thus leaving an excess of **negative** charges inside the cell.

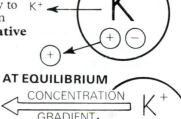

As K^+ moves out of the cell and the inside becomes more **negative** this **negativity**, by its attraction force, begins to oppose further outward movement of **positively** charged K^+. Positive charge **outside** also repels the outward movement of K^+.

When the electric force *opposing* the outward movement of K^+ *equals* the force due to the K^+ concentration difference the potential difference across the membrane is said to be at the **equilibrium potential** for **potassium**. In a nerve or muscle cell this will occur when the *inside* is about 100 mV **negative** to the *outside*. Both greater permeability of the membrane to K^+ and a larger **concentration difference** cause more K^+ to leave the cell. The inside of the cell would then become **more negative** i.e. the potential *difference* would be *larger*.

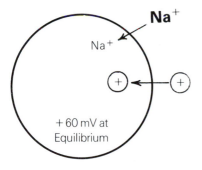

There is a *high* concentration of Na^+ *outside* a cell and a *low* concentration *inside*. If the membrane was **permeable** only to Na^+, diffusion of Na^+ into the cell carrying **positive** charge would make the inside **positive** with respect to the outside. At the **equilibrium potential** for Na^+ (about $+60$ mV) the force due to the concentration difference moving Na^+ inward equals the electrical force repelling Na^+ from the inside of the cell.

RESTING POTENTIAL

In a nerve or muscle cell at rest (not stimulated) there is a **resting membrane potential**, the size of which depends mainly on the permeability of the cell membrane to K^+ and Na^+ and to the concentration of these two ions inside and outside the cell.

In the membrane there are separate **channel proteins** (page 52) for each type of ion. These may be *open* or *closed*. The more channels that are open the greater will be the permeability of the membrane.

The resting membrane is about 75 times more permeable to K^+ than to Na^+, so K^+ diffuses *out* of the cell making the inside **negative** with respect to the outside. At the same time a small amount of Na^+ diffuses *into* the cell cancelling the effect of an equivalent small number of K^+ ions. Because of this effect of Na^+ the resting membrane potential is *not* equal to the **equilibrium potential** for K^+ (100 mV) but is much closer to that value than it is to the equilibrium potential for Na^+ (+60 mV). The resting membrane potential of a nerve cell is about -70 mV and in a skeletal muscle cell about -90 mV.

The concentrations of K^+ and Na^+ inside the cell are kept **constant** by the Na^+, K^+ ATPase pump which actively transports K^+ into the cell and Na^+ out of the cell to balance exactly the **diffusion** of Na^+ into and K^+ out of the cell.

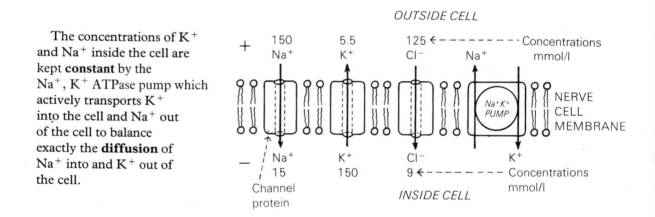

Most but not all cells are relatively permeable to Cl^-. They have Cl^- channels but do not have Cl^- pumps in their membrane. In such cells Cl^- does not help to establish the resting membrane potential. However the electrical force of the membrane potential moves Cl^- to the **positive** outside of the membrane until a concentration gradient of Cl^- builds up which has a diffusion force which equals the electrical force of the membrane potential and stops further net Cl^- movement. If the **permeability** of the membrane to Cl^- *increases*, more Cl^- will move into the cell making the cell **more negative**.

The membrane potential of an excitable cell can be changed by stimulation. If the inside of the cell becomes **less negative** (i.e. the potential difference is decreased) the membrane is said to be **depolarized**. If the inside of the cell becomes **more negative** the membrane is said to be **hyperpolarized** (i.e. the potential difference is increased).

A small localized change in the potential of a membrane can occur with **subthreshold** stimuli. This change spreads only a few millimetres from the point of stimulation and quickly dies out. This is called an **electrotonic potential** or the local response.

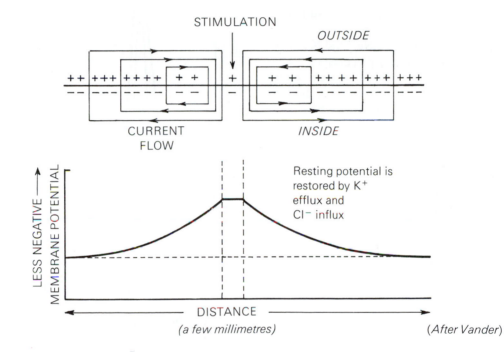

(a few millimetres)   *(After Vander)*

When such a potential change occurs, current flows through the extracellular and intracellular fluids between the stimulated and unstimulated parts of the membrane. The direction of current flow is conventionally regarded as the direction in which **positive** ions move.

At the point of stimulation, current will flow in the extracellular fluid, *outside* the membrane, *towards* the stimulation site (which is less positive), and in the intracellular fluid, *inside* the membrane, *away from* the stimulation site (which is less negative, i.e. more positive).

Action potentials occur only when the membrane potential reaches a level at which the depolarizing forces are *greater than* the repolarizing forces. The membrane potential at which this occurs is called the **threshold potential** or the **firing level**. A stimulus which is just sufficiently large to produce this change is a **threshold stimulus**.

THE ACTION POTENTIAL

An **action potential** is a rapid **reversal** of the resting membrane potential (inside becomes **positive** with respect to the outside). This is followed by a return to the resting membrane potential.

The action potential is the result of changes in the permeability of the membrane, mainly to Na^+ and K^+ ions. In the resting membrane, most of the Na^+ channels are closed. Stimulation of the membrane causes opening of the Na^+ channels, followed slightly later by opening of the K^+ channels. Both these channel types are voltage-gated. More channels open as the membrane potential changes.

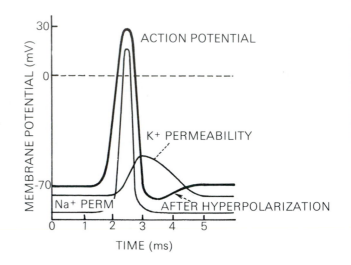

On stimulation, permeability of the membrane to Na^+ increases several hundred fold. Na^+ rushes into the cell and the inside rapidly becomes less and less negative; it reaches **zero** then becomes **positive** as Na^+ continues to enter the cell. The potential does *not* reach the **equilibrium potential** for Na^+ ($+60$ mV) because *some* K^+ channels are *open* (tending to make the inside **negative**).

Closure of the Na^+ channels now occurs, followed by rapid outflow of **positively** charged K^+ **ions**. The membrane potential is thus returned to its resting level. Indeed it may pass the resting level and produce an after **hyperpolarization** due to slow closure of some voltage-gated K^+ channels. The action potential in nerve axons lasts about 1 ms.

Only a minute fraction of the Na^+ and K^+ in the cell is involved in the action potential changes hence many action potentials can occur without a significant change in intracellular ion concentrations. These are continuously being restored by Na^+, K^+ ATPase pumps.

Ca^{2+} is involved in transporting **positive** charge through slow-conducting channels during the action potential in cardiac and smooth muscle.

Local anaesthetics can block nerve conduction by preventing the opening of fast Na^+ channels.

PROPAGATION OF THE NERVE IMPULSE

The **threshold** potential for most excitable cells is about 15 mV **less negative** than the **resting** membrane potential. In a nerve, if the membrane potential decreases from -70 mV to -55 mV the cell fires an action potential which **propagates** along the axon.

An action potential is propagated (i.e. 'handed on') with the same shape and size along the whole length of the axon or muscle cell.

One particular action potential does not itself travel along the membrane. Each action potential **activates** voltage-gated channels in the adjacent membrane and a *new* action potential occurs

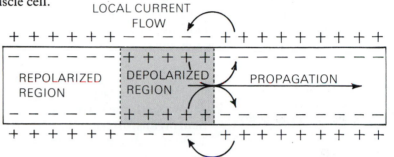

there. This triggers the next region of the membrane and the process is repeated again and again right along the nerve.

The **velocity** of propagation depends on the **diameter** of the fibre and whether or not the fibre is **myelinated**. The *larger* the fibre the *faster* is the propagation.

In **MYELINATED NERVE FIBRES**:
Myelin makes it difficult for currents to flow between intracellular and extracellular fluid. Consequently action potentials only occur where the myelin is interrupted, i.e. at the **nodes of Ranvier**. Thus the nerve impulse is propagated by leaping from **node** to **node**. This method of propagation is called **saltatory conduction**.

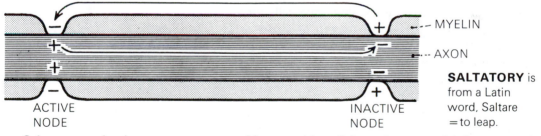

ACTIVE NODE

INACTIVE NODE

SALTATORY is from a Latin word, Saltare = to leap.

Saltatory conduction causes a more rapid propagation of the action potential than occurs in **non-myelinated** fibres of the same axon diameter.

COMMUNICATION BETWEEN CELLS

Cells communicate with one another by sending *messages* in the form of *chemicals* to bring about a change in the activity of the target cell.

A variety of cell types including cardiac muscle and smooth muscle have small channels linking their membranes at **gap junctions**. Small molecules and ions can pass through these gap junctions allowing the spread of electrical activity between the cells.

GAP JUNCTIONS

NEUROTRANSMITTERS

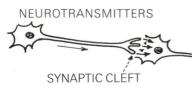

SYNAPTIC CLEFT

Messages can be passed long distances via impulses in nerve axons. The nerve impulses liberate a **neurotransmitter** at the nerve endings which diffuses across a small **synaptic cleft** and activates either another nerve or some other post-synaptic cell.

Other nerves secrete **neurohormones** from their endings into the blood stream. The blood conveys the hormone to cells elsewhere in the body. Such a mechanism is used, e.g. by neurons from the hypothalamus to deliver messages to the pituitary gland.

NEUROHORMONE

BLOOD VESSEL

HORMONE

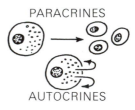

Endocrine glands secrete **hormones** directly into the blood stream which then delivers the message to a large number of cells which are widely distributed. Chemical messengers may be released by cells and diffuse to neighbouring cells through the **interstitial fluid**. Such messengers are called **para-crines**.

PARACRINES

AUTOCRINES

Similarly other chemical messengers act on the cell that secreted them. These are **autocrines**.

Chemical messages activate the correct target cells because the target cells have specific protein molecules called **receptors** (NB *not sensory* receptors, page 235) to which the chemical messenger binds. Many receptors are located in the plasma membrane but some are on the nucleus and some are elsewhere in the cell.

The number of receptors in the membrane is not constant. Excess messenger often causes the number of receptors for that messenger to decrease. This is **down regulation**. A deficiency of chemical messenger can increase the number of receptors. This is **up regulation**.

The binding of chemical messengers to membrane receptors is an initial step which leads ultimately to changed cellular activity, e.g. contraction, secretion, change in electrical potential, etc. The chemical messenger (or **ligand**) which binds to the receptor is called a **first messenger**. This binding triggers changes in the concentration of a mediator *inside* the cell which is vital to produce the change in cellular activity. The **intracellular** mediator is called a **second messenger**. The identity of all second messengers is not yet known but they include Ca^{2+}, cyclic 3', 5'-adenosine monophosphate (cAMP), inositol triphosphate (IP_3), diglyceride and probably cyclic 3', 5'-guanosine monophosphate (cGMP).

Ca^{2+} SYSTEM

The intracellular Ca^{2+} concentration can be increased by diffusion of Ca^{2+} into the cell through ligand-gated or voltage-gated channels (page 52). It may also be released from intracellular Ca^{2+} stores by IP_3 which is formed in the membrane by (a) ligand-receptor binding; (b) activation of phospholipase C via a regulatory protein (G protein) and (c) hydrolysis of phosphatidyl-inositol 4, 5-diphosphate (PIP_2). Ca^{2+} often binds to a Ca^{2+} binding protein **calmodulin** or **tropinin** and this complex changes cellular activity.

cAMP SYSTEM

Intracellular cAMP is increased by ligand-receptor binding which activates **adenylate cyclase** via a G protein. Adenylate cyclase then catalyses the conversion of ATP to cyclic AMP.

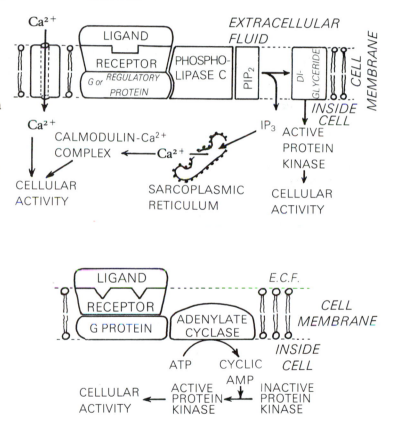

Ca^{2+}, cAMP and diglyceride can activate a class of enzymes called **protein kinases**. These enzymes **phosphorylate**, and by so doing activate a cascade of enzymes, during which the amounts of substances in the cascade are amplified. The end product brings about changed cellular activity, secretion, contraction, etc.

The Ca^{2+} and cAMP systems can interact with one another to stimulate or inhibit each other.

cGMP, another second messenger, has actions opposite to those of cAMP.

CHAPTER 4

DIGESTIVE SYSTEM

DIGESTIVE SYSTEM

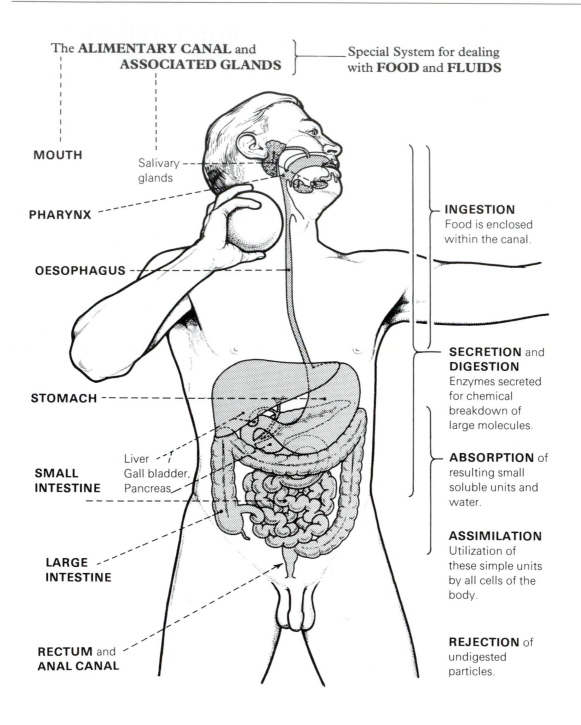

The **ALIMENTARY CANAL** and **ASSOCIATED GLANDS**

Special System for dealing with **FOOD** and **FLUIDS**

MOUTH

Salivary glands

PHARYNX

OESOPHAGUS

STOMACH

Liver
Gall bladder,
Pancreas

SMALL INTESTINE

LARGE INTESTINE

RECTUM and **ANAL CANAL**

INGESTION
Food is enclosed within the canal.

SECRETION and **DIGESTION**
Enzymes secreted for chemical breakdown of large molecules.

ABSORPTION of resulting small soluble units and water.

ASSIMILATION
Utilization of these simple units by all cells of the body.

REJECTION of undigested particles.

PROGRESS OF FOOD ALONG ALIMENTARY CANAL

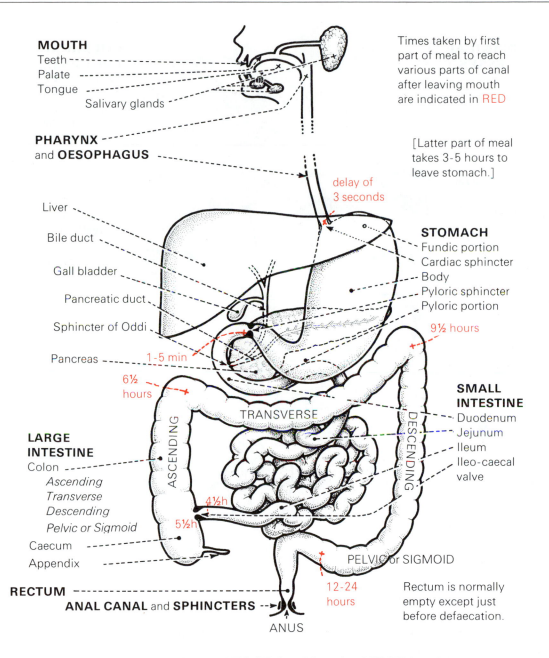

MOUTH
Teeth
Palate
Tongue
Salivary glands

PHARYNX
and **OESOPHAGUS**

Liver
Bile duct
Gall bladder
Pancreatic duct
Sphincter of Oddi
Pancreas

LARGE INTESTINE
Colon
Ascending
Transverse
Descending
Pelvic or Sigmoid
Caecum
Appendix

RECTUM
ANAL CANAL and **SPHINCTERS**
ANUS

Times taken by first part of meal to reach various parts of canal after leaving mouth are indicated in RED

[Latter part of meal takes 3-5 hours to leave stomach.]

delay of 3 seconds

STOMACH
Fundic portion
Cardiac sphincter
Body
Pyloric sphincter
Pyloric portion

9½ hours

1-5 min

6½ hours

TRANSVERSE
ASCENDING
DESCENDING

SMALL INTESTINE
Duodenum
Jejunum
Ileum
Ileo-caecal valve

4½h
5½h

PELVIC or SIGMOID

12-24 hours

Rectum is normally empty except just before defaecation.

During its progress along the canal **FOOD** is subjected to **MECHANICAL** as well as **CHEMICAL** changes to render it suitable for absorption and assimilation.

DIGESTION IN THE MOUTH

MECHANICAL PROCESSES

CHEMICAL PROCESSES

Saliva (1-1½ litres per day) is a slightly acid solution of salts and organic substances secreted mainly by 3 pairs of **salivary glands.**

PAROTID SALIVARY GLAND (serous)

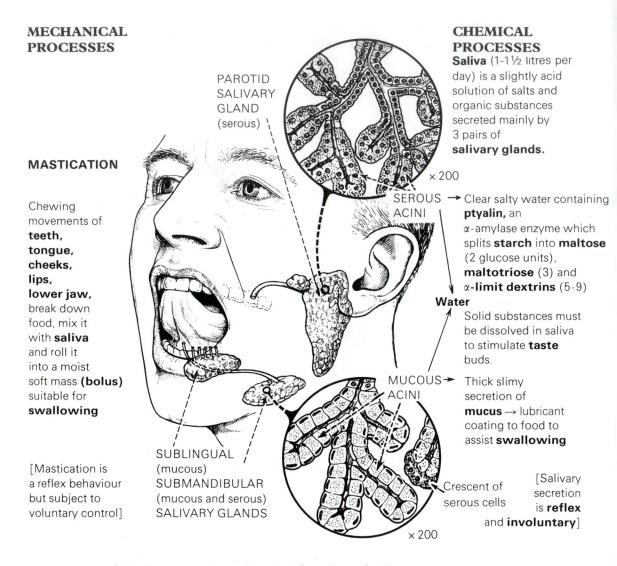

× 200

SEROUS ACINI → Clear salty water containing **ptyalin,** an α-amylase enzyme which splits **starch** into **maltose** (2 glucose units), **maltotriose** (3) and α-**limit dextrins** (5-9)

Water

Solid substances must be dissolved in saliva to stimulate **taste** buds.

MUCOUS → Thick slimy ACINI secretion of **mucus** → lubricant coating to food to assist **swallowing**

MASTICATION

Chewing movements of **teeth, tongue, cheeks, lips, lower jaw,** break down food, mix it with **saliva** and roll it into a moist soft mass **(bolus)** suitable for **swallowing**

[Mastication is a reflex behaviour but subject to voluntary control]

SUBLINGUAL (mucous) SUBMANDIBULAR (mucous and serous) SALIVARY GLANDS

Crescent of serous cells

[Salivary secretion is **reflex** and **involuntary**]

× 200

Other important (non-digestive) functions of **saliva**:-

CLEANSING – Mouth and teeth kept free of debris, etc., to inhibit bacteria.

MOISTENING and **LUBRICATING** – Soft parts of mouth kept pliable for **speech**.

EXCRETORY – Many organic substances (e.g. urea, sugar) and inorganic substances (e.g. mercury, lead) can be excreted in saliva.

Increased secretion at mealtimes is **reflex** (involuntary).
Salivary reflexes are of two types:-

(a) **UNCONDITIONED** (inborn)

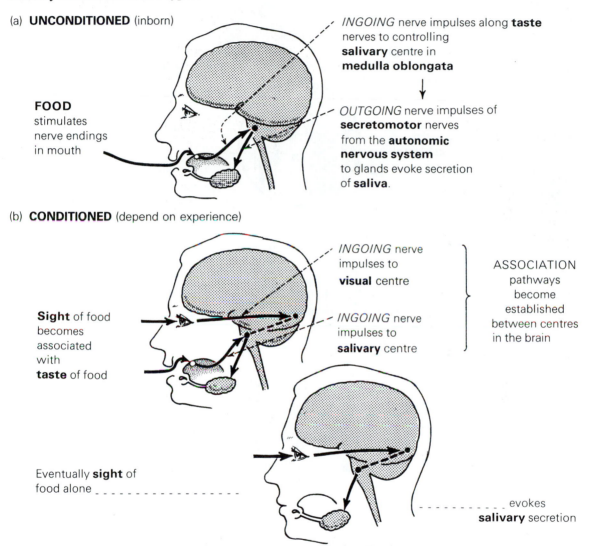

INGOING nerve impulses along **taste** nerves to controlling **salivary** centre in **medulla oblongata**

OUTGOING nerve impulses of **secretomotor** nerves from the **autonomic nervous system** to glands evoke secretion of **saliva**.

FOOD stimulates nerve endings in mouth

(b) **CONDITIONED** (depend on experience)

INGOING nerve impulses to **visual** centre

INGOING nerve impulses to **salivary** centre

ASSOCIATION pathways become established between centres in the brain

Sight of food becomes associated with **taste** of food

Eventually **sight** of food alone _ evokes **salivary** secretion

Similar conditioned reflexes are established by smell, by thought of food, and even by the sounds of its preparation.

Secretomotor, parasympathetic nerves release (a) **acetylcholine** which greatly increases salivary secretion and (b) **vasoactive intestinal polypeptide** (VIP) which dilates the salivary gland blood vessels.

OESOPHAGUS

The oesophagus is a muscular tube about 25 cm long which conveys ingested food and fluid from the mouth to the stomach.

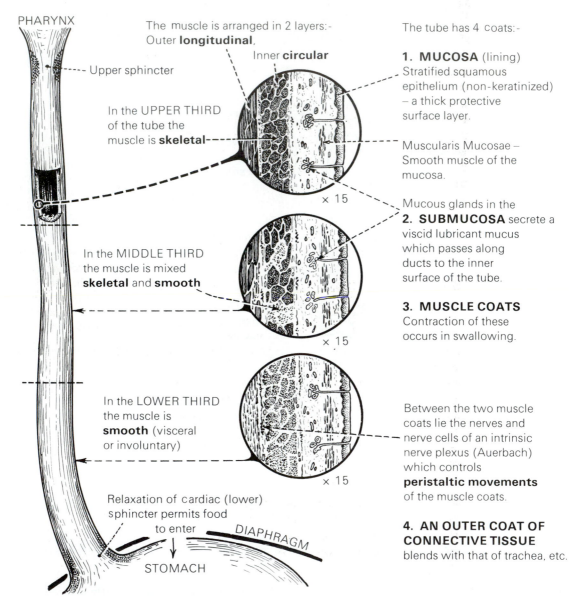

PHARYNX

The muscle is arranged in 2 layers:-
Outer **longitudinal**,
Inner **circular**

---- Upper sphincter

In the UPPER THIRD of the tube the muscle is **skeletal**----

× 15

In the MIDDLE THIRD the muscle is mixed **skeletal** and **smooth**

× 15

In the LOWER THIRD the muscle is **smooth** (visceral or involuntary)

× 15

Relaxation of cardiac (lower) sphincter permits food to enter

DIAPHRAGM

STOMACH

The tube has 4 coats:-

1. MUCOSA (lining)
Stratified squamous epithelium (non-keratinized) – a thick protective surface layer.

Muscularis Mucosae – Smooth muscle of the mucosa.

Mucous glands in the
2. SUBMUCOSA secrete a viscid lubricant mucus which passes along ducts to the inner surface of the tube.

3. MUSCLE COATS
Contraction of these occurs in swallowing.

Between the two muscle coats lie the nerves and nerve cells of an intrinsic nerve plexus (Auerbach) which controls **peristaltic movements** of the muscle coats.

4. AN OUTER COAT OF CONNECTIVE TISSUE
blends with that of trachea, etc.

Except during passage of food the oesophagus is flattened and closed; its mucosa thrown into several longitudinal folds.

Swallowing is a complex act initiated **voluntarily** and completed **involuntarily** (or **reflexly**).

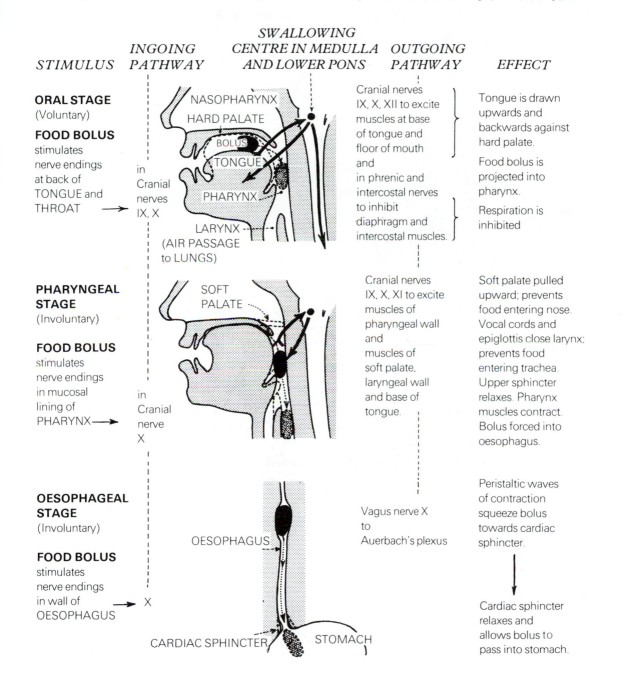

STOMACH

STRUCTURE

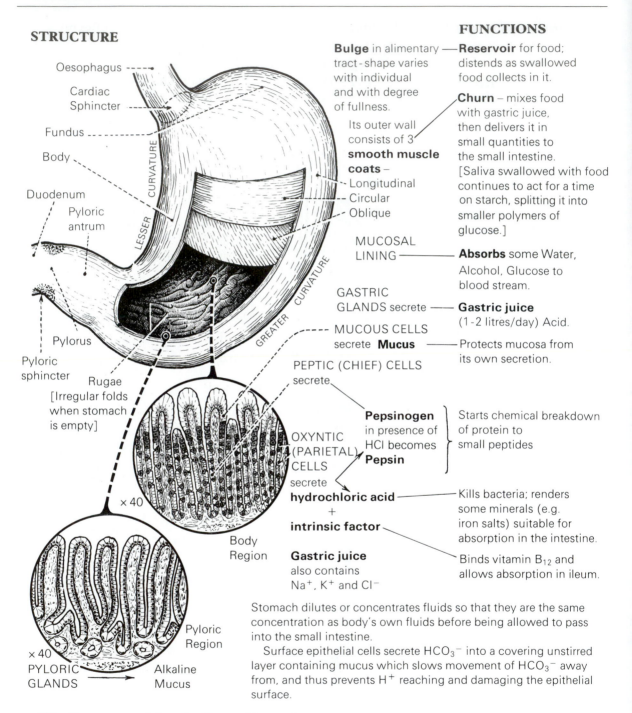

Oesophagus

Cardiac Sphincter

Fundus

Body

Duodenum

Pyloric antrum

LESSER CURVATURE

GREATER CURVATURE

Pylorus

Pyloric sphincter

Rugae [Irregular folds when stomach is empty]

× 40

Body Region

× 40
PYLORIC GLANDS → Alkaline Mucus

Pyloric Region

FUNCTIONS

Bulge in alimentary tract - shape varies with individual and with degree of fullness. —— **Reservoir** for food; distends as swallowed food collects in it.

Its outer wall consists of 3 **smooth muscle coats –**
– Longitudinal
– Circular
– Oblique

Churn – mixes food with gastric juice, then delivers it in small quantities to the small intestine. [Saliva swallowed with food continues to act for a time on starch, splitting it into smaller polymers of glucose.]

MUCOSAL LINING —— **Absorbs** some Water, Alcohol, Glucose to blood stream.

GASTRIC GLANDS secrete —— **Gastric juice** (1-2 litres/day) Acid.

MUCOUS CELLS secrete **Mucus** —— Protects mucosa from its own secretion.

PEPTIC (CHIEF) CELLS secrete

Pepsinogen in presence of HCl becomes **Pepsin** } Starts chemical breakdown of protein to small peptides

OXYNTIC (PARIETAL) CELLS secrete **hydrochloric acid** —— Kills bacteria; renders some minerals (e.g. iron salts) suitable for absorption in the intestine.

+

intrinsic factor —— Binds vitamin B_{12} and allows absorption in ileum.

Gastric juice also contains Na^+, K^+ and Cl^-

Stomach dilutes or concentrates fluids so that they are the same concentration as body's own fluids before being allowed to pass into the small intestine.

Surface epithelial cells secrete HCO_3^- into a covering unstirred layer containing mucus which slows movement of HCO_3^- away from, and thus prevents H^+ reaching and damaging the epithelial surface.

'G' cells secrete **gastrin;** stimulates acid secretion.

Secretion of gastric juice is under 2 types of control:-

(a) **NERVOUS** – Messages are conveyed rapidly from brain centre by nerve fibres of the **autonomic nervous system** for *immediate* effect, e.g. stimulation of parasympathetic (vagus) nerves to **gastric glands** → secretion of **highly acid juice** containing **enzymes**.

(b) **HUMORAL** – Message is **chemical** and carried in **blood stream** for *slower* and longer-lasting control. [*Note:*- Chemical messengers (hormones) travel all over the body even though the message stimulates only one part – e.g. in this case the gastric glands.]

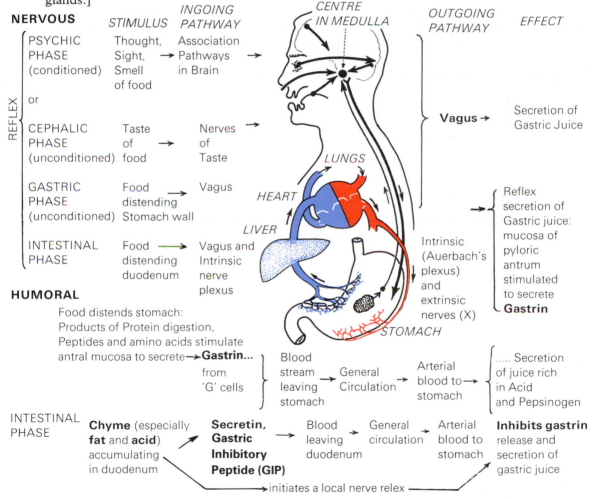

NERVOUS

	STIMULUS	INGOING PATHWAY	CENTRE IN MEDULLA	OUTGOING PATHWAY	EFFECT

REFLEX

PSYCHIC PHASE (conditioned) — Thought, Sight, Smell of food → Association Pathways in Brain →

or

CEPHALIC PHASE (unconditioned) — Taste of food → Nerves of Taste →

Vagus → Secretion of Gastric Juice

GASTRIC PHASE (unconditioned) — Food distending Stomach wall → Vagus

LUNGS

HEART

LIVER

INTESTINAL PHASE — Food distending duodenum → Vagus and Intrinsic nerve plexus

Intrinsic (Auerbach's plexus) and extrinsic nerves (X) → Reflex secretion of Gastric juice: mucosa of pyloric antrum stimulated to secrete **Gastrin**

STOMACH

HUMORAL

Food distends stomach: Products of Protein digestion, Peptides and amino acids stimulate antral mucosa to secrete → **Gastrin...** from 'G' cells → Blood stream leaving stomach → General Circulation → Arterial blood to stomach → ..... Secretion of juice rich in Acid and Pepsinogen

INTESTINAL PHASE

Chyme (especially **fat** and **acid**) accumulating in duodenum ⬈ **Secretin, Gastric Inhibitory Peptide (GIP)** → Blood leaving duodenum → General circulation → Arterial blood to stomach → **Inhibits gastrin** release and secretion of gastric juice

→ initiates a local nerve relex →

Cells in mucosa of stomach secrete **Histamine**; acts on H_2 receptors on parietal cells; causes acid secretion. Blocked by H_2 receptor blocking agent e.g. drug Cimetidine.

MOVEMENTS OF THE STOMACH

Very little movement is seen in empty stomach until onset of **hunger**.

FILLING

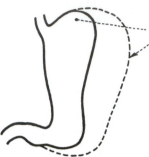

FUNDUS and
GREATER CURVATURE bulge and lengthen as stomach fills with food. This is 'receptive relaxation' – the smooth muscle cells relax so that the pressure in the stomach does not rise until the organ is fairly large. This is a **nervous reflex** involving vagal fibres which release an inhibitory neurotransmitter which is neither noradrenaline nor acetylcholine.

TWO TYPES OF MOVEMENT occur while food is in stomach.

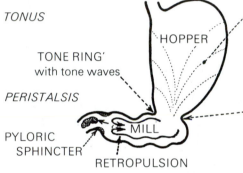

TONUS

HOPPER

TONE RING'
with tone waves

PERISTALSIS

PYLORIC
SPHINCTER

MILL

RETROPULSION

BODY – Food lies in layers and the walls exert a slight but steady pressure on it. This squeezes food steadily towards the PYLORUS even when stomach is becoming relatively empty near the end of a meal.

From INCISURA ANGULARIS vigorous waves of contraction carry chyme through the PYLORUS in small squirts. The sphincter then contracts forcing chyme back by **retropulsion** and mixing the food with digestive juices.

EMPTYING

Control of **motility** and **emptying**

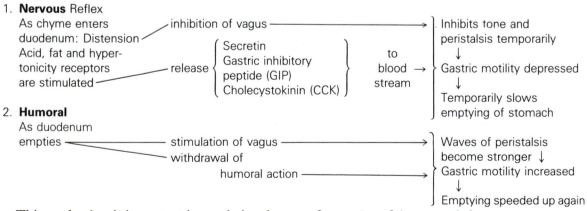

1. **Nervous** Reflex

As chyme enters duodenum: Distension Acid, fat and hypertonicity receptors are stimulated

inhibition of vagus ⟶ . Inhibits tone and peristalsis temporarily

release { Secretin / Gastric inhibitory peptide (GIP) / Cholecystokinin (CCK) } to blood → stream } Gastric motility depressed ↓ Temporarily slows emptying of stomach

2. **Humoral**

As duodenum empties ⟶ stimulation of vagus ⟶ withdrawal of humoral action ⟶ } Waves of peristalsis become stronger ↓ Gastric motility increased ↓ Emptying speeded up again

This mechanism is important in regulating the rate of emptying of the stomach from moment to moment during the digestion of a meal.

[**Starchy foods** leave the stomach quickly: **meat** leaves relatively slowly: **fatty foods** pass through most slowly of all.]

72

Vomiting is a **reflex** action.
Nerve pathways involved:-

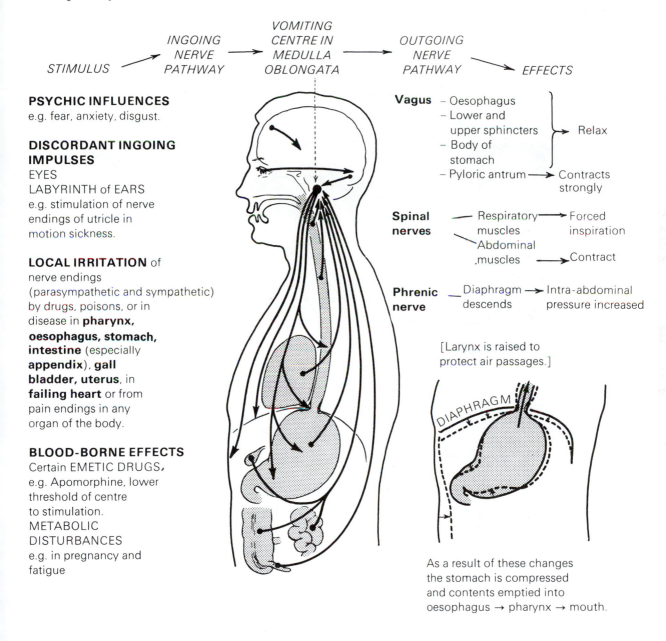

STIMULUS → INGOING NERVE PATHWAY → VOMITING CENTRE IN MEDULLA OBLONGATA → OUTGOING NERVE PATHWAY → EFFECTS

PSYCHIC INFLUENCES
e.g. fear, anxiety, disgust.

DISCORDANT INGOING IMPULSES
EYES
LABYRINTH of EARS
e.g. stimulation of nerve endings of utricle in motion sickness.

LOCAL IRRITATION of nerve endings (parasympathetic and sympathetic) by drugs, poisons, or in disease in **pharynx, oesophagus, stomach, intestine** (especially **appendix**), **gall bladder, uterus**, in **failing heart** or from pain endings in any organ of the body.

BLOOD-BORNE EFFECTS
Certain EMETIC DRUGS,
e.g. Apomorphine, lower threshold of centre to stimulation.
METABOLIC DISTURBANCES
e.g. in pregnancy and fatigue

Vagus
– Oesophagus
– Lower and upper sphincters
– Body of stomach
} Relax
– Pyloric antrum → Contracts strongly

Spinal nerves
– Respiratory muscles → Forced inspiration
– Abdominal muscles → Contract

Phrenic nerve
– Diaphragm descends → Intra-abdominal pressure increased

[Larynx is raised to protect air passages.]

DIAPHRAGM

As a result of these changes the stomach is compressed and contents emptied into oesophagus → pharynx → mouth.

In **retching**, gastric contents are forced into the oesophagus; they do not enter pharynx.

PANCREAS

The pancreas is a large gland lying across the posterior abdominal wall. It has 2 secretions – a digestive secretion poured into the duodenum, and a hormonal secretion passed into the blood stream.

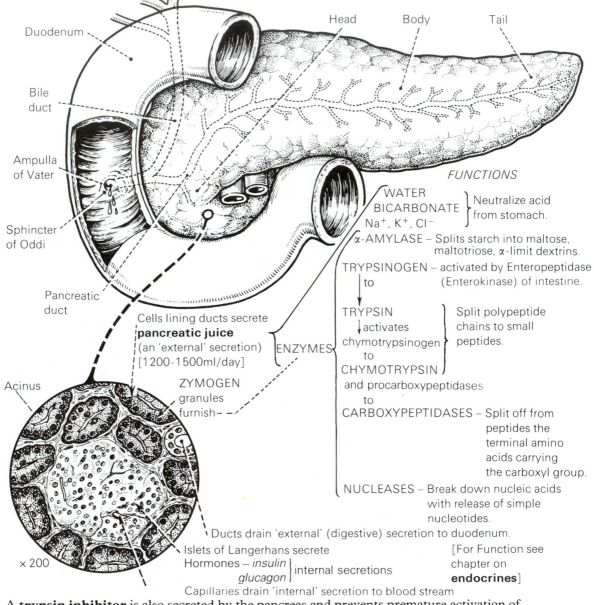

Duodenum

Head   Body   Tail

Bile duct

Ampulla of Vater

Sphincter of Oddi

Pancreatic duct

Acinus

Cells lining ducts secrete **pancreatic juice** (an 'external' secretion) [1200-1500ml/day]

ENZYMES

ZYMOGEN granules furnish– – –

× 200

FUNCTIONS

WATER
BICARBONATE
Na^+, K^+, Cl^- } Neutralize acid from stomach.

α-AMYLASE – Splits starch into maltose, maltotriose, α-limit dextrins.

TRYPSINOGEN – activated by Enteropeptidase (Enterokinase) of intestine.
to

TRYPSIN
↓activates
chymotrypsinogen
to
CHYMOTRYPSIN } Split polypeptide chains to small peptides.

and procarboxypeptidases
to
CARBOXYPEPTIDASES – Split off from peptides the terminal amino acids carrying the carboxyl group.

NUCLEASES – Break down nucleic acids with release of simple nucleotides.

Ducts drain 'external' (digestive) secretion to duodenum.

Islets of Langerhans secrete
Hormones – *insulin glucagon* } internal secretions

[For Function see chapter on **endocrines**]

Capillaries drain 'internal' secretion to blood stream

A **trypsin inhibitor** is also secreted by the pancreas and prevents premature activation of proteolytic enzymes in pancreatic ducts.

Secretion of pancreatic juice is under 2 types of control:- **nervous** and **humoral**. The humoral mechanism is the more important.

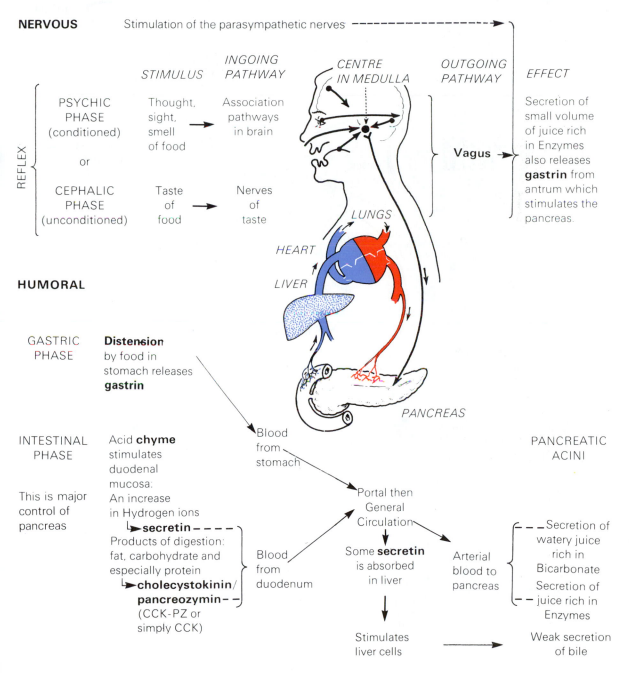

NERVOUS Stimulation of the parasympathetic nerves

	STIMULUS	INGOING PATHWAY	CENTRE IN MEDULLA	OUTGOING PATHWAY	EFFECT

REFLEX

PSYCHIC PHASE (conditioned) — Thought, sight, smell of food → Association pathways in brain

or

CEPHALIC PHASE (unconditioned) — Taste of food → Nerves of taste

Vagus → Secretion of small volume of juice rich in Enzymes also releases **gastrin** from antrum which stimulates the pancreas.

HUMORAL

GASTRIC PHASE — **Distension** by food in stomach releases **gastrin**

LUNGS
HEART
LIVER
PANCREAS

Blood from stomach

INTESTINAL PHASE — Acid **chyme** stimulates duodenal mucosa: An increase in Hydrogen ions

This is major control of pancreas

→ **secretin** - - - -
Products of digestion: fat, carbohydrate and especially protein
→ **cholecystokinin/ pancreozymin** - - -
(CCK-PZ or simply CCK)

Blood from duodenum

Portal then General Circulation

Some **secretin** is absorbed in liver

Arterial blood to pancreas

Stimulates liver cells

PANCREATIC ACINI

- - - Secretion of watery juice rich in Bicarbonate
Secretion of juice rich in Enzymes

Weak secretion of bile

LIVER AND GALL BLADDER

The liver is a large highly complex organ with many functions. One of these is the production of **bile** (500-1000 ml/day)

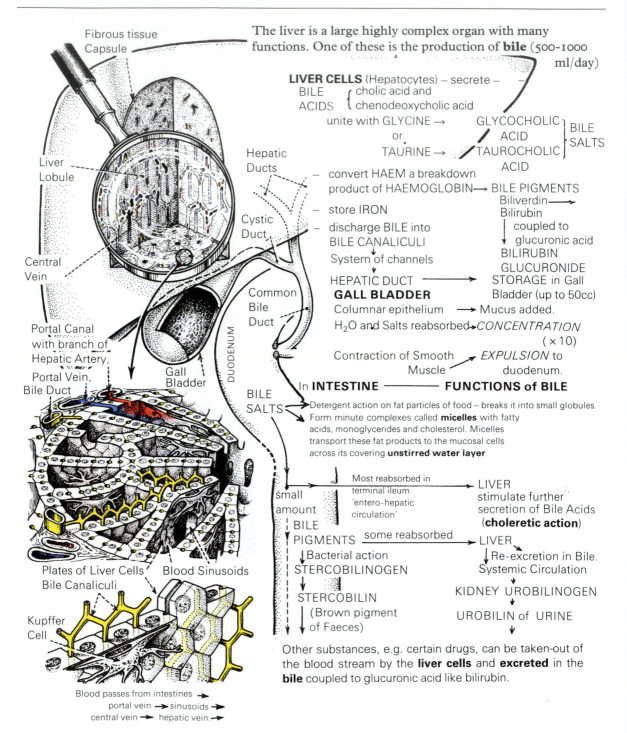

Fibrous tissue Capsule

Liver Lobule

Central Vein

Portal Canal with branch of Hepatic Artery, Portal Vein, Bile Duct

Gall Bladder

Hepatic Ducts

Cystic Duct

Common Bile Duct

DUODENUM

BILE SALTS

Plates of Liver Cells
Bile Canaliculi   Blood Sinusoids

Kupffer Cell

LIVER CELLS (Hepatocytes) – secrete –
BILE { cholic acid and
ACIDS { chenodeoxycholic acid
unite with GLYCINE → GLYCOCHOLIC ACID } BILE
or
TAURINE → TAUROCHOLIC ACID } SALTS

– convert HAEM a breakdown product of HAEMOGLOBIN → BILE PIGMENTS
  Biliverdin →
  Bilirubin

– store IRON | coupled to
– discharge BILE into | glucuronic acid
  BILE CANALICULI   BILIRUBIN
  System of channels GLUCURONIDE
                      STORAGE in Gall
  HEPATIC DUCT →   Bladder (up to 50cc)
  GALL BLADDER
  Columnar epithelium → Mucus added.
  H_2O and Salts reabsorbed → *CONCENTRATION* ($\times 10$)
  Contraction of Smooth → *EXPULSION* to
                 Muscle   duodenum.

In **INTESTINE** ———— **FUNCTIONS of BILE**

Detergent action on fat particles of food – breaks it into small globules. Form minute complexes called **micelles** with fatty acids, monoglycerides and cholesterol. Micelles transport these fat products to the mucosal cells across its covering **unstirred water layer**

small amount
↓ BILE
PIGMENTS

Most reabsorbed in terminal ileum 'entero-hepatic circulation' → LIVER stimulate further secretion of Bile Acids (**choleretic action**)

some reabsorbed → LIVER
↓ Re-excretion in Bile.
Systemic Circulation

↓ Bacterial action
STERCOBILINOGEN

↓
STERCOBILIN
↓ (Brown pigment
  of Faeces)

KIDNEY UROBILINOGEN
↓
UROBILIN of URINE
↓

Other substances, e.g. certain drugs, can be taken-out of the blood stream by the **liver cells** and **excreted** in the **bile** coupled to glucuronic acid like bilirubin.

Blood passes from intestines →
portal vein → sinusoids →
central vein → hepatic vein. →

EXPULSION OF BILE (FROM THE GALL BLADDER)

Bile acid secretion by the liver is stimulated by **bile salts** in portal vein blood. The volume of bile is increased by **secretin** released by acid in the duodenum causing a bicarbonate-rich watery secretion from the epithelium of small bile ducts.

Bile is *stored* and *concentrated* in the **gall bladder**. Periodically (e.g. during a meal) it is discharged into the duodenum. **Nervous** and **humoral** factors influence this expulsion:

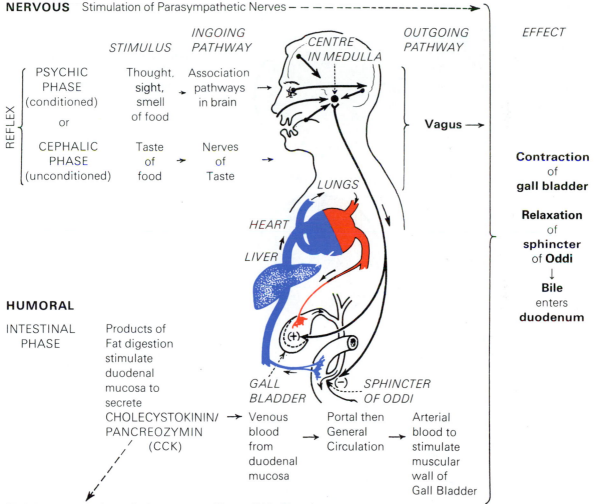

NERVOUS   Stimulation of Parasympathetic Nerves – – – – – – – – – – – →

| | STIMULUS | INGOING PATHWAY | CENTRE IN MEDULLA | OUTGOING PATHWAY | EFFECT |

REFLEX
- PSYCHIC PHASE (conditioned) — Thought, sight, smell of food → Association pathways in brain
- or
- CEPHALIC PHASE (unconditioned) — Taste of food → Nerves of Taste

Vagus →

Contraction of **gall bladder**

Relaxation of **sphincter** of **Oddi**
↓
Bile enters **duodenum**

LUNGS   HEART   LIVER   GALL BLADDER   SPHINCTER OF ODDI

HUMORAL

INTESTINAL PHASE — Products of Fat digestion stimulate duodenal mucosa to secrete CHOLECYSTOKININ/PANCREOZYMIN (CCK) → Venous blood from duodenal mucosa → Portal then General Circulation → Arterial blood to stimulate muscular wall of Gall Bladder

[This hormone is the main factor controlling gall bladder contraction in man.]

Nerve fibres containing **vasoactive intestinal peptide** (VIP) in the wall of the gall bladder **inhibit** its contraction.

SMALL INTESTINE

The small intestine is a long muscular tube – approximately 275 cm long in life and 700 cm after death. It receives **chyme** in small quantities from the stomach; **pancreatic juice** from the pancreas; **bile** from the gall bladder.

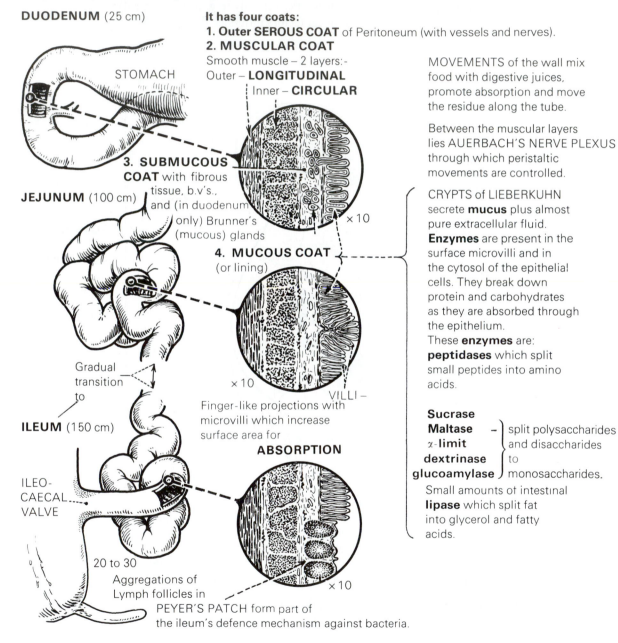

DUODENUM (25 cm)

STOMACH

JEJUNUM (100 cm)

Gradual transition to

ILEUM (150 cm)

ILEO-CAECAL VALVE

20 to 30 Aggregations of Lymph follicles in PEYER'S PATCH form part of the ileum's defence mechanism against bacteria.

It has four coats:
1. Outer SEROUS COAT of Peritoneum (with vessels and nerves).
2. MUSCULAR COAT
Smooth muscle – 2 layers:-
Outer – **LONGITUDINAL**
Inner – **CIRCULAR**

× 10

3. SUBMUCOUS COAT with fibrous tissue, b.v's., and (in duodenum only) Brunner's (mucous) glands

4. MUCOUS COAT (or lining)

× 10

VILLI – Finger-like projections with microvilli which increase surface area for

ABSORPTION

× 10

MOVEMENTS of the wall mix food with digestive juices, promote absorption and move the residue along the tube.

Between the muscular layers lies AUERBACH'S NERVE PLEXUS through which peristaltic movements are controlled.

CRYPTS of LIEBERKUHN secrete **mucus** plus almost pure extracellular fluid.
Enzymes are present in the surface microvilli and in the cytosol of the epithelial cells. They break down protein and carbohydrates as they are absorbed through the epithelium.
These **enzymes** are:
peptidases which split small peptides into amino acids.

Sucrase
Maltase – split polysaccharides
α-**limit** and disaccharides
dextrinase to
glucoamylase monosaccharides.
Small amounts of intestinal **lipase** which split fat into glycerol and fatty acids.

78

THE BASIC PATTERN OF THE GUT WALL

The wall of the digestive tube shows a basic structural pattern. *FOUR coats* are seen in transverse section.

STRUCTURE

1. MUCOUS COAT (or MUCOSA)
SURFACE EPITHELIUM / type varies with / with its **glands** / site and function /

LOOSE FIBROUS TISSUE (lamina propria)
with capillaries and lymphatic vessels.

MUSCULARIS MUCOSAE
(thin layers of smooth muscle)

LYMPHATIC TISSUE

2. SUBMUCOUS COAT
(or SUBMUCOSA)
DENSE FIBROUS TISSUE
in which lie blood vessels,
lymphatic vessels and
MEISSNER'S NERVE PLEXUS
[Glands in oesophagus
and 1st part of duodenum]
LYMPHATIC TISSUE

3. MUSCULAR COAT
(or MUSCULARIS EXTERNA)
SMOOTH MUSCLE LAYERS
Inner – Circular arrangement
Outer – Longitudinal
arrangement
Between them, blood- and
lymphatic vessels and
AUERBACH'S NERVE PLEXUS
[Some striated muscle
in oesophagus]

(After GARVEN)

4. SEROUS COAT (or SEROSA)
FIBROUS TISSUE
with fat, blood- and lymphatic vessels

MESOTHELIUM
Where tube is suspended by a **mesentery**
the serosa is formed by visceral layer of Peritoneum.

[Where there is no mesentery – replaced by
fibrous tissue which merges with surrounding fibrous tissue]

FUNCTIONS

Layer in close contact with gut
contents: specialized for –
SECRETION
ABSORPTION – nutrients and
hormones to blood stream.
MOBILITY (can continually change
degree of folding and the surface
of contact with gut contents).

DEFENCE against bacteria.

STRONG LAYER of SUPPLY
to specialized mucosa
(Blood vessels supply
needs and remove
absorbed materials).

COORDINATION of motor
and secretory activities
of mucosa.

MOVEMENT

Controls diameter of tube.
Mixes contents.
Propagates contents
along tube.

Nervous elements
coordinate secretory and
muscular activities.

Carries nerves, blood- and
lymphatic vessels to and
from mesentery.

Forms smooth, moist
membrane which reduces
friction between contacting
surfaces in the peritoneal cavity.

INTESTINAL SECRETIONS

The mucosa of the intestine secretes 1800 ml/day of **mucus, electrolytes** and **water**. Mucus protects the mucosa from mechanical damage. The *nature* and *control* of secretions differ in each part of the intestine.

DUODENUM First part is at risk from gastric acid and liable to peptic ulceration. Small coiled BRUNNER'S GLANDS in the submucosa secrete mucus to protect this region.

Stimuli for secretion:
(a) Tactile or irritating stimuli of the overlying mucosa act via Intrinsic Nerve Plexuses.
(b) Vagal stimulation.
(c) Intestinal hormones, especially **secretin**.
Brunner's glands are **inhibited** by sympathetic stimulation.

SMALL INTESTINE Mucus from GOBLET CELLS is secreted, in response to tactile stimuli, throughout the intestine. Mucus from CRYPTS OF LIEBERKÜHN is secreted in response to local nerve reflexes. Also from crypts a fluid like pure extracellular fluid is secreted. Provides a medium for absorption of food products. Mechanism involves active Cl^- and HCO_3^- transport into the lumen. Na^+ follows these anions then water follows by osmosis. Cholera toxin somehow increases Cl^- transport and thus a severe watery diarrhoea is produced.

COLON There are no villi here. Secretion is rich in **mucus, bicarbonate** and **potassium**. Mechanical irritation and cholinergic nerves stimulate secretion. Sympathetic nerves inhibit it.

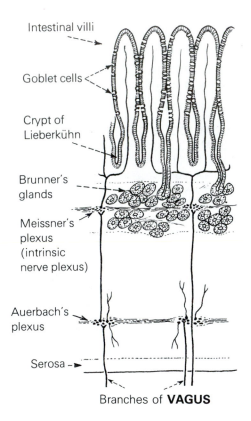

Intestinal villi

Goblet cells

Crypt of Lieberkühn

Brunner's glands

Meissner's plexus (intrinsic nerve plexus)

Auerbach's plexus

Serosa

Branches of **VAGUS**

The duodenum receives food in small quantities from the stomach. The mixture of food and digestive juices – chyme – is passed along the length of the small intestine.

TWO TYPES of MOVEMENT

SEGMENTATION
Rhythmical alternating
contractions and relaxations

— These 'shuttling' movements serve to mix **chyme** and to bring it into contact with the absorptive mucosa, i.e. **digestion** and **absorption** are promoted.

This type of movement is **myogenic**, i.e. it is the property of the smooth muscle cells. It does not depend on a nervous mechanism.

PERISTALSIS
Food probably acts as stimulus
to stretch receptors in muscle
and perhaps in mucous membrane

— Waves of this contraction move the food along the canal.

The contraction behind the bolus sweeps it into the portion of the tube ahead.

Circular muscle
behind bolus **contracts**

Muscle in front of
bolus may **relax**

BOLUS

This type of movement is **neurogenic**, i.e. it is carried out through a ' local' reflex mediated through **intrinsic** nerve plexuses within the wall of the tube.

The extrinsic nerve supply influences it. { Parasympathetic stimulation → ↑ contractions. Sympathetic stimulation → ↓ motility.

EMPTYING

ILEO-CAECAL VALVE ------- opens and closes during
Formed by thickening of
circular muscle layer
digestion to allow spurts of fluid material from the Ileum to enter the large intestine.

ILEUM

CAECUM

Relaxed by
gastrin from
stomach

Constricted by distension of caecum. **Relaxed** by **gastro-ileal** reflex produced by entry of food into stomach.

Meals of different composition travel along the intestine at different rates. **Digestion** and **absorption** of food are usually complete by the time the residue reaches the ileo-caecal valve.
Contractions of the small intestine are coordinated by slow waves of depolarization which travel in the smooth muscle from the duodenum to the ileum at a frequency of 9-12/minute.

ABSORPTION IN SMALL INTESTINE

Absorption of most digested foodstuffs occurs in the small intestine through the striated border epithelium covering the villi.

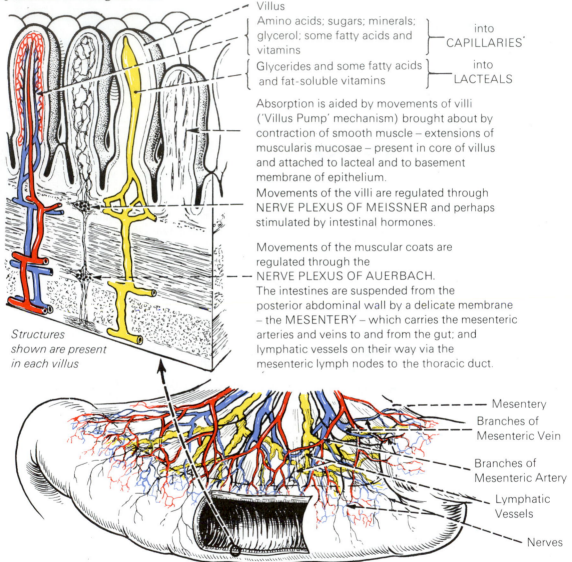

Villus

Amino acids; sugars; minerals; glycerol; some fatty acids and vitamins — into CAPILLARIES

Glycerides and some fatty acids and fat-soluble vitamins — into LACTEALS

Absorption is aided by movements of villi ('Villus Pump' mechanism) brought about by contraction of smooth muscle – extensions of muscularis mucosae – present in core of villus and attached to lacteal and to basement membrane of epithelium.
Movements of the villi are regulated through NERVE PLEXUS OF MEISSNER and perhaps stimulated by intestinal hormones.

Movements of the muscular coats are regulated through the NERVE PLEXUS OF AUERBACH.
The intestines are suspended from the posterior abdominal wall by a delicate membrane – the MESENTERY – which carries the mesenteric arteries and veins to and from the gut; and lymphatic vessels on their way via the mesenteric lymph nodes to the thoracic duct.

Structures shown are present in each villus

Mesentery

Branches of Mesenteric Vein

Branches of Mesenteric Artery

Lymphatic Vessels

Nerves

Absorption is not just a process of simple diffusion of substances from areas of high to areas of low concentration. Movement of ions can take place against a concentration gradient. In other words – absorption is often an active process involving the use of energy by the epithelial cells.

FOOD ABSORPTION BY CELLS OF THE SMALL INTESTINE

Digestion of food continues as it is absorbed through the **brush border** and **cytosol** of the endothelial cells of the small intestine.

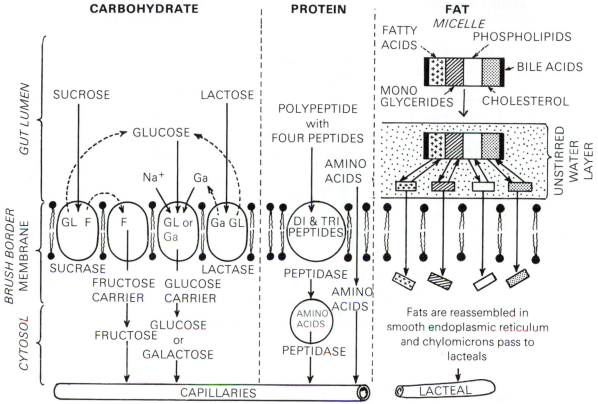

Small **carbohydrates** are broken down by enzymes (e.g. **sucrase, lactase**) in the membrane, then **fructose** (F) is transported by a fructose carrier into the cytosol. **Glucose** (GL) competes with **galactose** (Ga) for another carrier which requires Na^+ for its function. From the cytosol the monosaccharides pass by diffusion or facilitated diffusion into the capillaries.

    Polypeptides are broken down by a peptidase in the membrane, then further broken down to amino acids in the cytosol. Some amino acids simply diffuse through the brush border into the cytosol. From the cytosol amino acids either diffuse or are transported into the capillaries. Some of these amino acid transport systems require Na^+ to function.

    Micelles transport fatty acids, phospholipids, cholesterol, monoglycerides and sometimes fat-soluble vitamins across the **unstirred water layer** which lies next to the membrane. From there, these substances diffuse across the membrane into the cytosol where they are reprocessed into fats, packaged into **chylomicrons** and passed into the lacteals.

TRANSPORT OF ABSORBED FOODSTUFFS

After absorption the **nutrients** are transported in:-

(a) **BLOOD** through
Mesenteric Veins
to
Portal Vein
to
LIVER – ⎰ Many substances
⎱ undergo further
changes;
some are stored;
some are passed on
via ↓
Hepatic Veins → Inferior
Vena Cava →
Heart and
Systemic
Circulation

(b) **LYMPH** in
Lymphatic Vessels
to
THORACIC DUCT
and via a large vein
in the neck to the
Systemic Circulation
↓
which then distributes
Food to **all tissues**
of the **body**

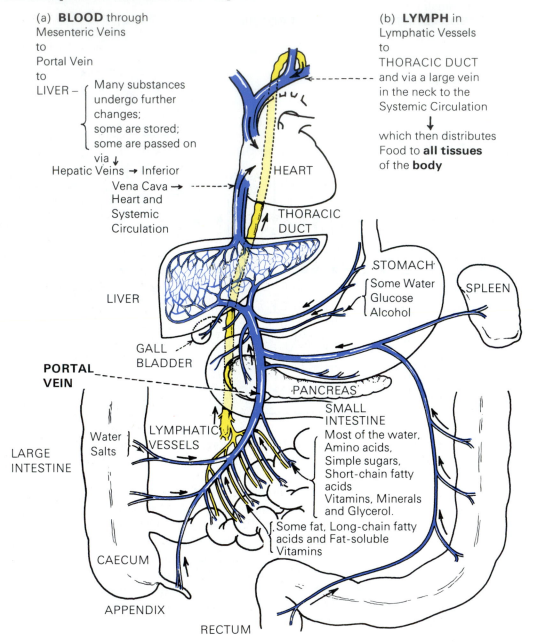

HEART

THORACIC DUCT

LIVER

GALL BLADDER

PORTAL VEIN

STOMACH
Some Water
Glucose
Alcohol

SPLEEN

PANCREAS

SMALL INTESTINE
Most of the water,
Amino acids,
Simple sugars,
Short-chain fatty
acids
Vitamins, Minerals
and Glycerol.

Water
Salts

LYMPHATIC VESSELS

LARGE INTESTINE

Some fat, Long-chain fatty
acids and Fat-soluble
Vitamins

CAECUM

APPENDIX

RECTUM

Total length about 1.6 metres

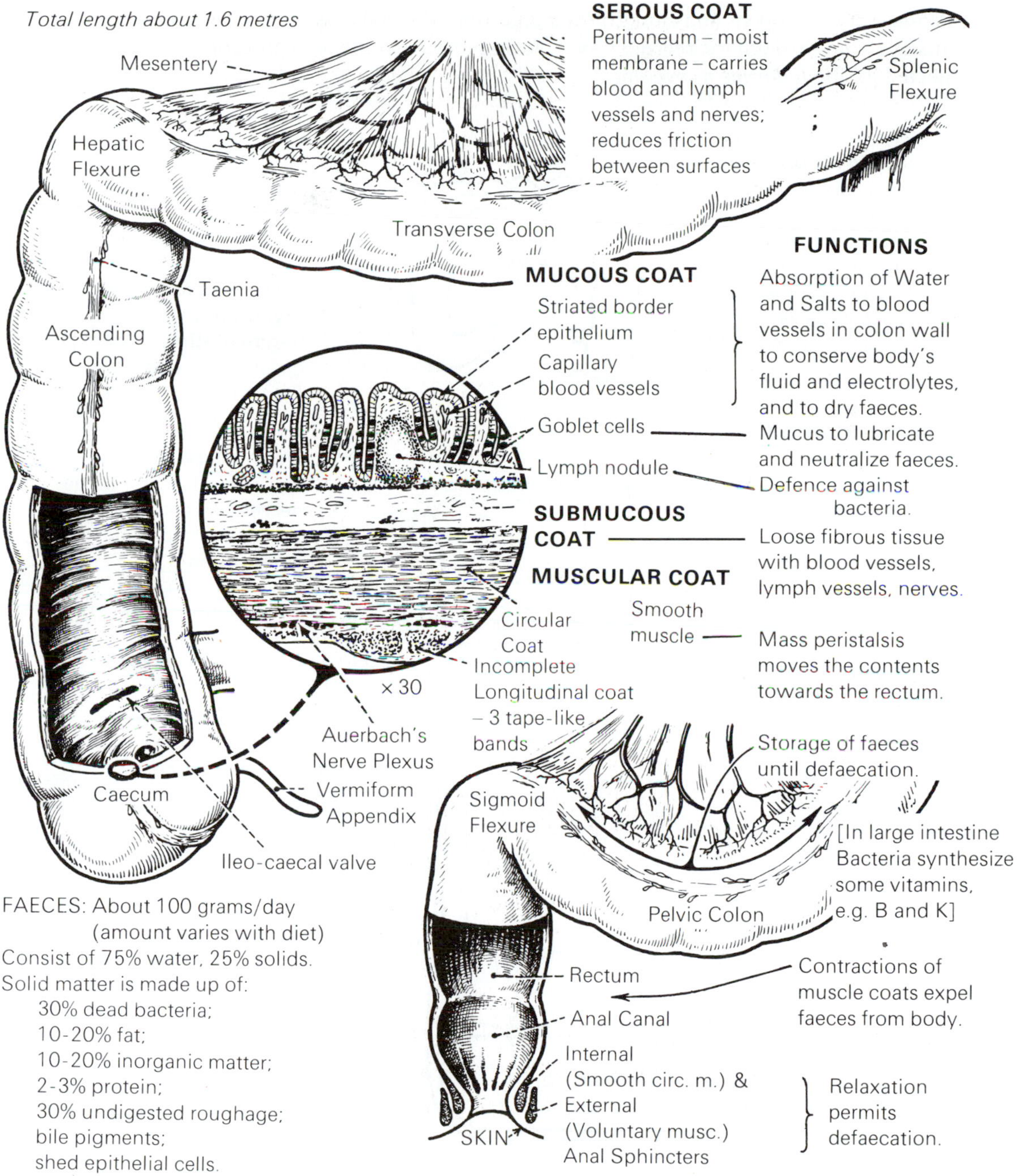

Mesentery

Hepatic Flexure

Ascending Colon

Taenia

Transverse Colon

SEROUS COAT
Peritoneum – moist membrane – carries blood and lymph vessels and nerves; reduces friction between surfaces

Splenic Flexure

MUCOUS COAT
Striated border epithelium

Capillary blood vessels

Goblet cells

Lymph nodule

SUBMUCOUS COAT

MUSCULAR COAT
Circular Coat Incomplete
Longitudinal coat – 3 tape-like bands

× 30

Smooth muscle

Auerbach's Nerve Plexus

Vermiform Appendix

Sigmoid Flexure

Caecum

Ileo-caecal valve

Pelvic Colon

Rectum

Anal Canal

Internal (Smooth circ. m.) & External (Voluntary musc.) Anal Sphincters

SKIN

FUNCTIONS

Absorption of Water and Salts to blood vessels in colon wall to conserve body's fluid and electrolytes, and to dry faeces.

Mucus to lubricate and neutralize faeces.

Defence against bacteria.

Loose fibrous tissue with blood vessels, lymph vessels, nerves.

Mass peristalsis moves the contents towards the rectum.

Storage of faeces until defaecation.

[In large intestine Bacteria synthesize some vitamins, e.g. B and K]

Contractions of muscle coats expel faeces from body.

Relaxation permits defaecation.

FAECES: About 100 grams/day
(amount varies with diet)
Consist of 75% water, 25% solids.
Solid matter is made up of:
   30% dead bacteria;
   10-20% fat;
   10-20% inorganic matter;
   2-3% protein;
   30% undigested roughage;
   bile pigments;
   shed epithelial cells.

MOVEMENTS OF THE LARGE INTESTINE

The ileo-caecal valve opens and closes during digestion. Peristaltic waves sweep semi-fluid contents of ileum through the relaxed valve. During its stay in the large intestine faecal matter is subjected to: the *following movements.*

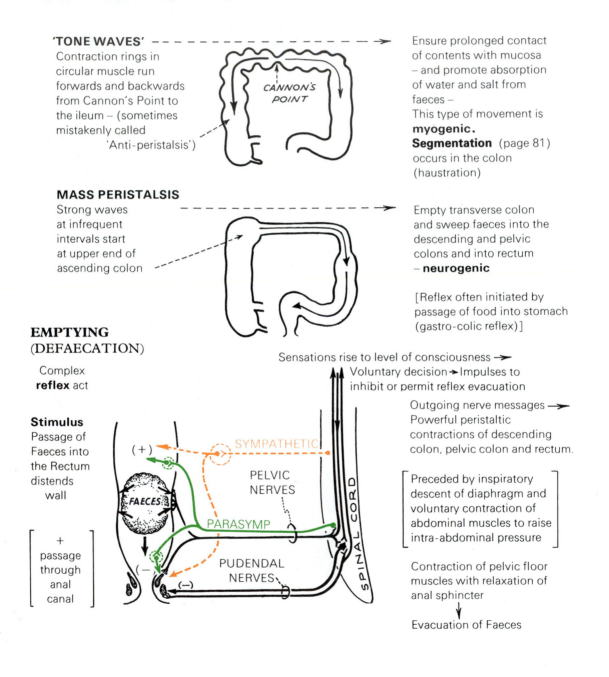

'TONE WAVES'
Contraction rings in circular muscle run forwards and backwards from Cannon's Point to the ileum – (sometimes mistakenly called 'Anti-peristalsis')

CANNON'S POINT

Ensure prolonged contact of contents with mucosa – and promote absorption of water and salt from faeces –
This type of movement is **myogenic.**
Segmentation (page 81) occurs in the colon (haustration)

MASS PERISTALSIS
Strong waves at infrequent intervals start at upper end of ascending colon

Empty transverse colon and sweep faeces into the descending and pelvic colons and into rectum – **neurogenic**

[Reflex often initiated by passage of food into stomach (gastro-colic reflex)]

EMPTYING (DEFAECATION)

Complex **reflex** act

Stimulus
Passage of Faeces into the Rectum distends wall

[
+
passage through anal canal
]

Sensations rise to level of consciousness →
Voluntary decision → Impulses to inhibit or permit reflex evacuation

Outgoing nerve messages →
Powerful peristaltic contractions of descending colon, pelvic colon and rectum.

(+)

SYMPATHETIC

PELVIC NERVES

FAECES

PARASYMP

SPINAL CORD

PUDENDAL NERVES

(−)

(−)

[
Preceded by inspiratory descent of diaphragm and voluntary contraction of abdominal muscles to raise intra-abdominal pressure
]

Contraction of pelvic floor muscles with relaxation of anal sphincter

↓

Evacuation of Faeces

In the wall of the gut there is an **intrinsic** nervous system called the **enteric nervous system** (i.e. Auerbach's plexus plus Meissner's plexus, (page 80). The enteric nervous system controls most gut movements and secretions. It is made up of (a) post-ganglionic parasympathetic nerves, (b) inhibitory nerves secreting **vasoactive intestinal polypeptide** (VIP) and (c) other nerves of unknown function containing e.g. ATP, substance P, enkephalins, serotonin (5-HT), somatostatin, gastrin releasing peptide (GRP) and neurotensin.

The degree of activity of the enteric nervous system can be altered by **extrinsic** nerves of the **autonomic nervous system**. This has *two* parts – **sympathetic** and **parasympathetic** which can respectively *decrease* and *increase* gut movements and secretions.

Sympathetic nerves when stimulated release at their ganglia a chemical substance **acetylcholine**, and at their post-ganglionic endings **noradrenaline**. Sympathetic post-ganglionic nerves to the gut terminate in the enteric nervous system and on blood vessels. The noradrenaline released inhibits ($-$) transmission in Auerbach's plexus thus decreasing gut movements. It constricts some sphincters ($+$) and the blood vessels ($+$).

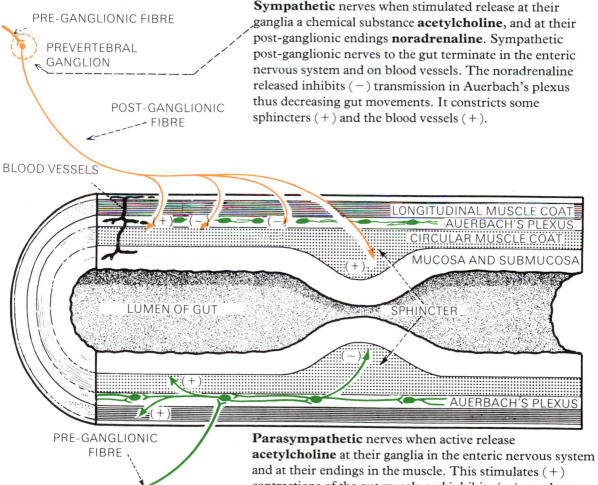

Parasympathetic nerves when active release **acetylcholine** at their ganglia in the enteric nervous system and at their endings in the muscle. This stimulates ($+$) contractions of the gut muscle and inhibits ($-$) or relaxes the sphincters.

NERVOUS CONTROL OF GUT MOVEMENTS

Movements in the wall of the gastro-intestinal tract are either

- (a) **myogenic** – a property of the smooth muscle, or
- (b) **neurogenic** – dependent on the enteric nervous System.

They can occur even after *extrinsic* nerves to the tract have been cut. Normally, however, impulses travelling in these nerves of the **sympathetic** and **parasympatheic** systems, from the **controlling centres** of the **autonomic nervous system** in the **brain** and spinal cord, influence and coordinate events in the whole tract.

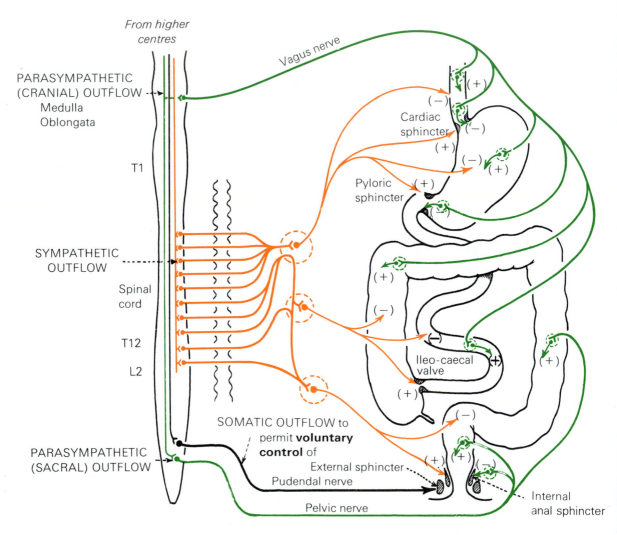

CHAPTER 5

TRANSPORT SYSTEM
THE HEART, BLOOD VESSELS AND BODY
FLUIDS: HAEMOPOIETIC SYSTEM

CARDIOVASCULAR SYSTEM

The **CIRCULATORY** System

Chief **TRANSPORT** System
of the body

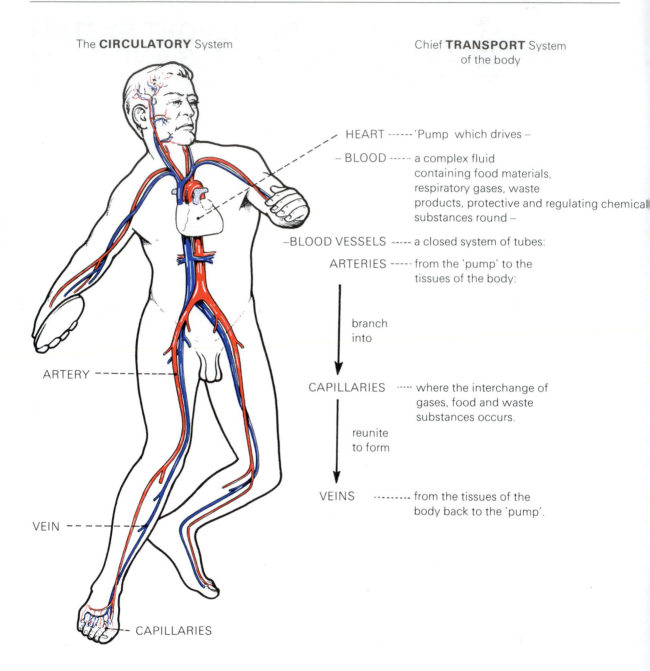

HEART ------ 'Pump which drives –

– BLOOD ----- a complex fluid
containing food materials,
respiratory gases, waste
products, protective and regulating chemical
substances round –

–BLOOD VESSELS ----- a closed system of tubes:

ARTERIES ----- from the 'pump' to the
tissues of the body:

branch
into

CAPILLARIES ----- where the interchange of
gases, food and waste
substances occurs.

reunite
to form

VEINS -------- from the tissues of the
body back to the 'pump'.

ARTERY

VEIN

CAPILLARIES

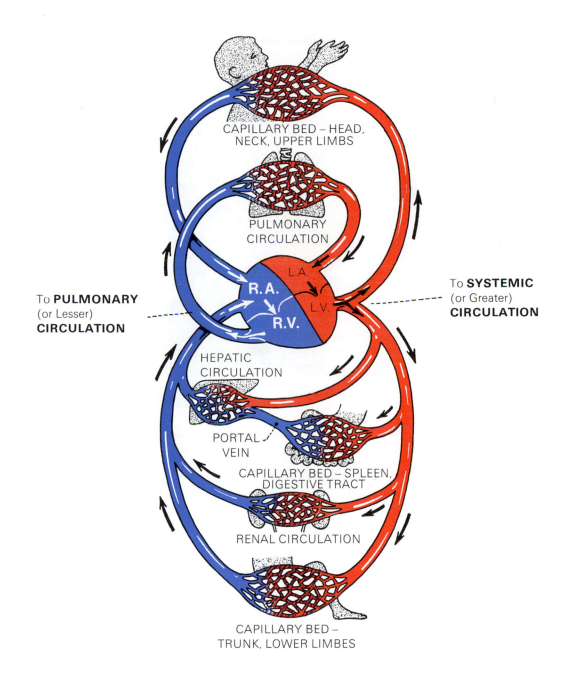

CAPILLARY BED – HEAD,
NECK, UPPER LIMBS

PULMONARY
CIRCULATION

L.A.

R.A.

L.V.

R.V.

To **PULMONARY**
(or Lesser)
CIRCULATION

To **SYSTEMIC**
(or Greater)
CIRCULATION

HEPATIC
CIRCULATION

PORTAL
VEIN

CAPILLARY BED – SPLEEN,
DIGESTIVE TRACT

RENAL CIRCULATION

CAPILLARY BED –
TRUNK, LOWER LIMBES

HEART

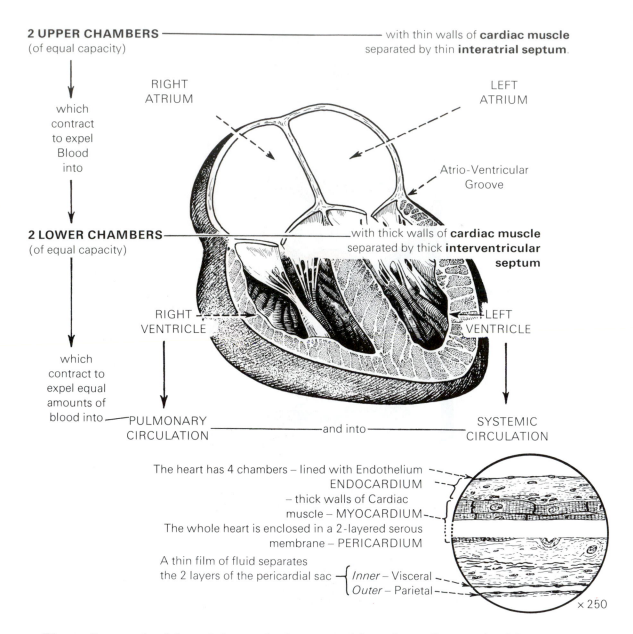

2 UPPER CHAMBERS ———————————————— with thin walls of **cardiac muscle**
(of equal capacity)   separated by thin **interatrial septum**.

↓

which
contract
to expel
Blood
into

RIGHT   LEFT
ATRIUM   ATRIUM

Atrio-Ventricular
Groove

2 LOWER CHAMBERS ————— with thick walls of **cardiac muscle**
(of equal capacity)   separated by thick **interventricular**
                                                              septum

↓

which
contract to
expel equal
amounts of
blood into —

RIGHT   LEFT
VENTRICLE   VENTRICLE

PULMONARY ————————————— and into ————————— SYSTEMIC
CIRCULATION   CIRCULATION

The heart has 4 chambers – lined with Endothelium
ENDOCARDIUM
– thick walls of Cardiac
muscle – MYOCARDIUM
The whole heart is enclosed in a 2-layered serous
membrane – PERICARDIUM

A thin film of fluid separates
the 2 layers of the pericardial sac ⎰ *Inner* – Visceral
                                                          ⎱ *Outer* – Parietal

× 250

The cardiac muscle of the atria is completely separated from the cardiac muscle of the
ventricles by a ring of fibrous tissue — at the atrio-ventricular groove.

This is a diagrammatic section through the heart.

HEART VALVES have a core of fibrous tissue
covered on both sides with **Endothelium**

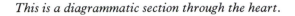

Extensions from
ATRIO-VENTRICULAR (A-V)
FIBROUS RING

Designed to
allow blood
to flow in
one direction
only –
from **atrium**
to **ventricle** –
and on into
arteries

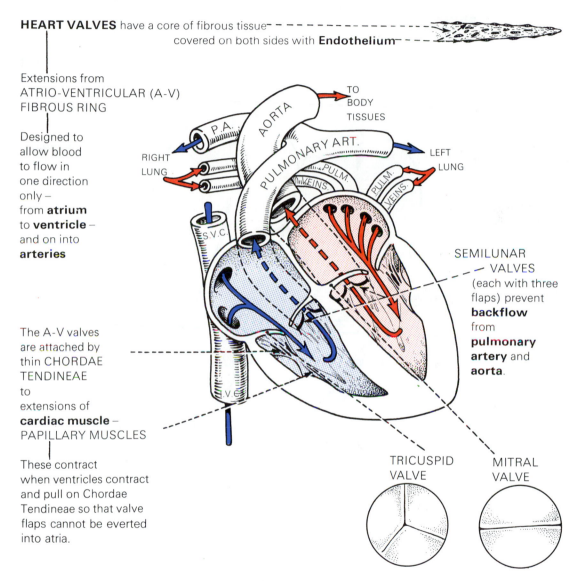

SEMILUNAR
VALVES
(each with three
flaps) prevent
backflow
from
**pulmonary
artery** and
aorta.

The A-V valves
are attached by
thin CHORDAE
TENDINEAE
to
extensions of
cardiac muscle –
PAPILLARY MUSCLES

These contract
when ventricles contract
and pull on Chordae
Tendineae so that valve
flaps cannot be everted
into atria.

TRICUSPID
VALVE

MITRAL
VALVE

The great veins do not have valves guarding their entrance to the heart.
Thickening and contraction of the muscle around their mouths prevent **backflow** of blood
from heart.

HEART

The human heart is really a *DOUBLE PUMP* – each pump quite separate from the other.

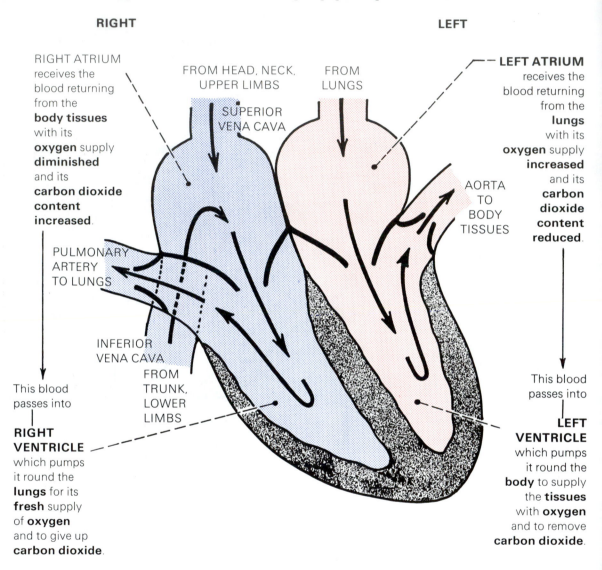

RIGHT

LEFT

RIGHT ATRIUM receives the blood returning from the **body tissues** with its **oxygen** supply **diminished** and its **carbon dioxide content increased**.

FROM HEAD, NECK, UPPER LIMBS

SUPERIOR VENA CAVA

FROM LUNGS

LEFT ATRIUM receives the blood returning from the **lungs** with its **oxygen** supply **increased** and its **carbon dioxide content reduced**.

AORTA TO BODY TISSUES

PULMONARY ARTERY TO LUNGS

INFERIOR VENA CAVA

FROM TRUNK, LOWER LIMBS

This blood passes into

RIGHT VENTRICLE which pumps it round the **lungs** for its **fresh** supply of **oxygen** and to give up **carbon dioxide**.

This blood passes into

LEFT VENTRICLE which pumps it round the **body** to supply the **tissues** with **oxygen** and to remove **carbon dioxide**.

This diagram simplifies the structure of the heart to make it easier to understand the function of its various parts.

Diagrammatic representation of the sequence of events in the heart during *one* heart beat.

DIASTOLE [Period of Relaxation – i.e. when heart is resting]

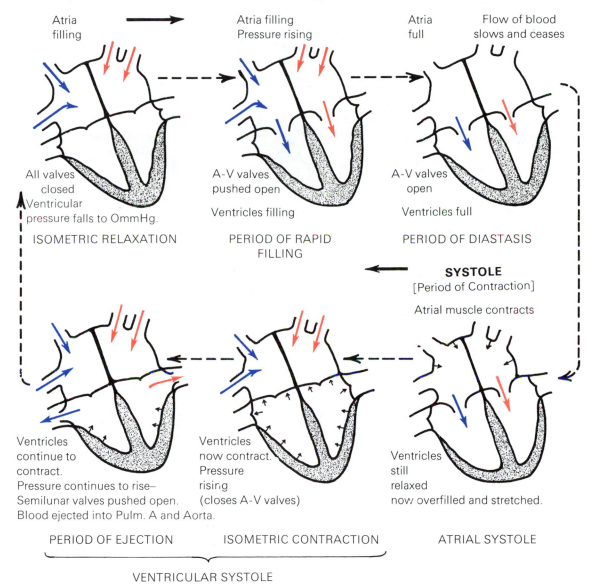

Atria
filling

Atria filling
Pressure rising

Atria
full

Flow of blood
slows and ceases

All valves
closed
Ventricular
pressure falls to 0mmHg.

ISOMETRIC RELAXATION

A-V valves
pushed open

Ventricles filling

PERIOD OF RAPID
FILLING

A-V valves
open

Ventricles full

PERIOD OF DIASTASIS

SYSTOLE
[Period of Contraction]

Atrial muscle contracts

Ventricles
continue to
contract.
Pressure continues to rise–
Semilunar valves pushed open.
Blood ejected into Pulm. A and Aorta.

PERIOD OF EJECTION

Ventricles
now contract.
Pressure
rising
(closes A-V valves)

ISOMETRIC CONTRACTION

Ventricles
still
relaxed
now overfilled and stretched.

ATRIAL SYSTOLE

VENTRICULAR SYSTOLE

The total cycle of events takes about 0.8 second when heart is beating 75 times per minute.

HEART SOUNDS

During each **cardiac cycle** *2 heart sounds* can be heard through a **stethoscope** applied to the **chest wall**.

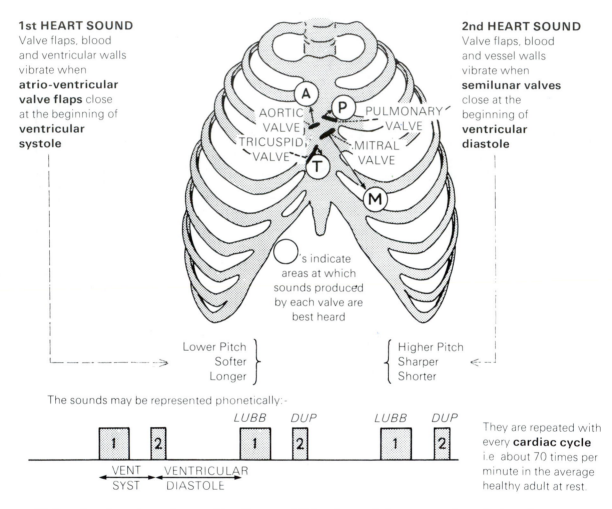

1st HEART SOUND
Valve flaps, blood and ventricular walls vibrate when **atrio-ventricular valve flaps** close at the beginning of **ventricular systole**

2nd HEART SOUND
Valve flaps, blood and vessel walls vibrate when **semilunar valves** close at the beginning of **ventricular diastole**

AORTIC VALVE
PULMONARY VALVE
TRICUSPID VALVE
MITRAL VALVE

○'s indicate areas at which sounds produced by each valve are best heard

Lower Pitch
Softer
Longer

Higher Pitch
Sharper
Shorter

The sounds may be represented phonetically:-

LUBB   DUP   LUBB   DUP

| 1 | 2 | 1 | 2 | 1 | 2 |

VENT SYST   VENTRICULAR DIASTOLE

They are repeated with every **cardiac cycle** i.e about 70 times per minute in the average healthy adult at rest.

If the valves have been damaged by disease additional sounds (**murmurs**) can be heard as the blood flows forwards through narrowed valves or leaks backwards through incompetent valves.

If the sounds of the heart are amplified, a third and a fourth heart sound can be detected. They are occasionally heard with a stethoscope over normal hearts.

ORIGIN AND CONDUCTION OF THE HEART BEAT

The rhythmic contraction of the heart is called the **heart beat**. The impulse to contract is generated in specialized cardiac muscle called the **sino-atrial node** lying in the wall of the right atrium.

Spontaneous impulses are discharged rhythmically from this **SINO-ATRIAL NODE** (The 'Pacemaker')

The wave of excitation spreads through three bundles of Purkinje-like tissue and the muscle of both ATRIA which are excited to contract.

The impulse is then conducted more slowly through another mass of NODAL TISSUE – the **ATRIO-VENTRICULAR NODE** and relayed by **PURKINJE TISSUE** (in Bundle of His and its branches) throughout the endocardium and then to the muscle of both ventricles. The impulse thus gets *very rapidly* to all the ventricular muscle so that all parts contract almost simultaneously.

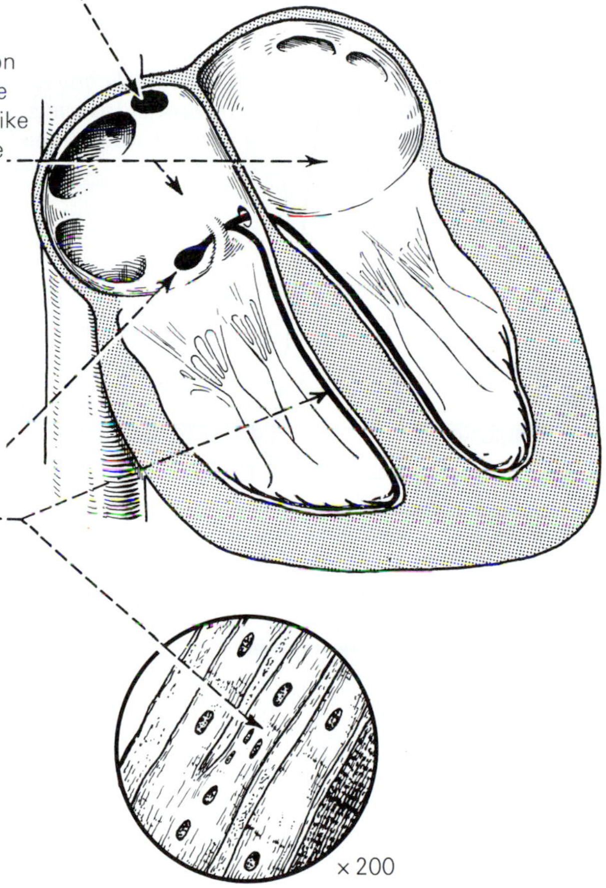

The right atrium starts contracting before left atrium.

A ring of fibrous tissue separates atria from ventricles. The heart beat is not transmitted from atria to ventricles directly by ordinary **cardiac** muscle.

Both ventricles contract together.

× 200

The wave of excitation is accompanied by an electrical change which is followed within 0.02 seconds by contraction of the cardiac muscle.

ELECTROCARDIOGRAM

The wave of excitation spreading through the heart wall is accompanied by electrical changes. Like nerve and skeletal muscle, the surface of **active** cardiac muscle is electrically negative relative to the **resting** cardiac muscle ahead of the zone of excitation. The electrical currents produced are conducted directly through the salty-water-like body fluids to the surface of the body and can be picked up, amplified and recorded by a special instrument – the **electrocardiograph**. The record obtained is the **electrocardiogram**. (ECG)

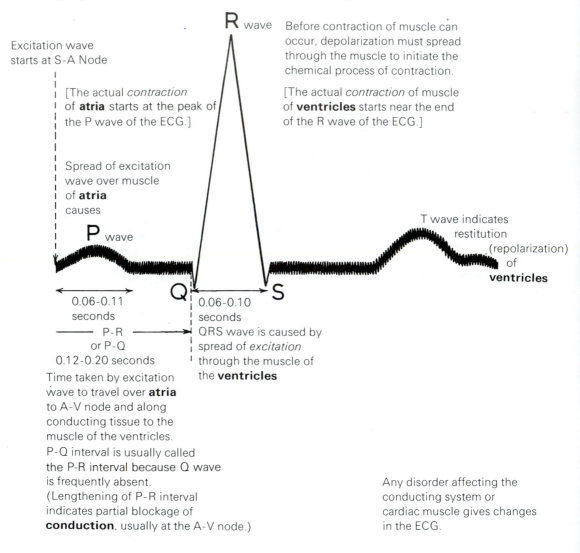

R wave

Excitation wave starts at S-A Node

Before contraction of muscle can occur, depolarization must spread through the muscle to initiate the chemical process of contraction.

[The actual *contraction* of **atria** starts at the peak of the P wave of the ECG.]

[The actual *contraction* of muscle of **ventricles** starts near the end of the R wave of the ECG.]

Spread of excitation wave over muscle of **atria** causes

P wave

T wave indicates restitution (repolarization) of **ventricles**

Q — S

0.06–0.11 seconds

0.06–0.10 seconds

P-R or P-Q 0.12–0.20 seconds

QRS wave is caused by spread of *excitation* through the muscle of the **ventricles**

Time taken by excitation wave to travel over **atria** to A-V node and along conducting tissue to the muscle of the ventricles. P-Q interval is usually called the P-R interval because Q wave is frequently absent. (Lengthening of P-R interval indicates partial blockage of **conduction**, usually at the A-V node.)

Any disorder affecting the conducting system or cardiac muscle gives changes in the ECG.

NERVOUS REGULATION OF ACTION OF HEART

Although the heart initiates its own impulse to contraction its activity is finely adjusted, to meet the body's constantly changing needs, by nervous impulses discharged from **controlling centres** in the **brain** and **spinal cord** along **parasympathetic** and **sympathetic outflows.**

ACTION of PARASYMPATHETIC –
Continuous stream of impulses to the **pacemaker** tends to restrain (−) Heart's Action:-
slows heart rate,
conduction at A-V node delayed,
force of contraction decreased,
excitability decreased.

[Variation in this
'vagal tone'
is chief factor
in giving
alteration
of Heart
Rate.
At rest the
Parasympathetic
influence
is dominant.]

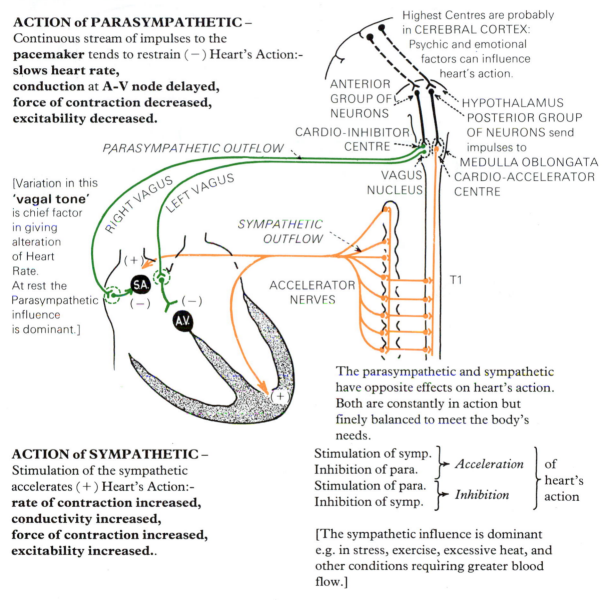

Highest Centres are probably in CEREBRAL CORTEX: Psychic and emotional factors can influence heart's action.

ANTERIOR GROUP OF NEURONS

CARDIO-INHIBITOR CENTRE

HYPOTHALAMUS
POSTERIOR GROUP OF NEURONS send impulses to

VAGUS NUCLEUS

MEDULLA OBLONGATA
CARDIO-ACCELERATOR CENTRE

PARASYMPATHETIC OUTFLOW

RIGHT VAGUS   LEFT VAGUS

SYMPATHETIC OUTFLOW

ACCELERATOR NERVES

T1

(+)

S.A.

(−)   (−)

A.V.

(+)

The parasympathetic and sympathetic have opposite effects on heart's action. Both are constantly in action but finely balanced to meet the body's needs.

ACTION of SYMPATHETIC –
Stimulation of the sympathetic accelerates (+) Heart's Action:-
rate of contraction increased,
conductivity increased,
force of contraction increased,
excitability increased..

Stimulation of symp.
Inhibition of para. } *Acceleration* } of
Stimulation of para. } } heart's
Inhibition of symp. } *Inhibition* } action

[The sympathetic influence is dominant e.g. in stress, exercise, excessive heat, and other conditions requiring greater blood flow.]

CARDIAC REFLEXES

There are **ingoing (sensory)** fibres travelling in the **parasympathetic** nerves which convey information to the **medulla** about events taking place in the heart.

    These afferent messages do not normally reach consciousness. They are important as the **afferent** pathways in **cardiac reflexes** by means of which heart's action is adjusted to body's needs.

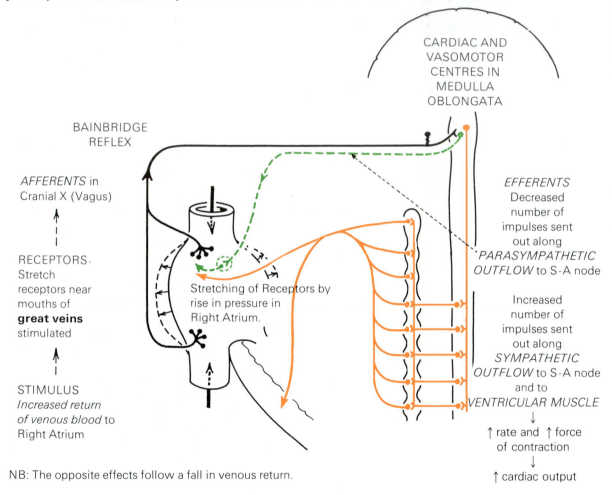

CARDIAC AND VASOMOTOR CENTRES IN MEDULLA OBLONGATA

BAINBRIDGE REFLEX

AFFERENTS in Cranial X (Vagus)

RECEPTORS. Stretch receptors near mouths of **great veins** stimulated

STIMULUS *Increased return of venous blood* to Right Atrium

Stretching of Receptors by rise in pressure in Right Atrium.

EFFERENTS Decreased number of impulses sent out along *PARASYMPATHETIC OUTFLOW* to S-A node

Increased number of impulses sent out along *SYMPATHETIC OUTFLOW* to S-A node and to *VENTRICULAR MUSCLE*
↓
↑ rate and ↑ force of contraction
↓
↑ cardiac output

NB: The opposite effects follow a fall in venous return.

    It is an important adaptive mechanism whereby heart rate and force of contraction are reflexly adjusted to match the quantity of venous blood returning to the heart.

One of the most important of these is the **baroreceptor reflex** for reflex control of arterial blood pressure.

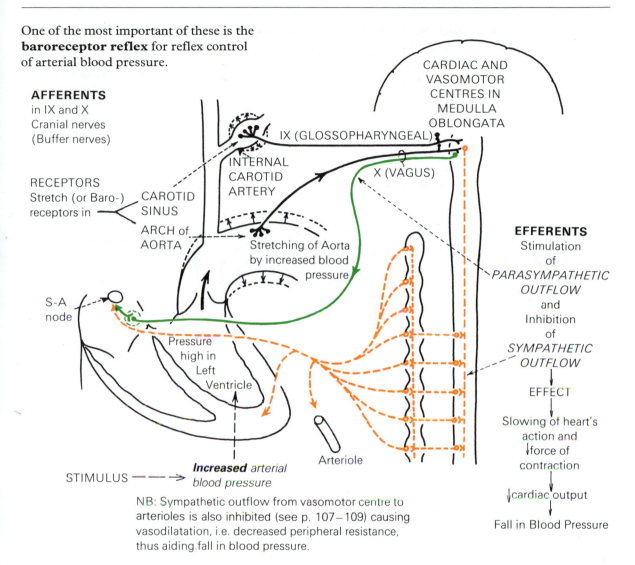

AFFERENTS
in IX and X
Cranial nerves
(Buffer nerves)

RECEPTORS
Stretch (or Baro-)
receptors in

CAROTID
SINUS

ARCH of
AORTA

INTERNAL
CAROTID
ARTERY

IX (GLOSSOPHARYNGEAL)

X (VAGUS)

CARDIAC AND
VASOMOTOR
CENTRES IN
MEDULLA
OBLONGATA

Stretching of Aorta
by increased blood
pressure

S-A
node

Pressure
high in
Left
Ventricle

Arteriole

STIMULUS - - - → **Increased** *arterial*
blood pressure

EFFERENTS
Stimulation
of
*PARASYMPATHETIC
OUTFLOW*
and
Inhibition
of
*SYMPATHETIC
OUTFLOW*

EFFECT

Slowing of heart's
action and
↓force of
contraction

↓cardiac output

Fall in Blood Pressure

NB: Sympathetic outflow from vasomotor centre to arterioles is also inhibited (see p. 107–109) causing vasodilatation, i.e. decreased peripheral resistance, thus aiding fall in blood pressure.

A **decrease** in arterial blood pressure has the opposite effects.

The baroreceptor reflex reduces the *short-term* variation in arterial pressure to about half that which would occur if there was no baroreceptor reflex present. The *long-term* control of blood pressure – over days or weeks – is determined by body fluid balance which is mainly controlled by the kidney.

CARDIAC OUTPUT

The volume of blood expelled from the heart can be measured using what is called the **Fick principle** as follows:

A semi-rigid tube is inserted into a **vein** in the arm and passed along into the **right atrium** of the heart – to obtain a sample of **mixed venous blood** – which has given up some of its oxygen to the tissues. The oxygen content is analysed.

100 ml VENOUS Blood hold 14 ml OXYGEN.

The amount of **oxygen** taken up by the lungs per minute is measured by a **spirometer**.

250 ml OXYGEN are removed from the lungs by blood each minute.

A needle is inserted into an **artery** in the leg and a sample of **arterial blood** – which has received its fresh oxygen supply in the lungs – is obtained. The oxygen content is analysed.

100 ml ARTERIAL Blood hold 19 ml OXYGEN.

(after wishart)

Each 100 ml blood gains 5 ml oxygen as it passes through the lungs. The blood in the lungs takes up 250 ml oxygen from the atmosphere per minute.

Therefore there must be $\left(\frac{250}{5} \times 100 \right)$ ml

[i.e. 5000 ml] of blood leaving the right ventricle and passing through the lungs to the left atrium per minute to take up this 250 ml oxygen.

The same volume of blood must leave the left ventricle and enter the aorta in the same time otherwise blood would soon be dammed back in the lungs; i.e. If heart contracts 72 times per minute

stroke volume $= \frac{5000}{72} \approx 70$ ml per beat for each ventricle.

A more popular method for measuring cardiac output is the **thermodilution** technique. A bolus of cold saline is injected into the **right atrium**. The temperature change that this causes in the **pulmonary artery** is recorded with a thermister. This change is proportional to the cardiac output which can be computed. Other techniques employ the use of computers along with **echocardiography** and **transthoracic impedance** measurements.

In **exercise**, cardiac output can be increased to 30 litres per minute mainly by increased **heart rate** but partly by increased stroke volume.

The system of tubes through which the heart pumps blood.

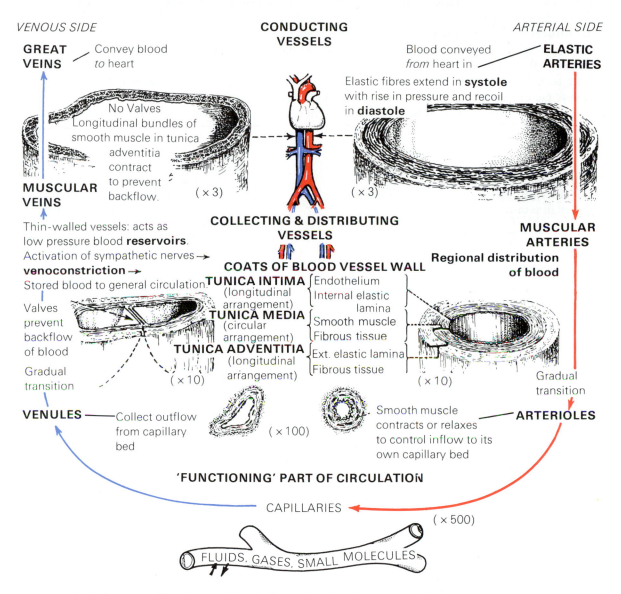

VENOUS SIDE

CONDUCTING VESSELS

ARTERIAL SIDE

GREAT VEINS — Convey blood *to* heart

Blood conveyed *from* heart in — **ELASTIC ARTERIES**

No Valves
Longitudinal bundles of smooth muscle in tunica adventitia contract to prevent backflow.

(× 3)

Elastic fibres extend in **systole** with rise in pressure and recoil in **diastole**

(× 3)

MUSCULAR VEINS

Thin-walled vessels: acts as low pressure blood **reservoirs**.
Activation of sympathetic nerves →
venoconstriction →
Stored blood to general circulation

COLLECTING & DISTRIBUTING VESSELS

MUSCULAR ARTERIES

Regional distribution of blood

COATS OF BLOOD VESSEL WALL

TUNICA INTIMA (longitudinal arrangement)
Endothelium
Internal elastic lamina

TUNICA MEDIA (circular arrangement)
Smooth muscle
Fibrous tissue

TUNICA ADVENTITIA (longitudinal arrangement)
Ext. elastic lamina
Fibrous tissue

Valves prevent backflow of blood

Gradual transition

(× 10)

(× 10)

Gradual transition

VENULES — Collect outflow from capillary bed

(× 100)

Smooth muscle contracts or relaxes to control inflow to its own capillary bed

ARTERIOLES

'FUNCTIONING' PART OF CIRCULATION

CAPILLARIES

(× 500)

FLUIDS, GASES, SMALL MOLECULES

Only from **capillaries** can blood give up food and oxygen to tissues, and receive waste products and carbon dioxide from tissues.

BLOOD PRESSURE

Each ventricle at each heart beat ejects forcibly about 70 ml of blood into the blood vessels. All of this blood cannot pass through arterioles into capillaries and veins during one contraction of the heart. This means that roughly ⅚ of the **cardiac output** at each heart beat has to be stored during systole and passed on during diastole.

Conducting arteries are always more or less stretched.

Peripheral resistance is offered to the passage of blood from arterial to venous side of the system chiefly by partial constriction ('tone') of smooth muscle in walls of arterioles. (The calibre is regulated mainly by action of the sympathetic nervous system – see pages 107–109).

These factors are largely responsible for the considerable pressure of the blood in the arterial system. The pressure is highest at the height of the heart's contraction, i.e. **systolic blood pressure**, and lowest when the heart is relaxing, i.e. **diastolic blood pressure**.

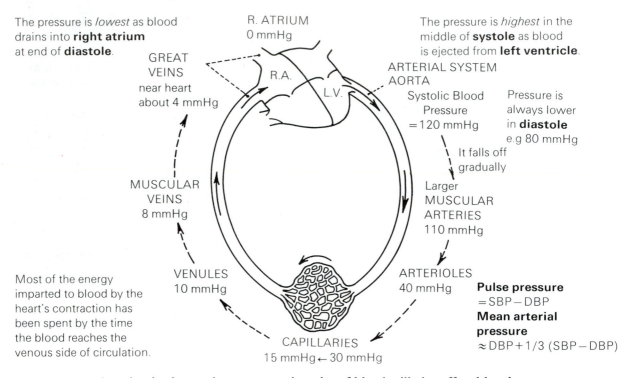

The pressure is *lowest* as blood drains into **right atrium** at end of **diastole**.

R. ATRIUM
0 mmHg

The pressure is *highest* in the middle of **systole** as blood is ejected from **left ventricle**.

GREAT VEINS
near heart
about 4 mmHg

R.A.

L.V.

ARTERIAL SYSTEM
AORTA
Systolic Blood
Pressure
= 120 mmHg

Pressure is always lower in **diastole** e.g 80 mmHg

It falls off gradually

MUSCULAR VEINS
8 mmHg

Larger
MUSCULAR
ARTERIES
110 mmHg

Most of the energy imparted to blood by the heart's contraction has been spent by the time the blood reaches the venous side of circulation.

VENULES
10 mmHg

ARTERIOLES
40 mmHg

Pulse pressure
= SBP − DBP
Mean arterial pressure
≈ DBP + 1/3 (SBP − DBP)

CAPILLARIES
15 mmHg ← 30 mmHg

NOTE:- Any alteration in the **total amount** or **viscosity** of blood will also affect **blood pressure**.

When standing still, the effect of **gravity** on the venous blood in the legs increases the pressure in the veins of the feet to 90 mmHg. Muscle movements lower this pressure.

MEASUREMENT OF ARTERIAL BLOOD PRESSURE

The arterial blood pressure is measured in man by means of a **sphygmomanometer**.

This consists of a rubber bag (covered with a cloth envelope) which is wrapped round the upper arm over the **brachial artery**.
One tube connects the inside of the bag with a **manometer** containing **mercury**.

Another tube connects the inside of the bag to a hand operated **pump** with a release **valve**.

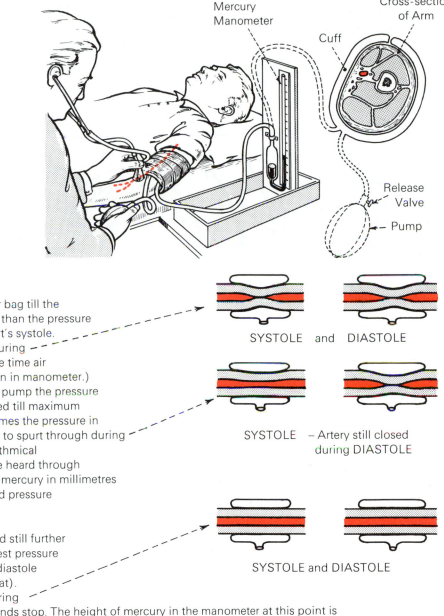

Mercury Manometer

Cross-section of Arm

Cuff

Release Valve

Pump

SYSTOLE   and   DIASTOLE

SYSTOLE   – Artery still closed
              during DIASTOLE

SYSTOLE and DIASTOLE

METHOD
Air is pumped into the rubber bag till the pressure in the cuff is greater than the pressure in the artery even during heart's systole. Artery is then closed down during systole and diastole. (At same time air is pushing up mercury column in manometer.) By releasing the valve on the pump the pressure in the cuff is gradually reduced till maximum pressure in artery just overcomes the pressure in the cuff – some blood begins to spurt through during systole At this point *faint* rhythmical **tapping sounds** begin to be heard through **stethoscope**. The height of mercury in millimetres is taken as the **systolic** blood pressure (e.g. 120 mmHg).

Pressure in the cuff is reduced still further till it is just less than the lowest pressure in artery towards the end of diastole (i.e. just before next heart beat). Blood flow is unimpeded during systole and diastole. The sounds stop. The height of mercury in the manometer at this point is taken as the **diastolic** Blood Pressure (e.g. about 80 mmHg)

These values differ with **sex, age, exercise, sleep**, etc.

ELASTIC ARTERIES

The large **conducting arteries** near the **heart** are **elastic arteries**.

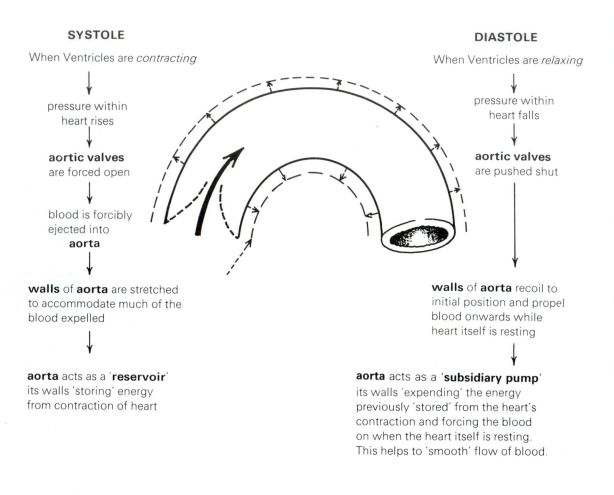

SYSTOLE

When Ventricles are *contracting*

↓

pressure within
heart rises

↓

aortic valves
are forced open

↓

blood is forcibly
ejected into
aorta

↓

walls of **aorta** are stretched
to accommodate much of the
blood expelled

↓

aorta acts as a 'reservoir'
its walls 'storing' energy
from contraction of heart

DIASTOLE

When Ventricles are *relaxing*

↓

pressure within
heart falls

↓

aortic valves
are pushed shut

↓

walls of **aorta** recoil to
initial position and propel
blood onwards while
heart itself is resting

↓

aorta acts as a '**subsidiary pump**'
its walls 'expending' the energy
previously 'stored' from the heart's
contraction and forcing the blood
on when the heart itself is resting.
This helps to 'smooth' flow of blood.

As blood is pumped from the heart during systole, this distension and increase in pressure which starts in the aorta passes along the whole arterial system as a wave – the **pulse wave**.

The expansion and subsequent relaxation of the wall of the radial artery can be felt as '**the pulse**' at the wrist.

A great increase in blood pressure can result if these walls lose some of their elasticity with age or disease and can no longer stretch readily to accommodate so much of the heart's output during systole: nor recoil so far in diastole.

The systolic and diastolic values may both be higher.

The smooth muscle in **arterioles** can contract or relax to alter calibre of the vessels, and thus adjust distribution of blood to meet the constantly changing needs of different parts of the body and also to maintain the normal blood pressure.

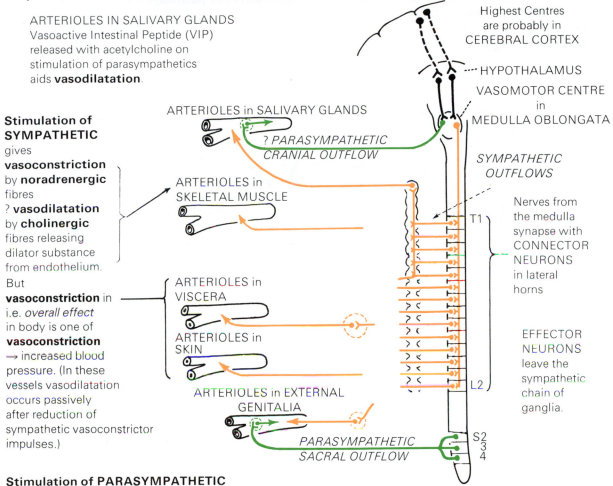

ARTERIOLES IN SALIVARY GLANDS
Vasoactive Intestinal Peptide (VIP) released with acetylcholine on stimulation of parasympathetics aids **vasodilatation**.

Stimulation of SYMPATHETIC
gives
vasoconstriction
by **noradrenergic**
fibres
? vasodilatation
by **cholinergic**
fibres releasing
dilator substance
from endothelium.
But
vasoconstriction in
i.e. *overall effect*
in body is one of
vasoconstriction
→ increased blood
pressure. (In these
vessels vasodilatation
occurs passively
after reduction of
sympathetic vasoconstrictor
impulses.)

ARTERIOLES in SALIVARY GLANDS

? PARASYMPATHETIC CRANIAL OUTFLOW

ARTERIOLES in SKELETAL MUSCLE

ARTERIOLES in VISCERA

ARTERIOLES in SKIN

ARTERIOLES in EXTERNAL GENITALIA

PARASYMPATHETIC SACRAL OUTFLOW

Highest Centres are probably in CEREBRAL CORTEX

---- HYPOTHALAMUS

VASOMOTOR CENTRE in MEDULLA OBLONGATA

SYMPATHETIC OUTFLOWS

T1

Nerves from the medulla synapse with CONNECTOR NEURONS in lateral horns

EFFECTOR NEURONS leave the sympathetic chain of ganglia.

L2

S2
3
4

Stimulation of PARASYMPATHETIC
appears to give **vasodilatation**
only in vessels to **external genitalia** and **salivary glands**.

Veins have a sympathetic nerve supply. When activated these cause generalized venoconstriction to mobilize the blood held in 'venous reservoirs'.

REFLEX AND CHEMICAL REGULATION OF ARTERIOLAR TONE – I

Afferent impulses are constantly reaching the **vasomotor centre** in the medulla oblongata in ingoing nerves from all parts of the body. These form the **afferent pathways** for **vasomotor reflexes**.

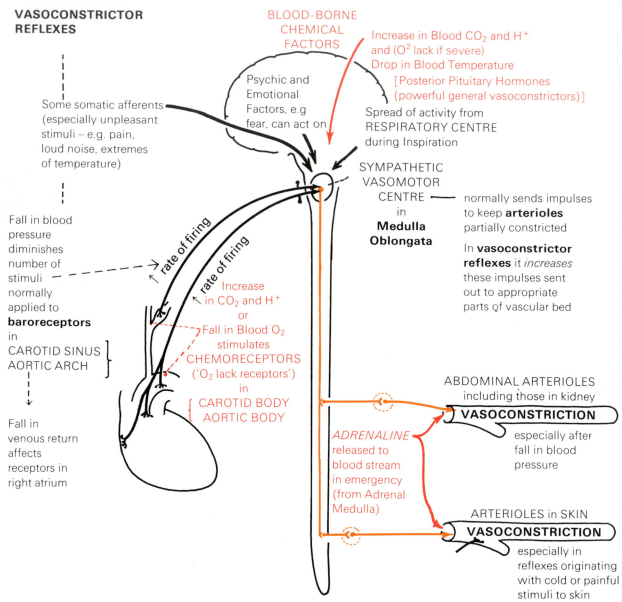

VASOCONSTRICTOR REFLEXES

BLOOD-BORNE CHEMICAL FACTORS

Increase in Blood CO_2 and H^+ and (O^2 lack if severe)
Drop in Blood Temperature
[Posterior Pituitary Hormones (powerful general vasoconstrictors)]

Psychic and Emotional Factors, e.g fear, can act on

Some somatic afferents (especially unpleasant stimuli – e.g. pain, loud noise, extremes of temperature)

Spread of activity from RESPIRATORY CENTRE during Inspiration

SYMPATHETIC VASOMOTOR CENTRE in **Medulla Oblongata**

normally sends impulses to keep **arterioles** partially constricted

In **vasoconstrictor reflexes** it *increases* these impulses sent out to appropriate parts of vascular bed

rate of firing

rate of firing

Fall in blood pressure diminishes number of stimuli normally applied to **baroreceptors** in CAROTID SINUS AORTIC ARCH

Increase in CO_2 and H^+ or Fall in Blood O_2 stimulates CHEMORECEPTORS ('O_2 lack receptors') in CAROTID BODY AORTIC BODY

Fall in venous return affects receptors in right atrium

ABDOMINAL ARTERIOLES including those in kidney
VASOCONSTRICTION
especially after fall in blood pressure

ADRENALINE released to blood stream in emergency (from Adrenal Medulla)

ARTERIOLES in SKIN
VASOCONSTRICTION
especially in reflexes originating with cold or painful stimuli to skin

Widespread **vasoconstriction** increases the peripheral resistance and gives rise in blood pressure.

REFLEX AND CHEMICAL REGULATION OF ARTERIOLAR TONE – II

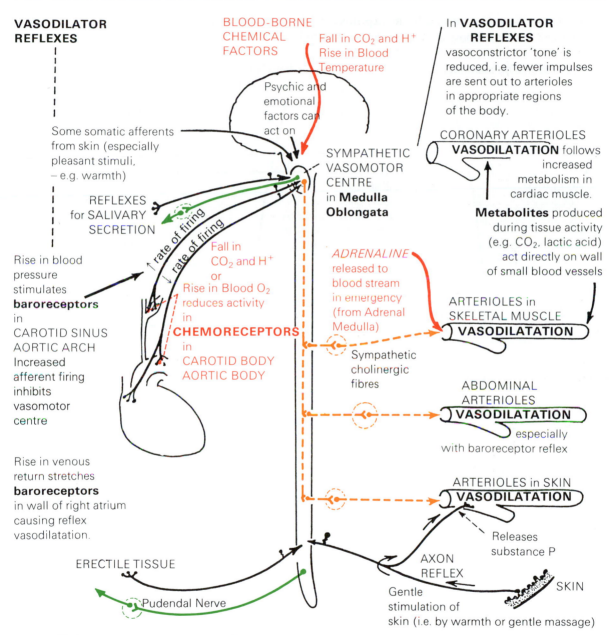

VASODILATOR REFLEXES

BLOOD-BORNE CHEMICAL FACTORS

Fall in CO_2 and H^+ Rise in Blood Temperature

Psychic and emotional factors can act on

In **VASODILATOR REFLEXES** vasoconstrictor 'tone' is reduced, i.e. fewer impulses are sent out to arterioles in appropriate regions of the body.

Some somatic afferents from skin (especially pleasant stimuli, – e.g. warmth)

REFLEXES for SALIVARY SECRETION

SYMPATHETIC VASOMOTOR CENTRE in **Medulla Oblongata**

CORONARY ARTERIOLES **VASODILATATION** follows increased metabolism in cardiac muscle.

Metabolites produced during tissue activity (e.g. CO_2, lactic acid) act directly on wall of small blood vessels

↑ rate of firing   ↓ rate of firing

Rise in blood pressure stimulates **baroreceptors** in CAROTID SINUS AORTIC ARCH Increased afferent firing inhibits vasomotor centre

Fall in CO_2 and H^+ or Rise in Blood O_2 reduces activity in **CHEMORECEPTORS** in CAROTID BODY AORTIC BODY

ADRENALINE released to blood stream in emergency (from Adrenal Medulla)

ARTERIOLES in SKELETAL MUSCLE **VASODILATATION**

Sympathetic cholinergic fibres

ABDOMINAL ARTERIOLES **VASODILATATION** especially with baroreceptor reflex

Rise in venous return stretches **baroreceptors** in wall of right atrium causing reflex vasodilatation.

ARTERIOLES in SKIN **VASODILATATION**

Releases substance P

ERECTILE TISSUE

Pudendal Nerve

AXON REFLEX

SKIN

Gentle stimulation of skin (i.e. by warmth or gentle massage)

Widespread **vasodilatation** decreases peripheral resistance and gives a fall in blood pressure.

CAPILLARIES

Blood is distributed by **arterioles** to **capillaries**. Through the semi-permeable walls of capillaries the sole purpose of the circulating system is achieved.

Forces determining **exchange** of **water** and **electrolytes** between
BLOOD———— and ————**INTERSTITIAL FLUID**

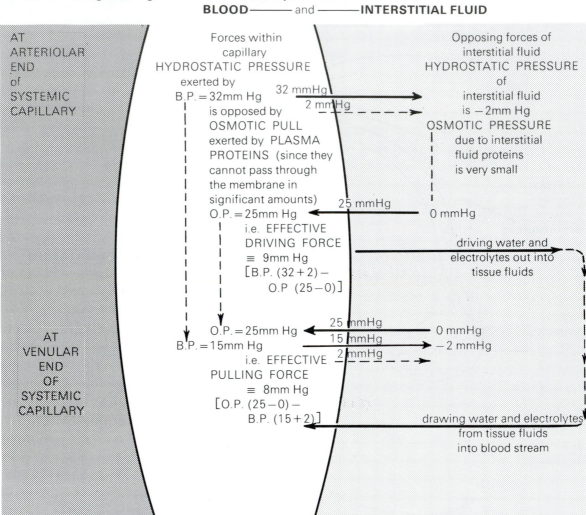

These exchange in systemic capillaries result in a continuous turnover and renewal of **interstitial fluid**.

    Electrolytes (Crystalloids, e.g. Na^+, Cl^-, etc.) in plasma and interstitial fluid also exert an **osmotic pressure** (OP) – this is huge (about 6000 mmHg). As the electrolyte concentration is the same on each side of the capillary membrane the **crystalloid OP** *does not* affect fluid movement. Protein is confined mainly to the plasma hence its OP *does* affect fluid movement.

Capillaries unite to form veins which convey blood back to the heart. By the time blood reaches the veins much of the force imparted to it by the heart's contraction has been spent but some remains.

VENOUS RETURN to the heart depends on various factors:-

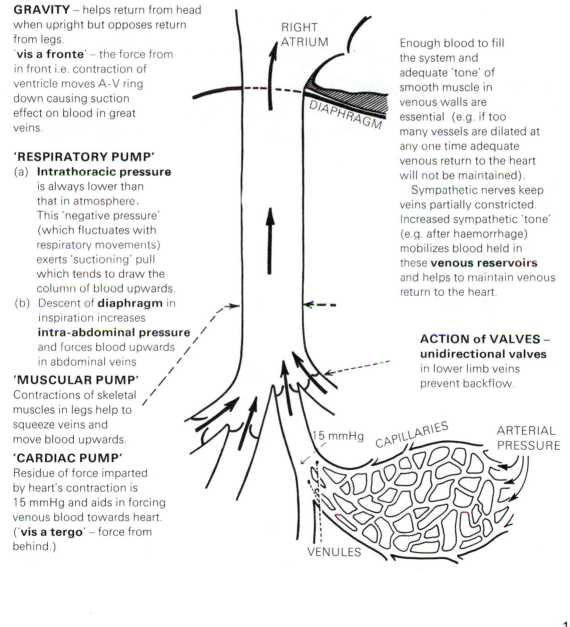

GRAVITY – helps return from head when upright but opposes return from legs.
'**vis a fronte**' – the force from in front i.e. contraction of ventricle moves A-V ring down causing suction effect on blood in great veins.

'RESPIRATORY PUMP'
(a) **Intrathoracic pressure** is always lower than that in atmosphere. This 'negative pressure' (which fluctuates with respiratory movements) exerts 'suctioning' pull which tends to draw the column of blood upwards.
(b) Descent of **diaphragm** in inspiration increases **intra-abdominal pressure** and forces blood upwards in abdominal veins

'MUSCULAR PUMP'
Contractions of skeletal muscles in legs help to squeeze veins and move blood upwards.

'CARDIAC PUMP'
Residue of force imparted by heart's contraction is 15 mmHg and aids in forcing venous blood towards heart. ('**vis a tergo**' – force from behind.)

RIGHT ATRIUM

DIAPHRAGM

Enough blood to fill the system and adequate 'tone' of smooth muscle in venous walls are essential (e.g. if too many vessels are dilated at any one time adequate venous return to the heart will not be maintained).
Sympathetic nerves keep veins partially constricted. Increased sympathetic 'tone' (e.g. after haemorrhage) mobilizes blood held in these **venous reservoirs** and helps to maintain venous return to the heart.

ACTION of VALVES – **unidirectional valves** in lower limb veins prevent backflow.

15 mmHg

CAPILLARIES

ARTERIAL PRESSURE

VENULES

111

BLOOD FLOW

The rate of blood flow varies in different parts of the vascular system. It is rapid in large vessels; slower in small vessels. [The larger the *total* cross-sectional area represented the slower the rate of flow.]

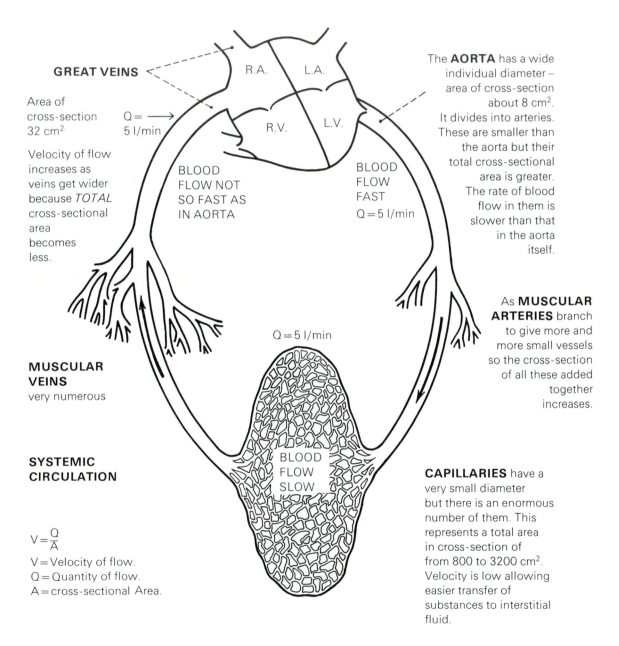

GREAT VEINS

Area of cross-section 32 cm^2

$Q = 5$ l/min

Velocity of flow increases as veins get wider because *TOTAL* cross-sectional area becomes less.

R.A.   L.A.

R.V.   L.V.

BLOOD FLOW NOT SO FAST AS IN AORTA

BLOOD FLOW FAST $Q = 5$ l/min

The **AORTA** has a wide individual diameter – area of cross-section about 8 cm^2. It divides into arteries. These are smaller than the aorta but their total cross-sectional area is greater. The rate of blood flow in them is slower than that in the aorta itself.

MUSCULAR VEINS
very numerous

$Q = 5$ l/min

As **MUSCULAR ARTERIES** branch to give more and more small vessels so the cross-section of all these added together increases.

SYSTEMIC CIRCULATION

BLOOD FLOW SLOW

$V = \dfrac{Q}{A}$

V = Velocity of flow.
Q = Quantity of flow.
A = cross-sectional Area.

CAPILLARIES have a very small diameter but there is an enormous number of them. This represents a total area in cross-section of from 800 to 3200 cm^2. Velocity is low allowing easier transfer of substances to interstitial fluid.

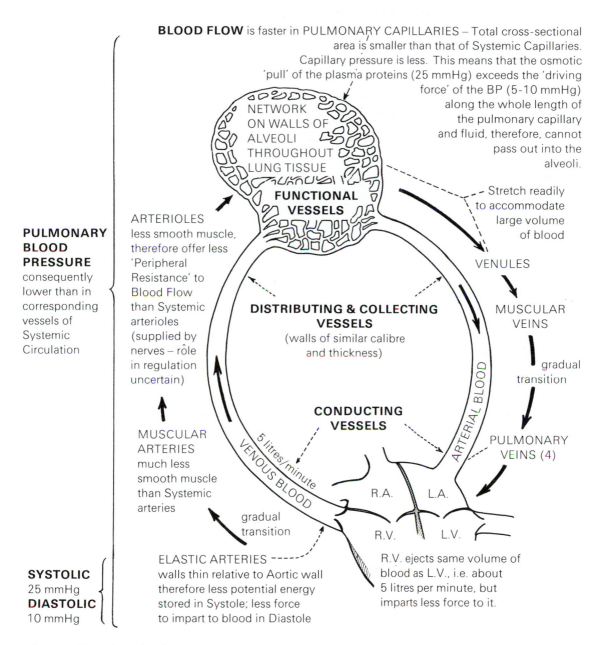

BLOOD FLOW is faster in PULMONARY CAPILLARIES – Total cross-sectional area is smaller than that of Systemic Capillaries. Capillary pressure is less. This means that the osmotic 'pull' of the plasma proteins (25 mmHg) exceeds the 'driving force' of the BP (5-10 mmHg) along the whole length of the pulmonary capillary and fluid, therefore, cannot pass out into the alveoli.

NETWORK ON WALLS OF ALVEOLI THROUGHOUT LUNG TISSUE

FUNCTIONAL VESSELS

Stretch readily to accommodate large volume of blood

ARTERIOLES less smooth muscle, therefore offer less 'Peripheral Resistance' to Blood Flow than Systemic arterioles (supplied by nerves – rôle in regulation uncertain)

VENULES

MUSCULAR VEINS

gradual transition

PULMONARY BLOOD PRESSURE consequently lower than in corresponding vessels of Systemic Circulation

DISTRIBUTING & COLLECTING VESSELS (walls of similar calibre and thickness)

ARTERIAL BLOOD

CONDUCTING VESSELS

PULMONARY VEINS (4)

5 litres/minute VENOUS BLOOD

MUSCULAR ARTERIES much less smooth muscle than Systemic arteries

R.A.   L.A.

R.V.   L.V.

gradual transition

SYSTOLIC 25 mmHg **DIASTOLIC** 10 mmHg

ELASTIC ARTERIES walls thin relative to Aortic wall therefore less potential energy stored in Systole; less force to impart to blood in Diastole

R.V. ejects same volume of blood as L.V., i.e. about 5 litres per minute, but imparts less force to it.

A great dilatation of pulmonary vessels occurs in exercise.

DISTRIBUTION OF WATER AND ELECTROLYTES IN BODY FLUIDS

Water makes up about 60% of adult human body, i.e. about 42 litres in a 70 kg man.

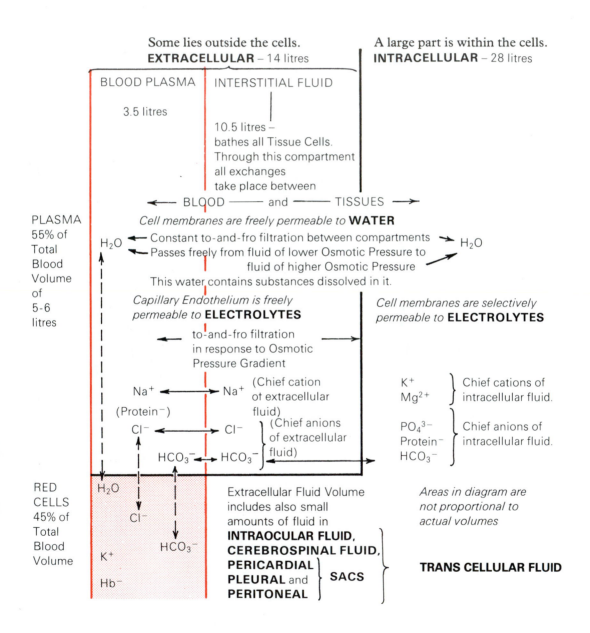

Some lies outside the cells.
EXTRACELLULAR – 14 litres

A large part is within the cells.
INTRACELLULAR – 28 litres

BLOOD PLASMA

3.5 litres

INTERSTITIAL FLUID

10.5 litres –
bathes all Tissue Cells.
Through this compartment
all exchanges
take place between

← BLOOD ——— and ——— TISSUES →

PLASMA
55% of
Total
Blood
Volume
of
5-6
litres

Cell membranes are freely permeable to **WATER**

H_2O ← Constant to-and-fro filtration between compartments → H_2O
← Passes freely from fluid of lower Osmotic Pressure to
fluid of higher Osmotic Pressure
This water contains substances dissolved in it.

*Capillary Endothelium is freely
permeable to* **ELECTROLYTES**

← to-and-fro filtration
in response to Osmotic
Pressure Gradient

*Cell membranes are selectively
permeable to* **ELECTROLYTES**

Na^+ ⟷ Na^+ (Chief cation
of extracellular
fluid)
(Protein⁻)
Cl^- ⟷ Cl^- (Chief anions
of extracellular
HCO_3^- ⟷ HCO_3^- fluid)

K^+
Mg^{2+} } Chief cations of
intracellular fluid.

PO_4^{3-}
Protein⁻
HCO_3^- } Chief anions of
intracellular fluid.

RED
CELLS
45% of
Total
Blood
Volume

H_2O
Cl^-
HCO_3^-
K^+
Hb^-

Extracellular Fluid Volume
includes also small
amounts of fluid in
**INTRAOCULAR FLUID,
CEREBROSPINAL FLUID,
PERICARDIAL
PLEURAL** and } **SACS**
PERITONEAL

*Areas in diagram are
not proportional to
actual volumes*

TRANS CELLULAR FLUID

In health the total amount of body water (and salt) is kept reasonably constant in spite of wide fluctuations in daily intake.

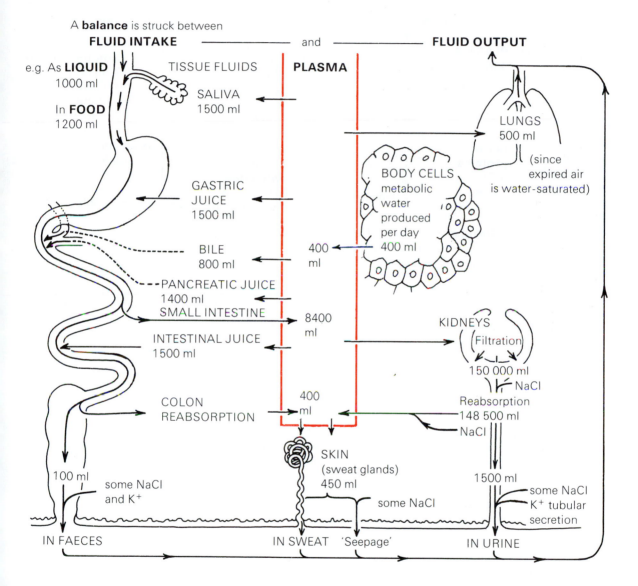

A **balance** is struck between

FLUID INTAKE ———————— and ———————— **FLUID OUTPUT**

e.g. As **LIQUID** 1000 ml

TISSUE FLUIDS   PLASMA

In **FOOD** 1200 ml

SALIVA 1500 ml

LUNGS 500 ml (since expired air is water-saturated)

GASTRIC JUICE 1500 ml

BODY CELLS metabolic water produced per day 400 ml

BILE 800 ml   400 ml ← 400 ml

PANCREATIC JUICE 1400 ml
SMALL INTESTINE   8400 ml

INTESTINAL JUICE 1500 ml

KIDNEYS (Filtration) 150 000 ml
NaCl Reabsorption 148 500 ml NaCl

COLON REABSORPTION   400 ml

100 ml some NaCl and K$^+$

SKIN (sweat glands) 450 ml   some NaCl

1500 ml some NaCl K$^+$ tubular secretion

IN FAECES   IN SWEAT   'Seepage'   IN URINE

Except in growth, convalescence or pregnancy, when new tissue is being formed, an *increase* or *decrease* in **intake** leads to an appropriate *increase* or *decrease* in **output** to maintain the **balance**.

BLOOD

Blood is the specialized fluid tissue of the transport system. (Specific Gravity, 1.055-1.065; pH, 7.3-7.4; average amount, 5.6 litres, about 8% of body weight.)

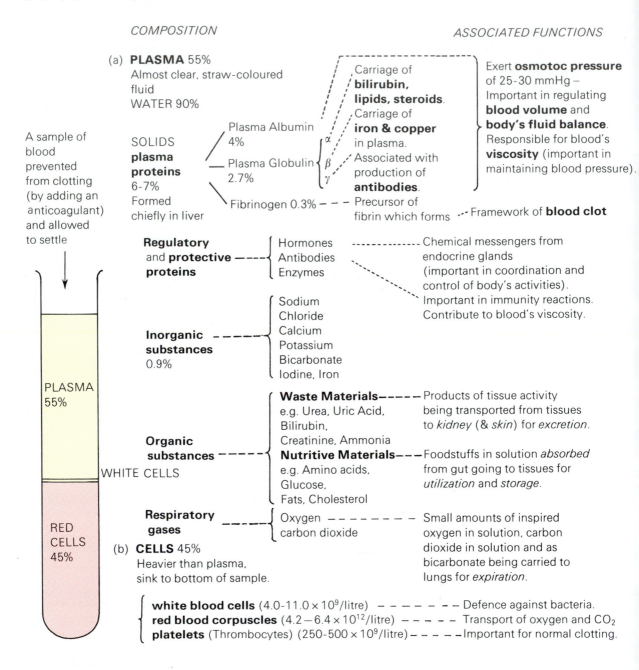

COMPOSITION

ASSOCIATED FUNCTIONS

A sample of blood prevented from clotting (by adding an anticoagulant) and allowed to settle

(a) **PLASMA** 55%
Almost clear, straw-coloured fluid
WATER 90%

SOLIDS
plasma proteins 6-7%
Formed chiefly in liver

Plasma Albumin 4%

Plasma Globulin 2.7% { α β γ

Fibrinogen 0.3%

Carriage of **bilirubin, lipids, steroids**.
Carriage of **iron & copper** in plasma.
Associated with production of **antibodies**.
Precursor of fibrin which forms

Exert **osmotoc pressure** of 25-30 mmHg –
Important in regulating **blood volume** and **body's fluid balance**. Responsible for blood's **viscosity** (important in maintaining blood pressure).

Framework of **blood clot**

Regulatory and **protective proteins** { Hormones, Antibodies, Enzymes

Chemical messengers from endocrine glands (important in coordination and control of body's activities).
Important in immunity reactions.
Contribute to blood's viscosity.

Inorganic substances 0.9% { Sodium, Chloride, Calcium, Potassium, Bicarbonate, Iodine, Iron

Organic substances {
Waste Materials e.g. Urea, Uric Acid, Bilirubin, Creatinine, Ammonia
Nutritive Materials e.g. Amino acids, Glucose, Fats, Cholesterol

Products of tissue activity being transported from tissues to *kidney* (& *skin*) for *excretion*.

Foodstuffs in solution *absorbed* from gut going to tissues for *utilization* and *storage*.

Respiratory gases { Oxygen, carbon dioxide

(b) **CELLS** 45%
Heavier than plasma, sink to bottom of sample.

Small amounts of inspired oxygen in solution, carbon dioxide in solution and as bicarbonate being carried to lungs for *expiration*.

PLASMA 55%

WHITE CELLS

RED CELLS 45%

{
white blood cells (4.0-11.0 × 10^9/litre) – – – – – – Defence against bacteria.
red blood corpuscles (4.2 – 6.4 × 10^{12}/litre) – – – – – Transport of oxygen and CO_2
platelets (Thrombocytes) (250-500 × 10^9/litre) – – – – – Important for normal clotting.

116

HAEMOSTASIS AND BLOOD COAGULATION

3 main processes are involved in **haemostasis**:-
On **Injury** – spasm of smooth muscle   1. **blood vessel wall constricts.**
   Platelets stick to collagen of damaged lining 2. **platelets plug gap**
   Serotonin – constricts vessels.
   ADP and **prostaglandins** attract more platelets.   3. **blood clots**

BLOOD COAGULATION is initiated when
PLASMA CONTACTS DAMAGED LINING
OF BLOOD VESSEL

(The roman numbers are clotting factors.
Activation is indicated by 'a'.)

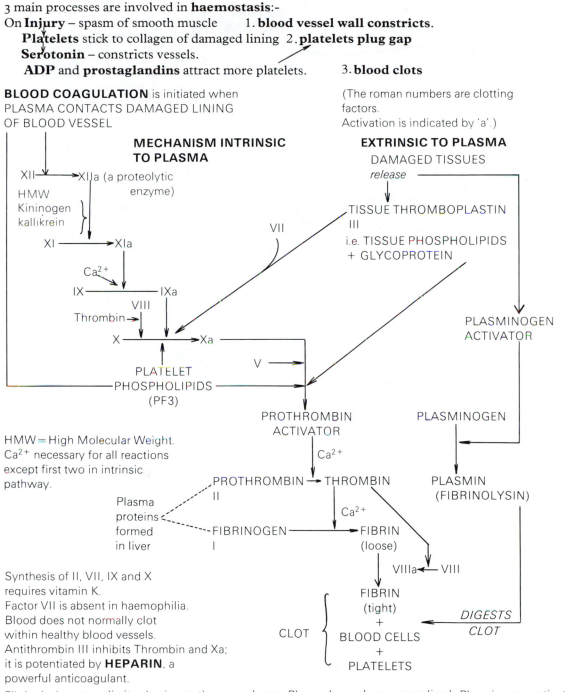

HMW = High Molecular Weight.
Ca^{2+} necessary for all reactions
except first two in intrinsic
pathway.

Synthesis of II, VII, IX and X
requires vitamin K.
Factor VII is absent in haemophilia.
Blood does not normally clot
within healthy blood vessels.
Antithrombin III inhibits Thrombin and Xa;
it is potentiated by **HEPARIN**, a
powerful anticoagulant.

Fibrinolysin system limits clotting to the wound area. Plugged vessels are recanalized. Plasminogen activators
include XIIa, streptokinase, urokinase.

FACTORS REQUIRED FOR NORMAL HAEMOPOIESIS

[Haemopoiesis: formation of normal blood cells.]

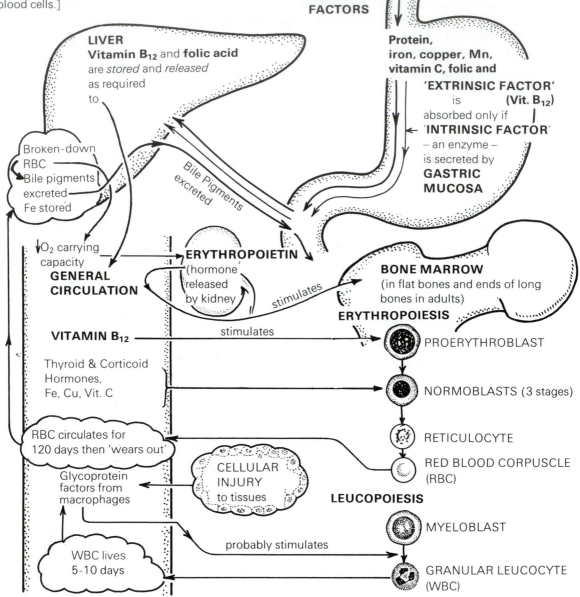

DIETARY FACTORS

LIVER
Vitamin B_{12} and **folic acid** are *stored* and *released* as required to

Protein, iron, copper, Mn, vitamin C, folic and

'EXTRINSIC FACTOR' is **(Vit. B_{12})** absorbed only if **'INTRINSIC FACTOR'** – an enzyme – is secreted by **GASTRIC MUCOSA**

Broken-down RBC
Bile pigments excreted
Fe stored

Bile Pigments excreted

↓O_2 carrying capacity
GENERAL CIRCULATION

ERYTHROPOIETIN (hormone released by kidney)   stimulates

BONE MARROW (in flat bones and ends of long bones in adults)
ERYTHROPOIESIS

VITAMIN B_{12}   stimulates

PROERYTHROBLAST

Thyroid & Corticoid Hormones, Fe, Cu, Vit. C

NORMOBLASTS (3 stages)

RETICULOCYTE

RBC circulates for 120 days then 'wears out'

RED BLOOD CORPUSCLE (RBC)

Glycoprotein factors from macrophages

CELLULAR INJURY to tissues

LEUCOPOIESIS

MYELOBLAST

WBC lives 5-10 days

probably stimulates

GRANULAR LEUCOCYTE (WBC)

In health the number of RBC and the amount of Hb in them remain fairly constant. Destruction of old red cells is balanced by formation of new.

In the adult the formed elements of the blood stream develop from primitive **reticular cells,** chiefly in **red bone marrow.**

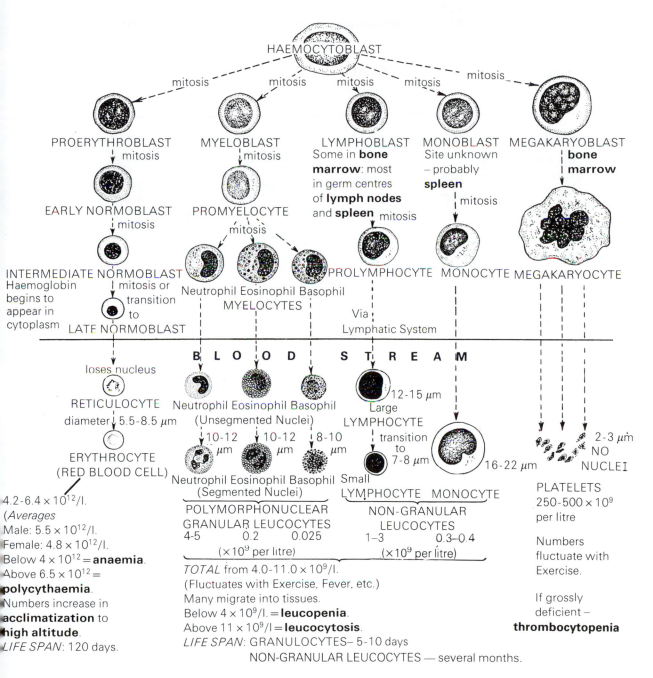

HAEMOCYTOBLAST

mitosis — mitosis — mitosis — mitosis — mitosis

PROERYTHROBLAST
mitosis

MYELOBLAST
mitosis

LYMPHOBLAST
Some in **bone marrow:** most in germ centres of **lymph nodes** and **spleen** — mitosis

MONOBLAST
Site unknown – probably **spleen** — mitosis

MEGAKARYOBLAST
bone marrow

EARLY NORMOBLAST
mitosis

PROMYELOCYTE
mitosis

INTERMEDIATE NORMOBLAST
Haemoglobin begins to appear in cytoplasm — mitosis or transition to LATE NORMOBLAST

Neutrophil Eosinophil Basophil
MYELOCYTES

PROLYMPHOCYTE

MONOCYTE

MEGAKARYOCYTE

Via Lymphatic System

loses nucleus

B L O O D   S T R E A M

RETICULOCYTE
diameter 5.5-8.5 μm

Neutrophil Eosinophil Basophil
(Unsegmented Nuclei)
10-12 μm 10-12 μm 8-10 μm

12-15 μm
Large LYMPHOCYTE
transition to 7-8 μm

16-22 μm

2-3 μm
NO NUCLEI

ERYTHROCYTE
(RED BLOOD CELL)

Neutrophil Eosinophil Basophil
(Segmented Nuclei)

Small LYMPHOCYTE   MONOCYTE

PLATELETS
250-500 $\times 10^9$ per litre

4.2-6.4 $\times 10^{12}$/l.
(*Averages*
Male: 5.5 $\times 10^{12}$/l.
Female: 4.8 $\times 10^{12}$/l.
Below 4 $\times 10^{12}$ = **anaemia.**
Above 6.5 $\times 10^{12}$ = **polycythaemia.**
Numbers increase in **acclimatization** to **high altitude.**
LIFE SPAN: 120 days.

POLYMORPHONUCLEAR
GRANULAR LEUCOCYTES
4-5   0.2   0.025
($\times 10^9$ per litre)

NON-GRANULAR
LEUCOCYTES
1–3   0.3–0.4
($\times 10^9$ per litre)

TOTAL from 4.0-11.0 $\times 10^9$/l.
(Fluctuates with Exercise, Fever, etc.)
Many migrate into tissues.
Below 4 $\times 10^9$/l. = **leucopenia.**
Above 11 $\times 10^9$/l = **leucocytosis.**
LIFE SPAN: GRANULOCYTES– 5-10 days
NON-GRANULAR LEUCOCYTES — several months.

Numbers fluctuate with Exercise.

If grossly deficient – **thrombocytopenia**

BLOOD GROUPS

There are present in the **plasma** of some individuals substances which can cause the **agglutination** (clumping together) and subsequent **haemolysis** (breakdown) of the **red blood cells** of some other individuals.

    If such reactions follow **blood transfusion** the two bloods are said to be **incompatible**.

Two **factors** *are involved in an* **agglutination** *reaction:-*

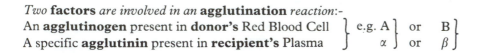

An **agglutinogen** present in **donor's** Red Blood Cell   e.g. A   or   B
A specific **agglutinin** present in **recipient's** Plasma α or β

Obviously no such combination occurs naturally otherwise auto-agglutination would result.

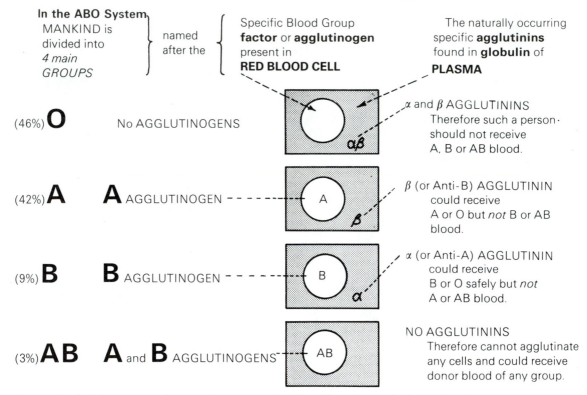

In practice it is important that the **donor's cells** should not be agglutinated by the **recipient's plasma**. Agglutination of recipient's cells by donor agglutinins is less likely to occur.

    Some individuals with A agglutinogen have an additional agglutinogen called A_1. Thus the A group is subdivided into types A_1 (those with both agglutinogens: 80%) and A_2 (those with only the A agglutinogen: 20%). Therefore there are really 6 ABO groups: O, A_1, A_2, B, A_1B and A_2B.

To determine the blood group to which an individual belongs **two test sera** only are required and the **red blood cells** to be grouped.

DROP of GROUP **A** SERUM & GROUP **B** SERUM
(on   [N.B. These droplets are of **plasma**
glass   – *NOT* red blood cells.
side)   i.e. only **agglutinins** are present.]

To each test serum a saline suspension of **red blood cells** is added, i.e. only **agglutinogens** are added.

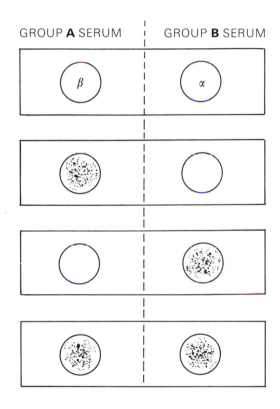

GROUP **A** SERUM   GROUP **B** SERUM

GROUP O blood cells give *no* agglutination since *no* agglutinogens are present in these cells to be clumped by test sera agglutinins.

GROUP B blood cells (B agglutinogen present) give agglutination with GROUP A serum – since the specific Anti-B (*β*) agglutinin is present in the first test serum.

GROUP A blood cells give agglutination with GROUP B serum – since the specific Anti-A (*α*) agglutinin is present in this serum.

GROUP AB blood cells give agglutination with both test sera.

    As well as determining blood group in this way, the blood of donor is always matched directly with blood of patient to avoid sub-group incompatibility.

    Agglutination is usually visible under the microscope within a few minutes. The clumped cells look like grains of cayenne pepper in a clear liquid. If no agglutination occurs the fluid remains uniformly pink.

    If the wrong blood is given to a patient, clumps of red blood cells may block small blood vessels in vital organs, e.g. lung or brain. The subsequent haemolysis (breakdown) of agglutinated cells may lead to haemoglobin in the urine and eventually to kidney failure and death.

RHESUS FACTOR

In addition to the antigens of the ABO blood group system there are innumerable other agglutinogens in red cells. Those of the **Rhesus (Rh) system** are clinically important. The **'Rh factor'** actually contains many antigens. By far the most important is the D agglutinogen.

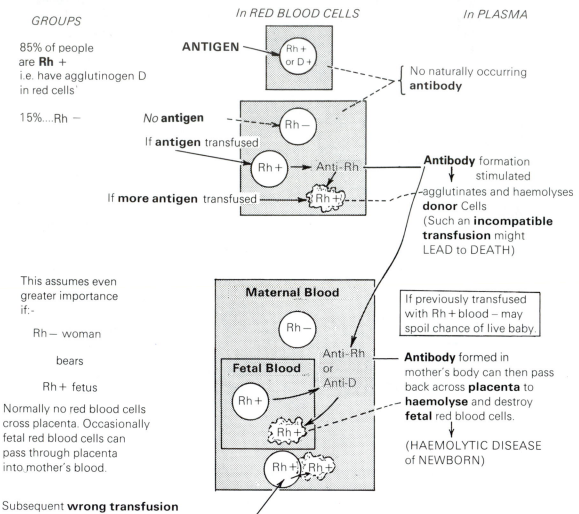

GROUPS   *In RED BLOOD CELLS*   *In PLASMA*

85% of people
are **Rh +**
i.e. have agglutinogen D
in red cells'

ANTIGEN — Rh + or D +

No naturally occurring **antibody**

15%....Rh —   *No* **antigen**   Rh —

If **antigen** transfused   Rh + → Anti-Rh

If **more antigen** transfused → Rh +

Antibody formation stimulated
↓
-agglutinates and haemolyses
donor Cells
(Such an **incompatible
transfusion** might
LEAD to DEATH)

This assumes even
greater importance
if:-

Rh — woman

bears

Rh + fetus

Normally no red blood cells
cross placenta. Occasionally
fetal red blood cells can
pass through placenta
into mother's blood.

Maternal Blood

Rh —

Fetal Blood   Anti-Rh
or
Anti-D

Rh +

Rh +

Rh + Rh +

If previously transfused
with Rh + blood — may
spoil chance of live baby.

Antibody formed in
mother's body can then pass
back across **placenta** to
haemolyse and destroy
fetal red blood cells.
↓
(HAEMOLYTIC DISEASE
of NEWBORN)

Subsequent **wrong transfusion**
of Rh + blood to mother could lead to dangerous **agglutination** and **haemolysis** of **donor cells**
within mother's own body.

A D-ve mother carrying a D + ve fetus is given anti-D antibodies during her pregnancy
and immediately after delivery of the child. This prevents the mother forming her own anti-D
antibodies.

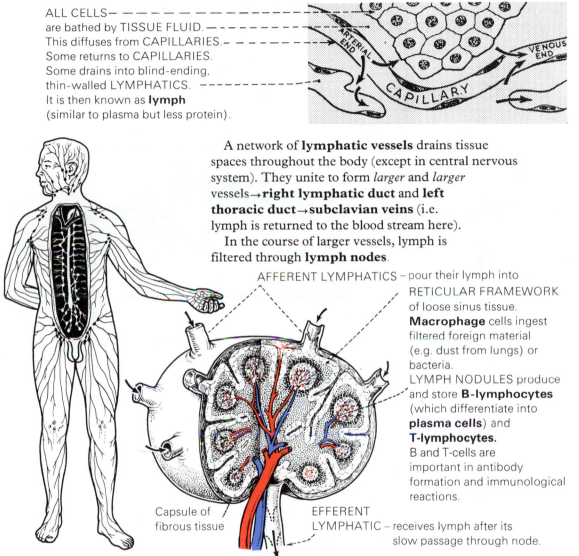

ALL CELLS— — — — — — — — — —
are bathed by TISSUE FLUID. — - - - - - - - -
This diffuses from CAPILLARIES. - — — — — —
Some returns to CAPILLARIES.
Some drains into blind-ending,
thin-walled LYMPHATICS. - - - - - - - - - -
It is then known as **lymph**
(similar to plasma but less protein).

A network of **lymphatic vessels** drains tissue spaces throughout the body (except in central nervous system). They unite to form *larger* and *larger* vessels→**right lymphatic duct** and **left thoracic duct**→**subclavian veins** (i.e. lymph is returned to the blood stream here).

In the course of larger vessels, lymph is filtered through **lymph nodes**.

AFFERENT LYMPHATICS – pour their lymph into RETICULAR FRAMEWORK of loose sinus tissue. **Macrophage** cells ingest filtered foreign material (e.g. dust from lungs) or bacteria. LYMPH NODULES produce and store **B-lymphocytes** (which differentiate into **plasma cells**) and **T-lymphocytes**. B and T-cells are important in antibody formation and immunological reactions.

Capsule of fibrous tissue

EFFERENT LYMPHATIC – receives lymph after its slow passage through node.

Movement of **lymph** towards **heart** depends partly on compression of lymphatic vessels by muscles of limbs and partly on 'suction' created by movements of respiration. Valves within the vessels prevent backflow. The lymphoid tissue of the body which includes lymph nodes, spleen, thymus, tonsils, etc., forms an important part of the body's defence against invading agents such as **protozoa, bacteria, viruses**, or other poisonous **toxins**. These act as **antigens** stimulating **antibody formation** – which can subsequently destroy or neutralize the antigen.

SPLEEN

The spleen is a vascular organ, weighing about 200 grams. It is situated in the left side of the abdomen behind the stomach and above the kidney.

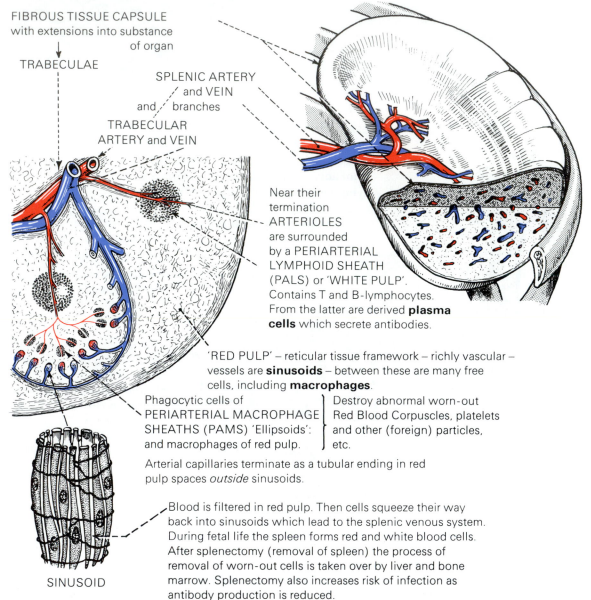

FIBROUS TISSUE CAPSULE
with extensions into substance
of organ

TRABECULAE

SPLENIC ARTERY
and VEIN
and branches

TRABECULAR
ARTERY and VEIN

Near their
termination
ARTERIOLES
are surrounded
by a PERIARTERIAL
LYMPHOID SHEATH
(PALS) or 'WHITE PULP'.
Contains T and B-lymphocytes.
From the latter are derived **plasma
cells** which secrete antibodies.

'RED PULP' – reticular tissue framework – richly vascular –
vessels are **sinusoids** – between these are many free
cells, including **macrophages**.

Phagocytic cells of
PERIARTERIAL MACROPHAGE
SHEATHS (PAMS) 'Ellipsoids':
and macrophages of red pulp.

Destroy abnormal worn-out
Red Blood Corpuscles, platelets
and other (foreign) particles,
etc.

Arterial capillaries terminate as a tubular ending in red
pulp spaces *outside* sinusoids.

Blood is filtered in red pulp. Then cells squeeze their way
back into sinusoids which lead to the splenic venous system.
During fetal life the spleen forms red and white blood cells.
After splenectomy (removal of spleen) the process of
removal of worn-out cells is taken over by liver and bone
marrow. Splenectomy also increases risk of infection as
antibody production is reduced.

SINUSOID

The thymus is an irregularly-shaped organ lying behind the breast bone. It is relatively large in the child and reaches its maximum size at puberty. It closely resembles a lymph node.

NEWBORN CHILD

× 50

At puberty weighs about 35 g

The thymus regresses after **puberty**

ADULT

× 50

NODULES or FOLLICLES contain large numbers of **lymphocytes**.
These are closely packed in the outer **cortex**,

less densely packed in the central **medulla** where occasional concentric **corpuscles** of **Hassall** are found.

Normally the active lymphatic tissue is replaced by **fatty tissue**.

T-LYMPHOCYTES

The thymus is important in **immunity**. Some lymphocytes undergo part maturation in the **thymus** and so they are called T-LYMPHOCYTES. They arise from lymphocyte precursor cells in fetal blood-forming tissue and are carried to the thymus by the blood. There they acquire the ability to recognize and react against foreign proteins, e.g. viruses, tissue transplants, etc., and an ability *not* to react against **self-antigens** (the person's own proteins).

Four types of T-lymphocyte develop:

Helper/Inducer T-cells   **Cytotoxic** T-cells (killer cells)
Suppressor T-cells   **Memory** T-cells (see page 126).

Development of these cells requires **thymosin**, a mixture of hormones secreted by thymic epithelium. T-lymphocytes are distributed throughout life to lymph nodes, spleen, etc. where they join B-lymphocytes which have developed in the liver and spleen.

Acquired immune deficiency syndrome (AIDS) is a disease caused by destruction of **T-helper** (T_4) cells by the virus **HTLV-III** or **HIV** (human immunodeficiency virus). Antibody formation is thus destroyed and the patient becomes vulnerable to infection and cancer.

IMMUNE SYSTEM

The **immune system** recognizes, remembers and produces **antibodies** against many millions of **antigens** (foreign agents) that invade the body. There are two types of immune system: (a) **humoral** – Protection is by **antibodies** – major defence against bacteria, and (b) **cell mediated** – Protection is by **T-lymphocytes** which react directly with foreign cells, e.g. tissue transplants and also cells infected by organisms.

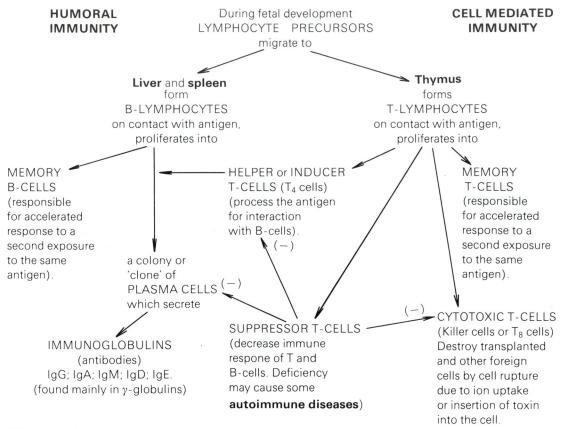

HUMORAL IMMUNITY

CELL MEDIATED IMMUNITY

During fetal development LYMPHOCYTE PRECURSORS migrate to

Liver and **spleen** form B-LYMPHOCYTES on contact with antigen, proliferates into

Thymus forms T-LYMPHOCYTES on contact with antigen, proliferates into

MEMORY B-CELLS (responsible for accelerated response to a second exposure to the same antigen).

HELPER or INDUCER T-CELLS (T$_4$ cells) (process the antigen for interaction with B-cells). (–)

MEMORY T-CELLS (responsible for accelerated response to a second exposure to the same antigen).

a colony or 'clone' of PLASMA CELLS (–) which secrete

SUPPRESSOR T-CELLS (decrease immune respone of T and B-cells. Deficiency may cause some **autoimmune diseases**)

(–) CYTOTOXIC T-CELLS (Killer cells or T$_8$ cells) Destroy transplanted and other foreign cells by cell rupture due to ion uptake or insertion of toxin into the cell.

IMMUNOGLOBULINS (antibodies) IgG; IgA; IgM; IgD; IgE. (found mainly in γ-globulins)

The complement system: A system of plasma enzymes. Activated by combination of antibody and antigen. Induces inflammatory response. Cells are lysed (ruptured). Bacteria are opsonized (made ready for eating). Leucocytes are attracted to antigen. Histamine and other substances are released from blood components causing increased capillary permeability and vasodilatation. If the reaction is severe bronchoconstriction can occur.

    Interleukins: Hormone substances produced by lymphocytes. **Interleukin-1** (IL-1) affects hypothalamus and produces fever. **Interleukin-2** (IL-2) stimulates clones of activated T and B-cells.

    Interferons (α, β and γ): Hormone-like proteins secreted by leucocytes, fibroblasts and lymphoid tissue. Inhibit virus multiplication and cell division. Activate natural killer cells.

    Natural killer cells (NK cells): Large lymphocytes – not thymic in origin. Kill virus-infected cells and malignant tumour cells; produce and are activated by interferon.

Cerebrospinal fluid (CSF) is like blood plasma but has very little protein, less K^+, slightly less HCO_3^-, but more Cl^- and Mg^{2+}. These differences indicate that active secretion is involved in its formation.

VOLUME:- 120-150 ml, in man. SPECIFIC GRAVITY:- 1.005-1.008.

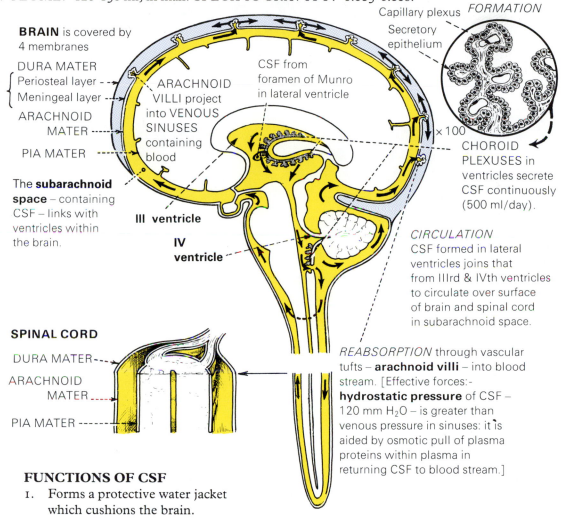

FORMATION

Capillary plexus

Secretory epithelium

BRAIN is covered by 4 membranes

DURA MATER
Periosteal layer - -
Meningeal layer - -

ARACHNOID MATER

PIA MATER - - - - -

ARACHNOID VILLI project into VENOUS SINUSES containing blood

CSF from foramen of Munro in lateral ventricle

The **subarachnoid space** – containing CSF – links with ventricles within the brain.

III ventricle

IV ventricle

×100

CHOROID PLEXUSES in ventricles secrete CSF continuously (500 ml/day).

CIRCULATION
CSF formed in lateral ventricles joins that from IIIrd & IVth ventricles to circulate over surface of brain and spinal cord in subarachnoid space.

SPINAL CORD

DURA MATER - - -

ARACHNOID MATER - - -

PIA MATER - - - - -

REABSORPTION through vascular tufts – **arachnoid villi** – into blood stream. [Effective forces:- **hydrostatic pressure** of CSF – 120 mm H_2O – is greater than venous pressure in sinuses: it is aided by osmotic pull of plasma proteins within plasma in returning CSF to blood stream.]

FUNCTIONS OF CSF
1. Forms a protective water jacket which cushions the brain.
2. Alteration of volume can compensate for fluctuations in amount of blood within skull and thus keep total volume of cranial contents constant.
3. Low K^+ concentration allows neurons to generate very high electrical potentials.

Endothelial cells of brain capillaries have tight junctions which prevent e.g. some drugs and transmitters passing from blood to brain interstitial fluid. They cannot cross this **blood-brain barrier**.

CHAPTER 6

RESPIRATORY SYSTEM

RESPIRATORY SYSTEM

All living cells require to get **oxygen** from the fluid around them and to get rid of **carbon dioxide** to it.

Internal respiration is the exchange of these gases between tissue cells and their fluid environment.

External respiration is the exchange of these gases (oxygen and carbon dioxide) between the body and the external environment.

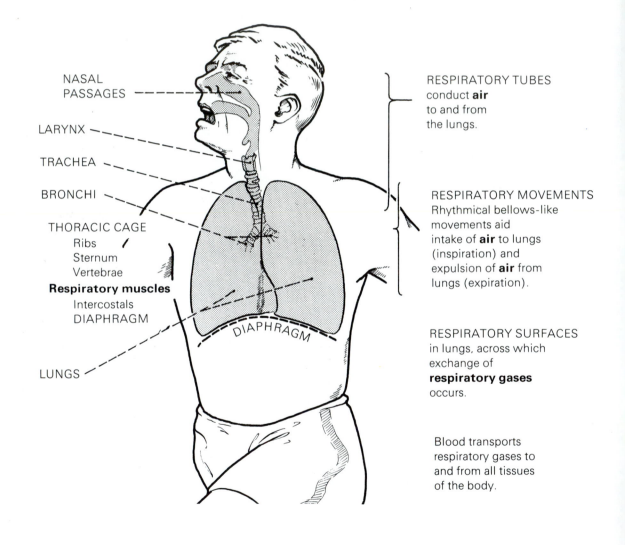

NASAL PASSAGES

LARYNX

TRACHEA

BRONCHI

THORACIC CAGE
Ribs
Sternum
Vertebrae
Respiratory muscles
Intercostals
DIAPHRAGM

LUNGS

DIAPHRAGM

RESPIRATORY TUBES conduct **air** to and from the lungs.

RESPIRATORY MOVEMENTS
Rhythmical bellows-like movements aid intake of **air** to lungs (inspiration) and expulsion of **air** from lungs (expiration).

RESPIRATORY SURFACES in lungs, across which exchange of **respiratory gases** occurs.

Blood transports respiratory gases to and from all tissues of the body.

AIR CONDUCTING PASSAGES

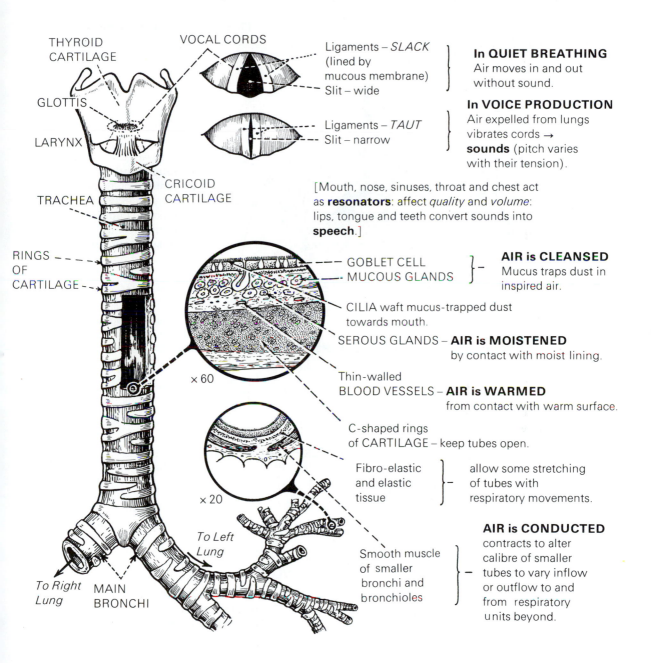

THYROID CARTILAGE

VOCAL CORDS

Ligaments – *SLACK*
(lined by
mucous membrane)
Slit – wide

In QUIET BREATHING
Air moves in and out
without sound.

GLOTTIS

LARYNX

Ligaments – *TAUT*
Slit – narrow

In VOICE PRODUCTION
Air expelled from lungs
vibrates cords →
sounds (pitch varies
with their tension).

CRICOID
CARTILAGE

[Mouth, nose, sinuses, throat and chest act
as **resonators**: affect *quality* and *volume*:
lips, tongue and teeth convert sounds into
speech.]

TRACHEA

RINGS
OF
CARTILAGE

GOBLET CELL
MUCOUS GLANDS

AIR is CLEANSED
Mucus traps dust in
inspired air.

CILIA waft mucus-trapped dust
towards mouth.

SEROUS GLANDS – **AIR is MOISTENED**
by contact with moist lining.

× 60

Thin-walled
BLOOD VESSELS – **AIR is WARMED**
from contact with warm surface.

C-shaped rings
of CARTILAGE – keep tubes open.

Fibro-elastic
and elastic
tissue

allow some stretching
of tubes with
respiratory movements.

× 20

To Left
Lung

AIR is CONDUCTED
contracts to alter
calibre of smaller
tubes to vary inflow
or outflow to and
from respiratory
units beyond.

Smooth muscle
of smaller
bronchi and
bronchioles

To Right
Lung

MAIN
BRONCHI

131

LUNGS: RESPIRATORY SURFACES

The trachea and the bronchial 'tree' conduct air down to the **respiratory surfaces**.

There is no exchange of gases in these tubes.

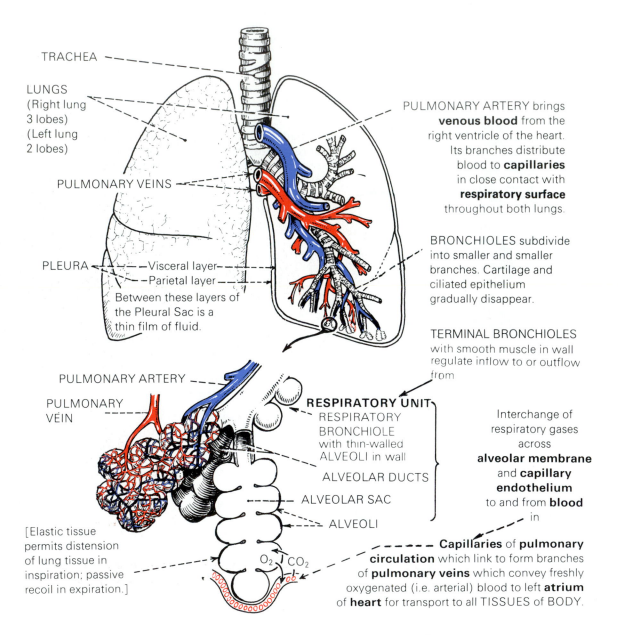

TRACHEA

LUNGS
(Right lung
3 lobes)
(Left lung
2 lobes)

PULMONARY VEINS

PLEURA — Visceral layer
— Parietal layer
Between these layers of
the Pleural Sac is a
thin film of fluid.

PULMONARY ARTERY

PULMONARY
VEIN

[Elastic tissue
permits distension
of lung tissue in
inspiration; passive
recoil in expiration.]

PULMONARY ARTERY brings
venous blood from the
right ventricle of the heart.
Its branches distribute
blood to **capillaries**
in close contact with
respiratory surface
throughout both lungs.

BRONCHIOLES subdivide
into smaller and smaller
branches. Cartilage and
ciliated epithelium
gradually disappear.

TERMINAL BRONCHIOLES
with smooth muscle in wall
regulate inflow to or outflow
from

RESPIRATORY UNIT
RESPIRATORY
BRONCHIOLE
with thin-walled
ALVEOLI in wall

ALVEOLAR DUCTS

ALVEOLAR SAC

ALVEOLI

O_2 CO_2

Interchange of
respiratory gases
across
alveolar membrane
and **capillary
endothelium**
to and from **blood**
in

Capillaries of **pulmonary
circulation** which link to form branches
of **pulmonary veins** which convey freshly
oxygenated (i.e. arterial) blood to left **atrium**
of **heart** for transport to all TISSUES of BODY.

The thorax (or chest) is the closed cavity which contains the **lungs, heart** and great vessels.

It is enclosed and
bounded:
ABOVE by the
   upper RIBS and
   tissues of the neck;
AT THE SIDES by
   the RIBS and
   INTERCOSTAL
   MUSCLES;
AT THE BACK by the
   RIBS and VERTEBRAL
   COLUMN (or back
   bone);
IN FRONT by the
   RIBS, COSTAL
   CARTILAGES and
   STERNUM (or
   breast bone);
BELOW by the
   DIAPHRAGM
   (a strong
   dome-shaped sheet
   of skeletal muscle
   with a central tendon
   which separates
   the thoracic cavity
   from the abdominal
   cavity

Parietal

Visceral

Crura

It is lined by a thin
moist membrane –
the PLEURA – the
inner layer of
which invests
the LUNGS.
In health
there is a
thin film of
fluid between these
two pleural layers
which causes adhesion
but allows them to slip
(like two glass sheets
with fluid between).
Elastic recoil of lungs
tends to pull visceral
layer away from
parietal layer. This
creates subatmospheric
or negative intrapleural
pressure (about – 2 mmHg).
In **quiet inspiration,** the
chest wall is **tending** to pull away
from lungs and the intrapleural
pressure increases to −6 mmHg.
With **forced inspiration** it can
become −30 mmHg.

Dimensions of thoracic cage and the **pressure** between pleural surfaces change rhythmically about 12-14 times a minute with the **movements** of **respiration** – air movement in and out of the lungs follows the dimension changes.

MECHANISM OF BREATHING

The rhythmical changes in the capacity of the thorax are brought about by muscular action. The changes in lung volume with intake or expulsion of air follow.

In NORMAL QUIET BREATHING

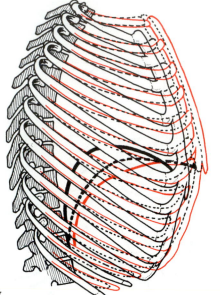

INSPIRATION

external intercostal muscles actively contract
– ribs and sternum move
   upwards and outwards
   because first rib is fixed
– width of chest increases
   from side to side and
   from front to back.

diaphragm contracts
– descends
– depth of chest increases.
capacity of **thorax** is
   increased
   ↓
pressure between **pleural surfaces** (already negative)
is *reduced* from -2 to -6 mmHg
(i.e. an increased 'suction pull'
is exerted on **lung tissue**)

elastic tissue of lungs is *stretched*

lungs *expand* to fill **thoracic cavity**

air pressure in alveoli is now -1.5 mmHg,
i.e. *less* than atmospheric pressure

air is sucked into **alveoli** from atmosphere
because of pressure difference.

EXPIRATION

external intercostal muscles relax
– ribs and sternum move
   downwards and inwards
– width of chest diminishes.

diaphragm relaxes–
   ascends
– depth of chest diminishes.
capacity of **thorax** is
   decreased
   ↓
pressure between **pleural surfaces** is *increased* from
-6 to -2 mmHg
(i.e. less pull is exerted on
lung tissue)

elastic tissue of lungs
   recoils

air pressure in alveoli is now
$+1.5$ mmHg.
i.e. *greater* than atmospheric pressure

air is forced out of **alveoli** to
atmosphere

In FORCED BREATHING

Muscles of nostrils and round glottis may contract to aid entrance of air to lungs. Extensors of vertebral column may aid inspiration. Muscles of neck contract
– move 1st rib upwards
   (and sternum upwards and forwards)

Internal intercostals may contract
– move ribs downwards more actively.
Abdominal muscles contract
– actively aid ascent of diaphragm.

If breathing has ceased in cases of drowning, electrocution, gas poisoning, etc., a life may be saved if artificial respiration is applied promptly. Respiration always ceases before the heart stops beating.

Mouth-to-mouth breathing is superior to all other methods.

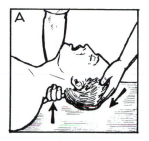

Applicator clears patient's mouth and throat of obstruction, then lays him on his back and positions himself at the side of the patient. He places one hand under his neck and the other on his forehead.

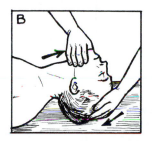

Applicator tilts the patient's head right back, raising his chin up. This causes the tongue to lift away from the back of the patient's throat and opens up airway.

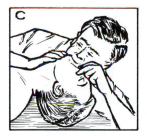

The applicator pinches shut the patient's nostrils, seals his lips round the patient's mouth and 12-14 times per minute blows in air until about twice the normal chest movement is observed.

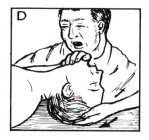

When he removes his mouth the patient breathes out passively. The applicator takes another breath. There is enough residual oxygen in the applicator's own expired air for the patient's needs.

VOLUMES AND CAPACITIES OF LUNGS

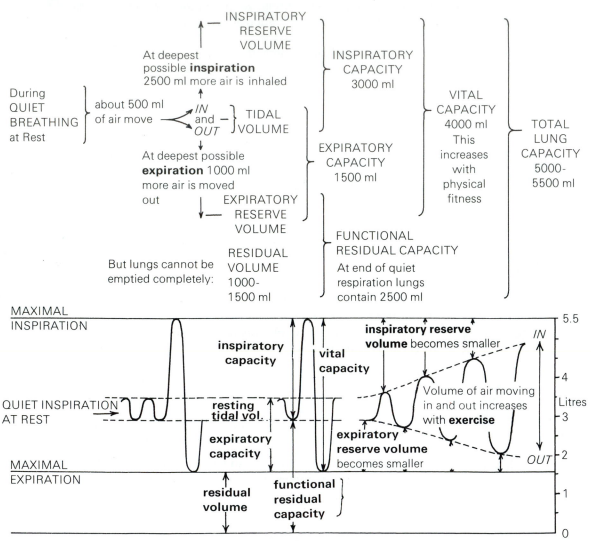

(after Pappenheimer, J.R., et al (1950) Fed. Proc., **9**,602)

At rest a normal male adult breathes in and out about 12 times per minute. The amount of air breathed in per minute is therefore 500 ml × 12 i.e. 6000 ml or 6 litres – this is the **respiratory minute volume** or **pulmonary ventilation**. In exercise it may go up to as much as 200 litres. These values are about 25% lower in women.

In deep breathing the volume of **atmospheric air inspired** with each inspiration and the amount which reaches the **alveoli** increase.

136

In QUIET BREATHING
Of the
500 ml **atmospheric air**
INSPIRED in a single inspiration

OXYGEN makes up	20.95%
NITROGEN	79.01%
CARBON DIOXIDE	0.04%

In most climates, water vapour in air will reduce these percentages slightly. There are also very small amounts of inert gases.

140 ml occupy the **conducting passages – 'dead space' air.** This remains unchanged in composition since it is not in contact with respiratory surfaces.

360ml reach the **respiratory units** and mix with 2- 2.5 litres alveolar air (Functional Residual Capacity). Alveolar air is saturated with WATER VAPOUR. It constantly gives up OXYGEN *to* the blood, and constantly takes up CARBON DIOXIDE *from* the blood.

Of the **air**
EXPIRED in a single expiration

OXYGEN makes up	15.7%
NITROGEN	74.5%
CARBON DIOXIDE	3.6%
WATER VAPOUR	6.2%

[NB: The percentage of nitrogen is changed because it is diluted by the addition of other gases, especially water vapour. Air is saturated with water vapour by the time it reaches the lungs.]

This represents a mixture of:
'dead space' air – air which has moved out *unchanged* from the **conducting passages**

and

alveolar air – air which has been in contact with respiratory surfaces and has given up some **oxygen** *to* the blood and taken up **carbon dioxide** *from* it.

OXYGEN	13.6%
NITROGEN	74.9%
CARBON DIOXIDE	5.3%
WATER VAPOUR	6.2%

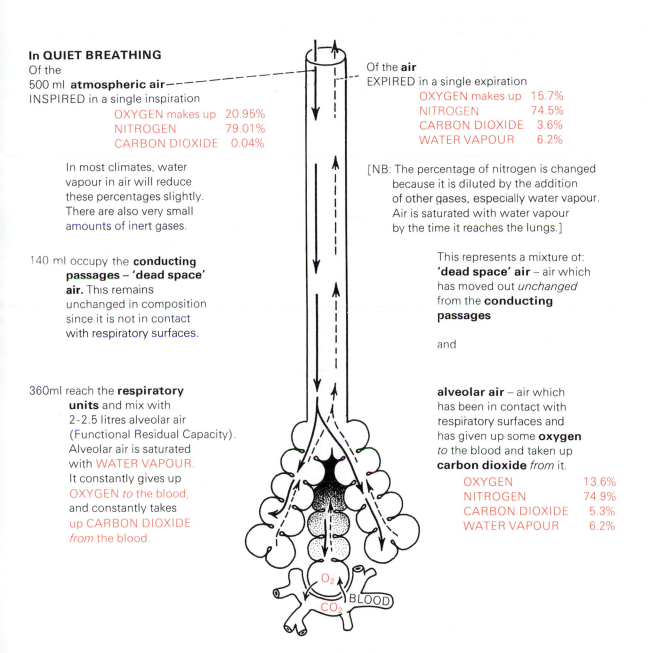

In DEEP BREATHING *at rest* (Hyperventilating) *more* new air exchanges with the alveolar air. Thus O_2 content of alveolar air will increase and the CO_2 content will decrease.

MOVEMENT OF RESPIRATORY GASES

A gas moves from an area where it is present at higher pressure to an area where it is present at lower pressure. The movement of gas molecules continues till the pressure exerted by them is the same throughout both areas. *Dry* atmospheric air (at sea level) has a pressure of 1 atmosphere = 760 mmHg = 101.3 kilopascals (kPa).

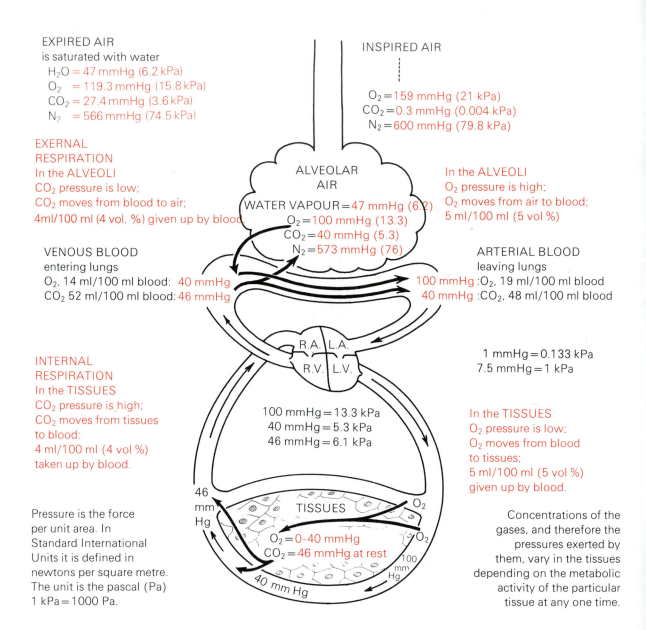

EXPIRED AIR
is saturated with water
H_2O = 47 mmHg (6.2 kPa)
O_2 = 119.3 mmHg (15.8 kPa)
CO_2 = 27.4 mmHg (3.6 kPa)
N_2 = 566 mmHg (74.5 kPa)

INSPIRED AIR

O_2 = 159 mmHg (21 kPa)
CO_2 = 0.3 mmHg (0.004 kPa)
N_2 = 600 mmHg (79.8 kPa)

EXERNAL
RESPIRATION
In the ALVEOLI
CO_2 pressure is low;
CO_2 moves from blood to air;
4ml/100 ml (4 vol. %) given up by blood

ALVEOLAR
AIR
WATER VAPOUR = 47 mmHg (6.2)
O_2 = 100 mmHg (13.3)
CO_2 = 40 mmHg (5.3)
N_2 = 573 mmHg (76)

In the ALVEOLI
O_2 pressure is high;
O_2 moves from air to blood;
5 ml/100 ml (5 vol %)

VENOUS BLOOD
entering lungs
O_2, 14 ml/100 ml blood: 40 mmHg
CO_2 52 ml/100 ml blood: 46 mmHg

ARTERIAL BLOOD
leaving lungs
100 mmHg : O_2, 19 ml/100 ml blood
40 mmHg : CO_2, 48 ml/100 ml blood

R.A. L.A.
R.V. L.V.

1 mmHg = 0.133 kPa
7.5 mmHg = 1 kPa

INTERNAL
RESPIRATION
In the TISSUES
CO_2 pressure is high;
CO_2 moves from tissues
to blood:
4 ml/100 ml (4 vol %)
taken up by blood.

100 mmHg = 13.3 kPa
40 mmHg = 5.3 kPa
46 mmHg = 6.1 kPa

In the TISSUES
O_2 pressure is low;
O_2 moves from blood
to tissues;
5 ml/100 ml (5 vol %)
given up by blood.

Pressure is the force
per unit area. In
Standard International
Units it is defined in
newtons per square metre.
The unit is the pascal (Pa)
1 kPa = 1000 Pa.

46
mm
Hg

TISSUES
O_2 = 0-40 mmHg
CO_2 = 46 mmHg at rest

O_2

O_2

100
mm
Hg

40 mm Hg

Concentrations of the
gases, and therefore the
pressures exerted by
them, vary in the tissues
depending on the metabolic
activity of the particular
tissue at any one time.

DISSOCIATION OF OXYGEN FROM HAEMOGLOBIN

The amount of O_2 taken up by **haemoglobin** in the **lungs** or given up by **oxyhaemoglobin** in the **tissues** depends on the **partial pressure** of the O_2 in the immediate environment.

It is also influenced by the **partial pressure** of CO_2, by **temperature**, by **acidity** and by the concentration of 2, 3 Diphosphoglycerate (DPG).

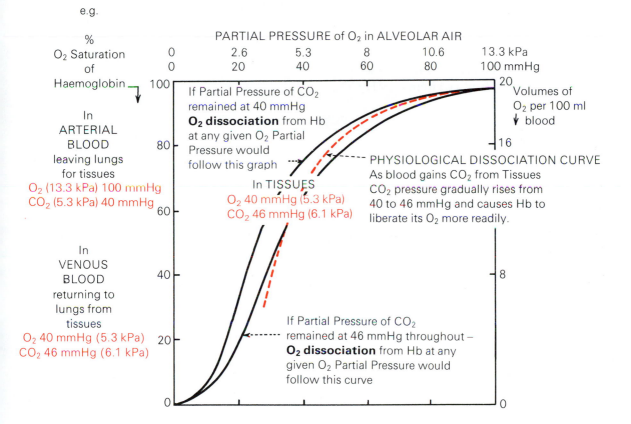

e.g.

%
O_2 Saturation of Haemoglobin

PARTIAL PRESSURE of O_2 in ALVEOLAR AIR

In ARTERIAL BLOOD leaving lungs for tissues
O_2 (13.3 kPa) 100 mmHg
CO_2 (5.3 kPa) 40 mmHg

In VENOUS BLOOD returning to lungs from tissues
O_2 40 mmHg (5.3 kPa)
CO_2 46 mmHg (6.1 kPa)

If Partial Pressure of CO_2 remained at 40 mmHg **O_2 dissociation** from Hb at any given O_2 Partial Pressure would follow this graph

In TISSUES
O_2 40 mmHg (5.3 kPa)
CO_2 46 mmHg (6.1 kPa)

Volumes of O_2 per 100 ml blood

PHYSIOLOGICAL DISSOCIATION CURVE
As blood gains CO_2 from Tissues CO_2 pressure gradually rises from 40 to 46 mmHg and causes Hb to liberate its O_2 more readily.

If Partial Pressure of CO_2 remained at 46 mmHg throughout – **O_2 dissociation** from Hb at any given O_2 Partial Pressure would follow this curve

This effect of CO_2 partial pressure on dissociation of O_2 from Hb (the Böhr effect) is advantageous.

E.g. an increase in CO_2 partial pressure locally during tissue activity causes Hb to part more readily with its O_2 to the active tissues.

Similarly, an increase in temperature, H^+ and DPG move the curve to the right. DPG is formed from glucose in the RBCs. Its presence favours the dissociation of oxygen from HbO_2

UPTAKE AND RELEASE OF CARBON DIOXIDE

CO_2 is carried by both red blood corpuscles and plasma. The bulk of it is carried as **bicarbonate** formed chiefly in red cells and carried largely by plasma. The amount of CO_2 taken up by the blood in the tissues or released by the blood in the lungs depends on the **partial pressures** of the CO_2 in the immediate environment and also on the state of the **haemoglobin** or on the amounts of **oxygen** linked to Hb at any one time.

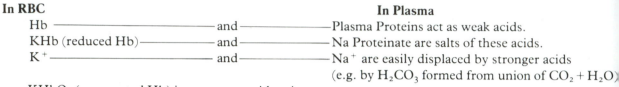

In RBC

Hb ——————————————— and ———————
KHb (reduced Hb) ——————— and ———————
K^+ ————————————————— and ———————

In Plasma

Plasma Proteins act as weak acids.
Na Proteinate are salts of these acids.
Na^+ are easily displaced by stronger acids
(e.g. by H_2CO_3 formed from union of $CO_2 + H_2O$)

$KHbO_2$ (oxygenated Hb) is a stronger acid and
K^+ is *less* readily displaced from it by H_2CO_3

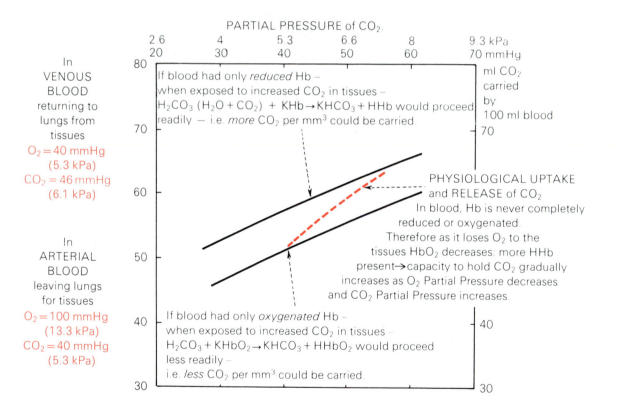

I.e. the more **oxygen** the blood holds the less **CO_2** it can hold and vice versa. This facilitates release of CO_2 in lungs and uptake of CO_2 in tissues.

CARRIAGE AND TRANSFER OF OXYGEN AND CARBON DIOXIDE

When **arterial blood** is delivered by **systemic capillaries** to the tissues it is exposed to:-

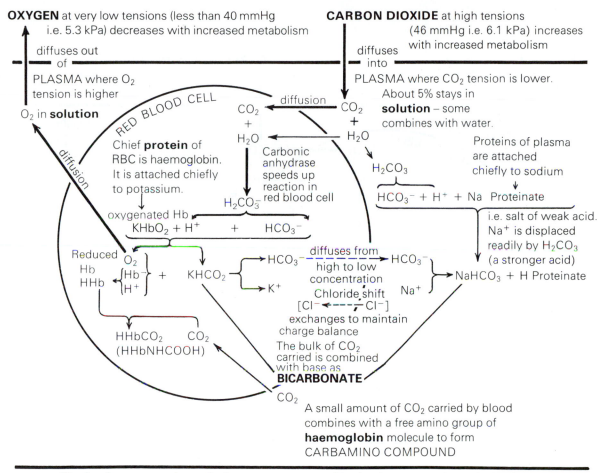

OXYGEN at very low tensions (less than 40 mmHg i.e. 5.3 kPa) decreases with increased metabolism

diffuses out of

PLASMA where O_2 tension is higher

O_2 in **solution**

RED BLOOD CELL

diffusion

Chief **protein** of RBC is haemoglobin. It is attached chiefly to potassium.

oxygenated Hb
$KHbO_2 + H^+$

Reduced Hb
HHb
O_2 $\begin{Bmatrix} Hb^- \\ H^+ \end{Bmatrix}$ + KHCO$_2$

$HHbCO_2$ CO_2
(HHbNHCOOH)

CARBON DIOXIDE at high tensions (46 mmHg i.e. 6.1 kPa) increases with increased metabolism

diffuses into

PLASMA where CO_2 tension is lower. About 5% stays in **solution** – some combines with water.

CO_2
+
H_2O

diffusion

CO_2
+
H_2O

Carbonic anhydrase speeds up reaction in red blood cell

H_2CO_3

H_2CO_3

$+ HCO_3^-$

HCO_3^- diffuses from high to low concentration HCO_3^-

K^+

Chloride shift
$[Cl^- \leftarrow - - \rightarrow Cl^-]$
exchanges to maintain charge balance

The bulk of CO_2 carried is combined with base as **BICARBONATE**

CO_2

Proteins of plasma are attached chiefly to sodium

$HCO_3^- + H^+ + Na$ Proteinate

i.e. salt of weak acid. Na^+ is displaced readily by H_2CO_3 (a stronger acid)

Na^+ $NaHCO_3 + H$ Proteinate

A small amount of CO_2 carried by blood combines with a free amino group of **haemoglobin** molecule to form CARBAMINO COMPOUND

During its passage through the tissues, blood gives up about 5 volumes % OXYGEN, i.e. its Hb is still up to 70% O_2 saturated.

The release of O_2 to tissues is speeded up by an increase in temperature, acidity or DPG such as occurs when tissues are active.

During its passage through the tissues, blood takes up about 4 volumes % CO_2.

CO_2 is carried, 5% in solution, 5% as carbamino compounds and 90% as HCO_3^-.

Carbonic anhydrase causes rapid formation of HCO_3^- inside the RBC and it then diffuses down a concentration gradient into the plasma.

CARRIAGE AND TRANSFER OF OXYGEN AND CARBON DIOXIDE

When **venous blood** flows through the **pulmonary capillaries** it is exposed to:-

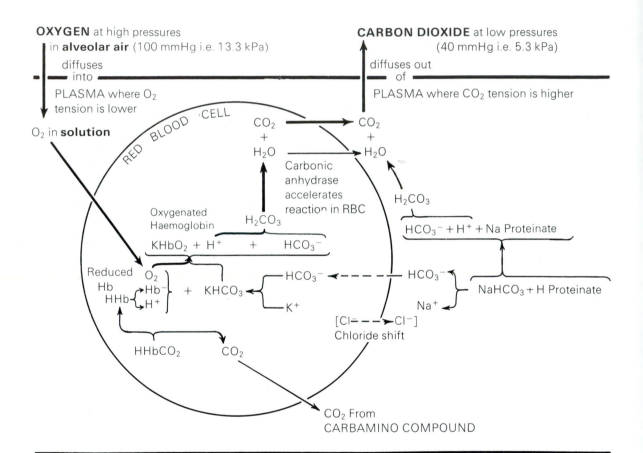

OXYGEN at high pressures in **alveolar air** (100 mmHg i.e. 13.3 kPa)

diffuses into

PLASMA where O_2 tension is lower

O_2 in **solution**

CARBON DIOXIDE at low pressures (40 mmHg i.e. 5.3 kPa)

diffuses out of

PLASMA where CO_2 tension is higher

RED BLOOD CELL

$CO_2 + H_2O$ → $CO_2 + H_2O$

Carbonic anhydrase accelerates reaction in RBC

H_2CO_3

Oxygenated Haemoglobin

H_2CO_3

$KHbO_2 + H^+ + HCO_3^-$

$HCO_3^- + H^+ + Na$ Proteinate

Reduced Hb HHb $\{$ Hb^- H^+ $\}$ $+ KHCO_3$

HCO_3^- ← — — HCO_3^-

K^+ Na$^+$

NaHCO$_3$ + H Proteinate

[Cl⁻ — — ➤ Cl⁻] Chloride shift

$HHbCO_2$ CO_2

CO_2 From CARBAMINO COMPOUND

As blood passes through capillaries of lungs it takes up approximately 5ml/100ml (5 vol. %) of **oxygen.** O_2 combines with haemoglobin (Hb) molecule. It becomes about 95-97% saturated with Oxygen.

As blood passes through capillaries of lungs it gives up approximately 4ml/100ml (4 vol. %) **carbon dioxide.** A small amount is released from combination with the free amino group in haemoglobin molecule – Carbamino compound. Most comes from **bicarbonate** in RBC and plasma by processes indicated in diagram.

NERVOUS CONTROL OF RESPIRATORY MOVEMENTS

Normal respiratory movements are involuntary. They are carried out automatically (i.e. without conscious control) through the rhythmical discharge of nerve impulses from **controlling centres** in the **brain**.

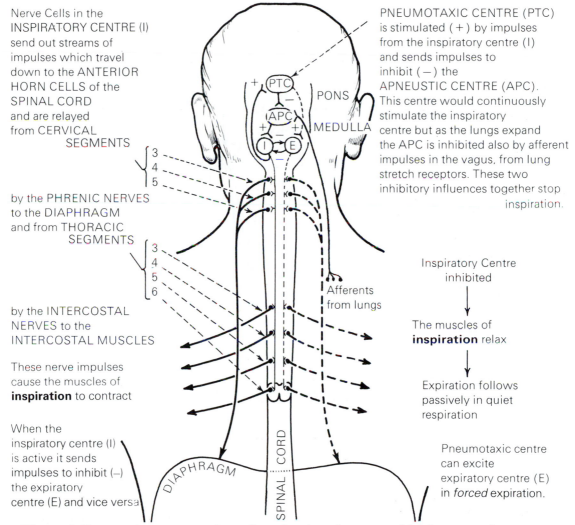

Nerve Cells in the INSPIRATORY CENTRE (I) send out streams of impulses which travel down to the ANTERIOR HORN CELLS of the SPINAL CORD and are relayed from CERVICAL SEGMENTS

by the PHRENIC NERVES to the DIAPHRAGM and from THORACIC SEGMENTS

by the INTERCOSTAL NERVES to the INTERCOSTAL MUSCLES

These nerve impulses cause the muscles of **inspiration** to contract

When the inspiratory centre (I) is active it sends impulses to inhibit (−) the expiratory centre (E) and vice versa

PNEUMOTAXIC CENTRE (PTC) is stimulated (+) by impulses from the inspiratory centre (I) and sends impulses to inhibit (−) the APNEUSTIC CENTRE (APC). This centre would continuously stimulate the inspiratory centre but as the lungs expand the APC is inhibited also by afferent impulses in the vagus, from lung stretch receptors. These two inhibitory influences together stop inspiration.

Inspiratory Centre inhibited
↓
The muscles of **inspiration** relax
↓
Expiration follows passively in quiet respiration

Pneumotaxic centre can excite expiratory centre (E) in *forced* expiration.

PTC
PONS
APC
MEDULLA
I E
3 4 5
3 4 5 6
Afferents from lungs
DIAPHRAGM
SPINAL CORD

The medullary respiratory centre has a **dorsal** group of neurons which innervate the diaphragm and a **ventral** group which innervate the accessory respiratory muscles and the intercostals. The dorsal group may drive the ventral group and may be switched on and off by the apneustic centre.

CHEMICAL REGULATION OF RESPIRATION

The activity of the respiratory centres is affected by the O_2, CO_2 and H^+ content of the blood. **Carbon dioxide** and H^+ are the most important. CO_2 dissolves in cerebrospinal fluid (CSF) which bathes receptors sensitive to H^+ on the ventral aspect of the medulla. Stimulation of these receptors leads to an increase in the rate and depth of respiration. Carotid bodies are stimulated if H^+ in blood rises; this also increases ventilation.

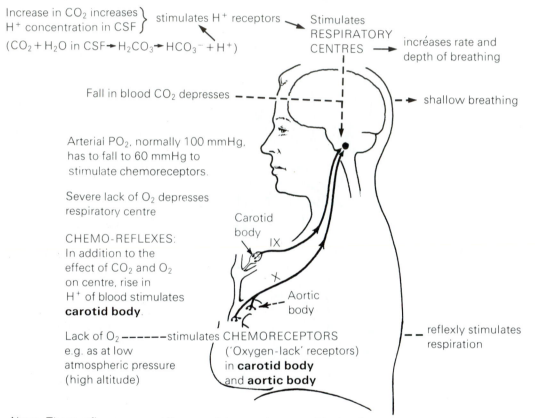

Increase in CO_2 increases H^+ concentration in CSF } stimulates H^+ receptors → Stimulates RESPIRATORY CENTRES → increases rate and depth of breathing

$(CO_2 + H_2O$ in CSF → H_2CO_3 → $HCO_3^- + H^+)$

Fall in blood CO_2 depresses --- → shallow breathing

Arterial PO_2, normally 100 mmHg, has to fall to 60 mmHg to stimulate chemoreceptors.

Severe lack of O_2 depresses respiratory centre

CHEMO-REFLEXES:
In addition to the effect of CO_2 and O_2 on centre, rise in H^+ of blood stimulates **carotid body**.

Lack of O_2 ------stimulates CHEMORECEPTORS e.g. as at low atmospheric pressure (high altitude)

Carotid body   IX

Aortic body

CHEMORECEPTORS ('Oxygen-lack' receptors) in **carotid body** and **aortic body**

reflexly stimulates respiration

Note:- These reflexes are usually powerful enough to override the direct depressant action of lack of O_2 on respiratory centres themselves.

The **chemical** and **nervous** means of regulating the activity of **respiratory centres** act together to adjust rate and depth of breathing to the needs of the body. E.g. **exercise** causes increased requirement for O_2 and production of more CO_2. Ventilation is increased to match this need for extra O_2 and get rid of the extra CO_2. However, the **alveolar** PO_2 and PCO_2 remain constant.

VOLUNTARY AND REFLEX FACTORS IN THE REGULATION OF RESPIRATION

Although fundamentally automatic and regulated by chemical factors in the blood, ingoing impulses from many parts of the body also modify the activity of the **respiratory centres** and consequently alter the outgoing impulses to the respiratory muscles to coordinate **rhythm, rate** or **depth** of breathing with other activities of the body.

Impulses from HIGHER CENTRES – PSYCHIC and EMOTIONAL INFLUENCES

{
Voluntary alterations in breathing.
Interruptions of expiration in **speech** and **singing**.
Deep inspiration then short spasmodic expirations in **laughter** and **weeping**.
Prolonged expiration in **sighing**.
Deep inspiration with mouth open in **yawning**.
Slow shallow breathing in **suspense** and **concentration**.
Rapid breathing in **fear** and **excitement**.
}

SENSORY STIMULI
e.g **pungent odours**
  irritating nerve endings in nasal mucosa.
bolus of **food**
  contacting pharynx.
irritant contacting larynx, trachea

painful, hot, cold
  stimuli to nerve endings in skin

Stretch-proprioceptors in INTERCOSTAL muscles, DIAPHRAGM ABDOMINAL muscles

Decrease (↓) in blood pressure
Increase (↑) in blood pressure
BARORECEPTORS in CAROTID sinus and AORTIC arch

↓O_2 as at high altitude
CHEMORECEPTORS in CAROTID BODY and AORTIC BODY

REFLEX alterations in respiratory movements

short inspiration, forced expirations with GLOTTIS open in **sneezing**.

inhibition of respiration during **swallowing**.

short inspiration; series of forced expirations with GLOTTIS closed (high pressure created in air passages); GLOTTIS opens suddenly; blast of air carries out irritant material in **coughing**.

sharp inspiration after sudden **pain** or **cold**; increasing rate and depth of breathing with **heat**.

spasmodic contractions of diaphragm with GLOTTIS closed in **hiccoughing**.

respiration stimulated.

respiration depressed.

respiration stimulated.

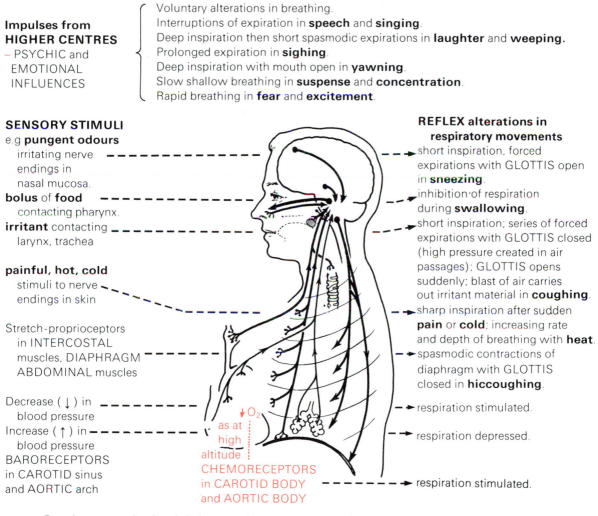

Proprioceptors stimulated during muscle movements send impulses to respiratory centre → ↑ rate and depth of breathing. (NB: This occurs with active or passive movements of limbs.)

    In normal breathing respiratory rate and rhythm are thought to be influenced rhythmically by the Hering-Breuer reflex.

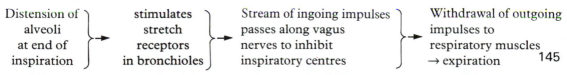

| Distension of alveoli at end of inspiration | → | stimulates stretch receptors in bronchioles | → | Stream of ingoing impulses passes along vagus nerves to inhibit inspiratory centres | → | Withdrawal of outgoing impulses to respiratory muscles → expiration |

145

CHAPTER 7

EXCRETORY SYSTEM

EXCRETORY SYSTEM

The respiratory system, the skin and the **kidneys** are the chief excretory organs of the body.

The kidneys adjust loss of water and electrolytes from the body to keep body fluids constant in amount and composition. They excrete waste products of metabolism, foreign chemicals such as drugs and food additives, secrete the hormones renin and erythropoietin and they activate vitamin D.

To understand the way in which the kidney carries out these functions, it is essential to understand first the way in which it is supplied with blood. About 25% of the left ventricle's output of blood in each cardiac cycle is distributed to the kidneys for filtration.

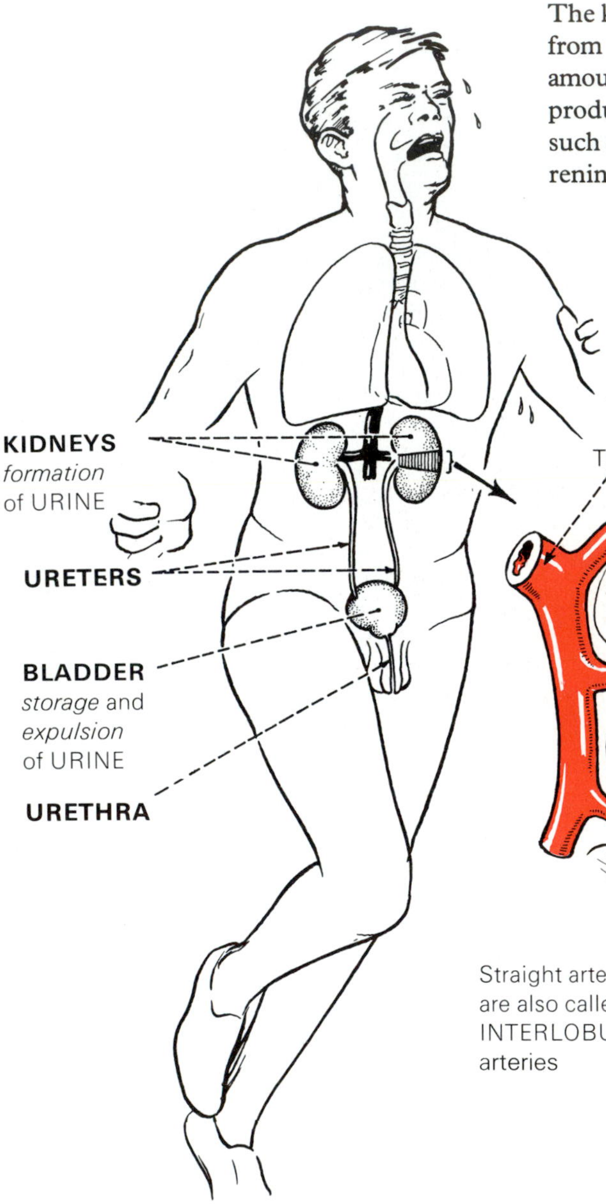

KIDNEYS
formation
of URINE

URETERS

BLADDER
storage and
expulsion
of URINE

URETHRA

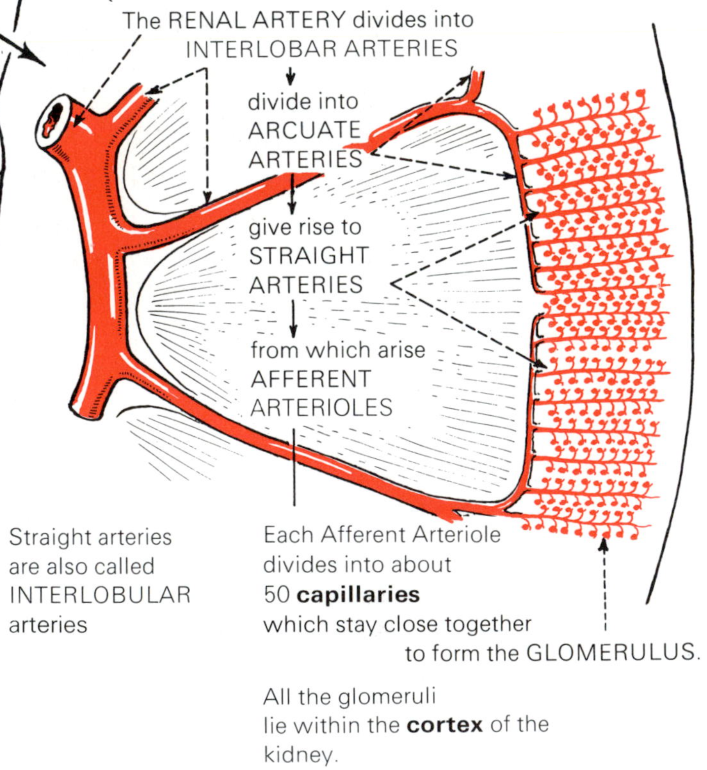

The RENAL ARTERY divides into
INTERLOBAR ARTERIES

divide into
ARCUATE
ARTERIES

give rise to
STRAIGHT
ARTERIES

from which arise
AFFERENT
ARTERIOLES

Straight arteries
are also called
INTERLOBULAR
arteries

Each Afferent Arteriole
divides into about
50 capillaries
which stay close together
to form the GLOMERULUS.

All the glomeruli
lie within the **cortex** of the
kidney.

Each kidney contains approximately one million functional units – **nephrons** – which form urine.

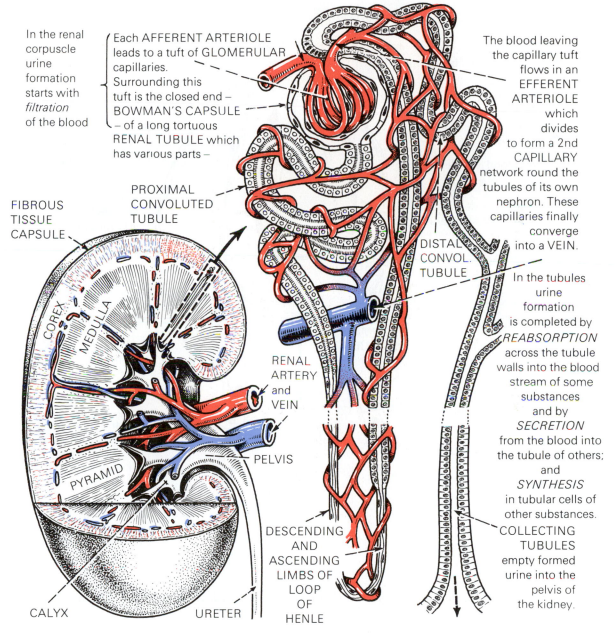

In the renal corpuscle urine formation starts with *filtration* of the blood

Each AFFERENT ARTERIOLE leads to a tuft of GLOMERULAR capillaries.
Surrounding this tuft is the closed end – BOWMAN'S CAPSULE – of a long tortuous RENAL TUBULE which has various parts –

The blood leaving the capillary tuft flows in an EFFERENT ARTERIOLE which divides to form a 2nd CAPILLARY network round the tubules of its own nephron. These capillaries finally converge into a VEIN.

FIBROUS TISSUE CAPSULE

PROXIMAL CONVOLUTED TUBULE

DISTAL CONVOL. TUBULE

In the tubules urine formation is completed by *REABSORPTION* across the tubule walls into the blood stream of some substances and by *SECRETION* from the blood into the tubule of others; and *SYNTHESIS* in tubular cells of other substances.

CORTEX

MEDULLA

RENAL ARTERY and VEIN

PYRAMID

PELVIS

COLLECTING TUBULES empty formed urine into the pelvis of the kidney.

CALYX

URETER

DESCENDING AND ASCENDING LIMBS OF LOOP OF HENLE

FORMATION OF URINE – 1. FILTRATION

About 25% of the left ventricle's total output of blood is distributed through the renal arteries to the kidneys where **filtration** of 20% of its plasma takes place.

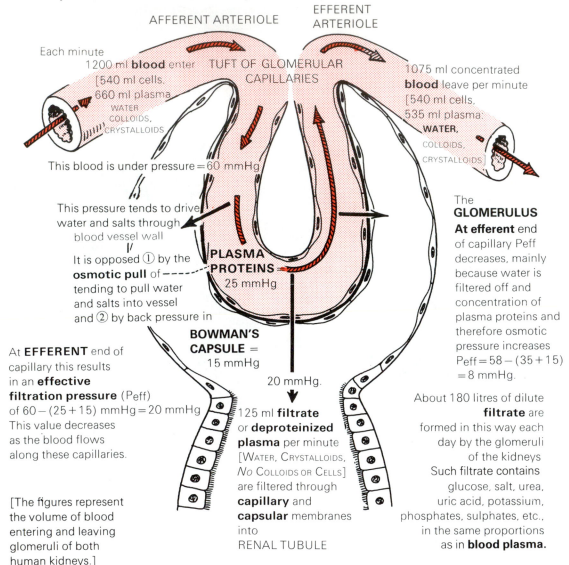

AFFERENT ARTERIOLE

EFFERENT ARTERIOLE

TUFT OF GLOMERULAR CAPILLARIES

Each minute
1200 ml **blood** enter
[540 ml cells,
660 ml plasma.
WATER
COLLOIDS,
CRYSTALLOIDS]

1075 ml concentrated
blood leave per minute
[540 ml cells,
535 ml plasma.
WATER,
COLLOIDS,
CRYSTALLOIDS]

This blood is under pressure = 60 mmHg

This pressure tends to drive water and salts through blood vessel wall

It is opposed ① by the **osmotic pull** of tending to pull water and salts into vessel and ② by back pressure in

PLASMA PROTEINS = 25 mmHg

The **GLOMERULUS**
At efferent end of capillary Peff decreases, mainly because water is filtered off and concentration of plasma proteins and therefore osmotic pressure increases
Peff = 58 − (35 + 15)
= 8 mmHg.

BOWMAN'S CAPSULE = 15 mmHg

20 mmHg.

At **EFFERENT** end of capillary this results in an **effective filtration pressure** (Peff) of 60 − (25 + 15) mmHg = 20 mmHg This value decreases as the blood flows along these capillaries.

125 ml **filtrate** or **deproteinized plasma** per minute [WATER, CRYSTALLOIDS, *No* COLLOIDS OR CELLS] are filtered through **capillary** and **capsular** membranes into RENAL TUBULE

About 180 litres of dilute **filtrate** are formed in this way each day by the glomeruli of the kidneys Such filtrate contains glucose, salt, urea, uric acid, potassium, phosphates, sulphates, etc., in the same proportions as in **blood plasma.**

[The figures represent the volume of blood entering and leaving glomeruli of both human kidneys.]

The glomerular membrane acts as a simple **filter** – i.e. no energy is used up by the cells in filtration.

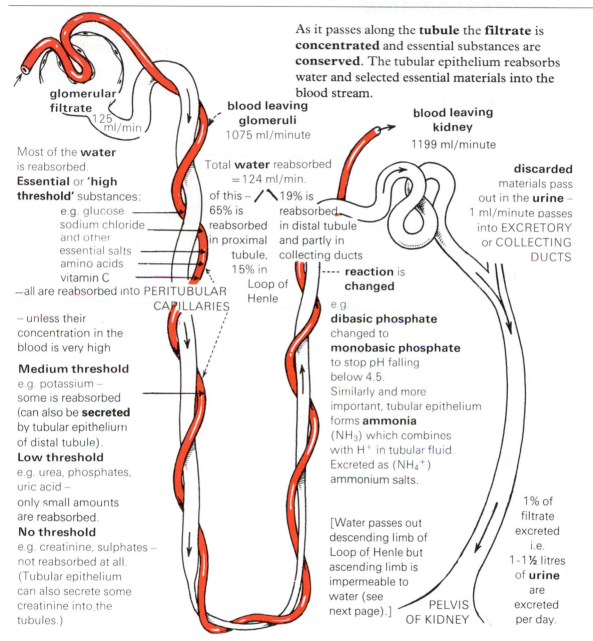

As it passes along the **tubule** the **filtrate** is **concentrated** and essential substances are **conserved**. The tubular epithelium reabsorbs water and selected essential materials into the blood stream.

glomerular filtrate
125 ml/min

blood leaving glomeruli
1075 ml/minute

blood leaving kidney
1199 ml/minute

Most of the **water** is reabsorbed.
Essential or 'high threshold' substances:
e.g. glucose
sodium chloride and other essential salts
amino acids
vitamin C
—all are reabsorbed into PERITUBULAR CAPILLARIES

– unless their concentration in the blood is very high

Medium threshold
e.g. potassium – some is reabsorbed (can also be **secreted** by tubular epithelium of distal tubule).

Low threshold
e.g. urea, phosphates, uric acid – only small amounts are reabsorbed.

No threshold
e.g. creatinine, sulphates – not reabsorbed at all.
(Tubular epithelium can also secrete some creatinine into the tubules.)

Total **water** reabsorbed = 124 ml/min.
of this – 65% is reabsorbed in proximal tubule, 15% in Loop of Henle

19% is reabsorbed in distal tubule and partly in collecting ducts

---- **reaction** is **changed**

e.g.
dibasic phosphate changed to **monobasic phosphate** to stop pH falling below 4.5.
Similarly and more important, tubular epithelium forms **ammonia** (NH_3) which combines with H^+ in tubular fluid. Excreted as (NH_4^+) ammonium salts.

[Water passes out descending limb of Loop of Henle but ascending limb is impermeable to water (see next page).]

PELVIS OF KIDNEY

discarded materials pass out in the **urine** – 1 ml/minute passes into EXCRETORY or COLLECTING DUCTS

1% of filtrate excreted i.e. 1-1½ litres of **urine** are excreted per day.

In concentrating the filtrate and in secretion of certain substances, the tubular epithelial cells use energy, i.e. they do work – much of it against an osmotic gradient.

FORMATION OF URINE: MECHANISM OF WATER REABSORPTION

Water is *not* actively reabsorbed by tubular cells. Its movements are determined passively by the **osmotic gradient** set up by solutes – chiefly by the **sodium** salts.

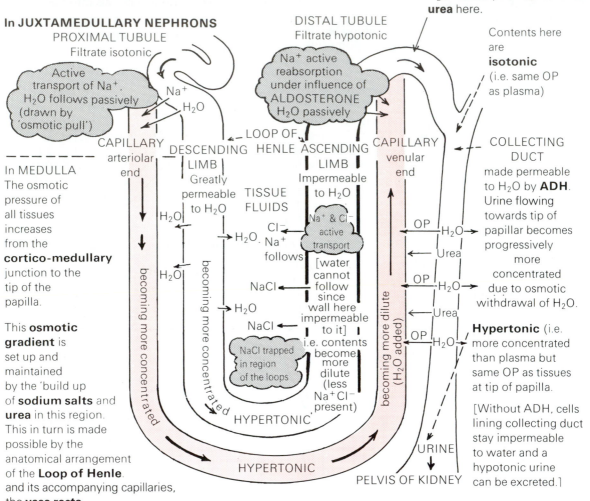

In JUXTAMEDULLARY NEPHRONS
PROXIMAL TUBULE
Filtrate isotonic

Active transport of Na^+. H_2O follows passively (drawn by 'osmotic pull')

Na^+
H_2O

DISTAL TUBULE
Filtrate hypotonic

Na^+ active reabsorption under influence of ALDOSTERONE H_2O passively

H_2O reabsorption concentrates **urea** here.

Contents here are **isotonic** (i.e. same OP as plasma)

CAPILLARY arteriolar end

LOOP OF HENLE
DESCENDING LIMB
ASCENDING LIMB

CAPILLARY venular end

COLLECTING DUCT

In MEDULLA
The osmotic pressure of all tissues increases from the **cortico-medullary** junction to the tip of the papilla.

DESCENDING LIMB
Greatly permeable to H_2O

H_2O

H_2O

TISSUE FLUIDS

Cl^-
$\rightarrow H_2O$. Na^+ follows

NaCl
$\rightarrow H_2O$
NaCl

NaCl trapped in region of the loops

ASCENDING LIMB
Impermeable to H_2O

Na^+ & Cl^- active transport

[water cannot follow since wall here impermeable to it] i.e. contents become more dilute (less Na^+Cl^- present)

HYPERTONIC

made permeable to H_2O by **ADH**. Urine flowing towards tip of papillar becomes progressively more concentrated due to osmotic withdrawal of H_2O.

OP $H_2O \rightarrow$
Urea
OP $H_2O \rightarrow$
Urea
OP $H_2O \rightarrow$

becoming more concentrated
becoming more concentrated
becoming more dilute (H_2O added)

This **osmotic gradient** is set up and maintained by the 'build up of **sodium salts** and **urea** in this region. This in turn is made possible by the anatomical arrangement of the **Loop of Henle**. and its accompanying capillaries, the **vasa recta**.

HYPERTONIC

URINE

PELVIS OF KIDNEY

Hypertonic (i.e. more concentrated than plasma but same OP as tissues at tip of papilla.

[Without ADH, cells lining collecting duct stay impermeable to water and a hypotonic urine can be excreted.]

These form a 'counter-current system' (i.e. *outflowing* fluid flows counter to (yet near) *inflowing*). Removal of water by osmosis from **descending** limb gives highly concentrated fluid at U bend. There is *passive* movement of $NA^+ + Cl^-$ out of the *thin* **ascending** limb and *active* transport of $Na^+ + Cl^-$ out of the *thick* part. Walls are 'water-tight' and no water follows so filtrate becomes hypotonic. **Urea** becomes concentrated in distal tubule. It diffuses out of collecting tubule to add to osmotic pressure of outer medulla.

65% of the **water** and **sodium** filtered into Bowman's capsule from the glomerular capillaries is reabsorbed in the **proximal convoluted tubule**.

Na$^+$ moves into the epithelial cells of the proximal tubule mainly by diffusion through channels. It is then actively transported into the **lateral intercellular spaces** by a Na$^+$, K$^+$ ATPase pump.

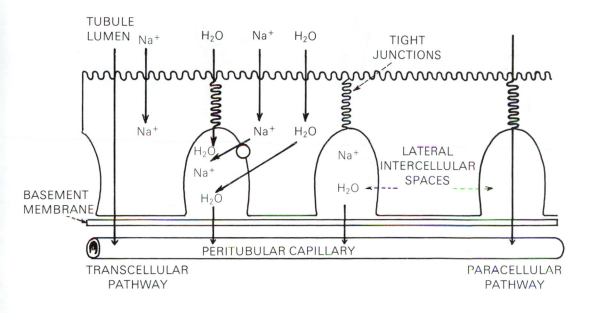

Accumulation of Na$^+$ in the lateral intercellular spaces creates an osmotic gradient across the epithelium. This osmotic gradient moves **water** into the lateral intercellular spaces either *through* the cells (i.e. via the **transcellular pathway**) or *across* the so-called **tight junctions** (i.e. via the **paracellular pathway**).

As fluid accumulates in the intercellular spaces the hydrostatic pressure increases and forces fluid across the basement membrane into the peritubular capillaries.

A similar mechanism exists for concentrating **bile** in the gall bladder by water reabsorption and for the reabsorption of water and electrolytes in the intestines.

This method of fluid absorption coupling water movement to sodium transport across tight-junctioned epithelia is called the **standing gradient mechanism**.

THE 'CLEARANCE' OF INULIN IN THE NEPHRON

The rate of glomerular filtration (GFR) can be found by measuring the 'plasma clearance' of a substance which is **filtered** by the renal corpuscle but **neither reabsorbed** nor **secreted** by the tubular epithelium. **Inulin** and **creatinine** are such substances. The use of inulin is more accurate but the technique using creatinine is simpler.

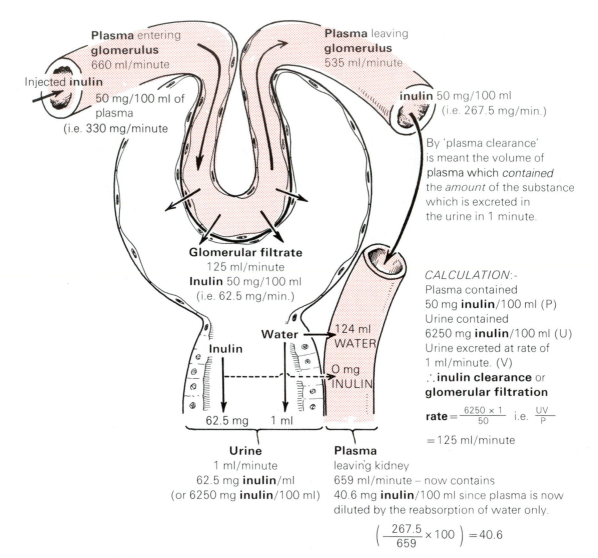

Plasma entering **glomerulus** 660 ml/minute

Injected **inulin** 50 mg/100 ml of plasma (i.e. 330 mg/minute)

Plasma leaving **glomerulus** 535 ml/minute

inulin 50 mg/100 ml (i.e. 267.5 mg/min.)

By 'plasma clearance' is meant the volume of plasma which *contained* the *amount* of the substance which is excreted in the urine in 1 minute.

Glomerular filtrate 125 ml/minute **Inulin** 50 mg/100 ml (i.e. 62.5 mg/min.)

Inulin **Water** 124 ml WATER

0 mg INULIN

62.5 mg 1 ml

Urine 1 ml/minute 62.5 mg **inulin**/ml (or 6250 mg **inulin**/100 ml)

Plasma leaving kidney 659 ml/minute – now contains 40.6 mg **inulin**/100 ml since plasma is now diluted by the reabsorption of water only.

$$\left(\frac{267.5}{659} \times 100 \right) = 40.6$$

CALCULATION:-
Plasma contained 50 mg **inulin**/100 ml (P)
Urine contained 6250 mg **inulin**/100 ml (U)
Urine excreted at rate of 1 ml/minute. (V)
∴ **inulin clearance** or **glomerular filtration**

$$\text{rate} = \frac{6250 \times 1}{50} \quad \text{i.e.} \quad \frac{UV}{P}$$

$$= 125 \text{ ml/minute}$$

This idea of clearance can be applied to other substances naturally present such as **urea**, or artificially introduced, such as **diodone**.

Urea, like inulin, is filtered by the renal corpuscle. Unlike inulin some urea is reabsorbed back into the blood stream from the tubules.

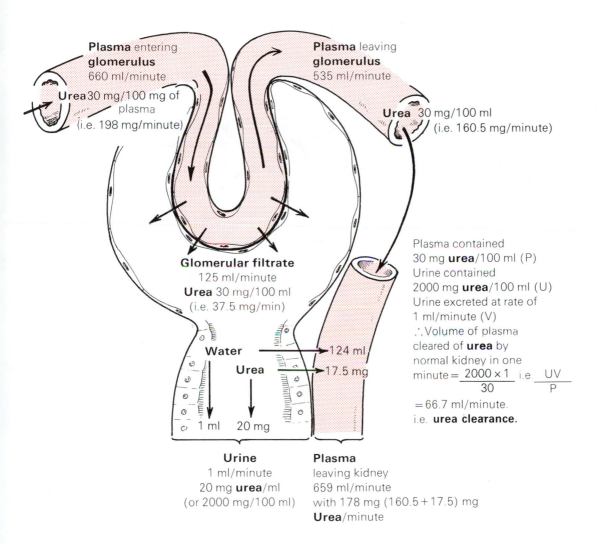

Plasma entering
glomerulus
660 ml/minute
Urea 30 mg/100 mg of
plasma
(i.e. 198 mg/minute)

Plasma leaving
glomerulus
535 ml/minute
Urea 30 mg/100 ml
(i.e. 160.5 mg/minute)

Glomerular filtrate
125 ml/minute
Urea 30 mg/100 ml
(i.e. 37.5 mg/min)

Water → 124 ml

Urea → 17.5 mg

1 ml   20 mg

Urine
1 ml/minute
20 mg **urea**/ml
(or 2000 mg/100 ml)

Plasma
leaving kidney
659 ml/minute
with 178 mg (160.5 + 17.5) mg
Urea/minute

Plasma contained
30 mg **urea**/100 ml (P)
Urine contained
2000 mg **urea**/100 ml (U)
Urine excreted at rate of
1 ml/minute (V)
∴ Volume of plasma
cleared of **urea** by
normal kidney in one

$$\text{minute} = \frac{2000 \times 1}{30} \quad \text{i.e.} \quad \frac{UV}{P}$$

= 66.7 ml/minute.
i.e. **urea clearance**.

Urea clearance is used as a test of renal function.

PAH 'CLEARANCE'

Certain special substances are filtered by the renal corpuscle and the rest that escapes filtration is then secreted totally from the peritubular blood into the tubule. Thus the renal artery contains the substance but the renal vein contains none. **Para-aminohippuric acid (PAH)** and **diodone** are such substances. The 'Plasma Clearance' of these substances measures the **renal plasma flow** rate.

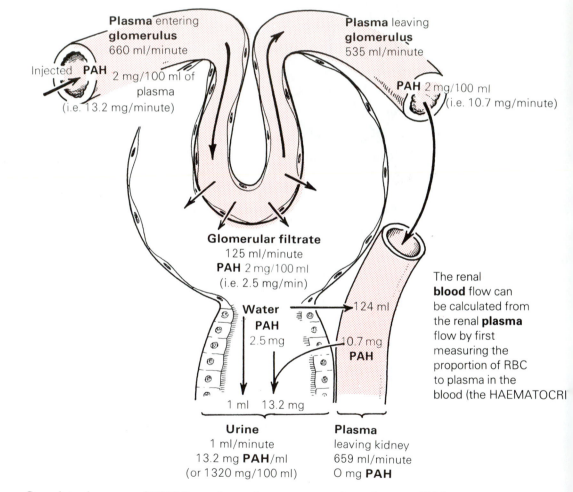

Plasma entering **glomerulus**
660 ml/minute

Injected **PAH** 2 mg/100 ml of plasma
(i.e. 13.2 mg/minute)

Plasma leaving **glomerulus**
535 ml/minute

PAH 2 mg/100 ml
(i.e. 10.7 mg/minute)

Glomerular filtrate
125 ml/minute
PAH 2 mg/100 ml
(i.e. 2.5 mg/min)

Water
PAH
2.5 mg

124 ml

10.7 mg
PAH

1 ml   13.2 mg

The renal **blood** flow can be calculated from the renal **plasma** flow by first measuring the proportion of RBC to plasma in the blood (the HAEMATOCRI

Urine
1 ml/minute
13.2 mg **PAH**/ml
(or 1320 mg/100 ml)

Plasma
leaving kidney
659 ml/minute
O mg **PAH**

Complete clearance of **PAH** from plasma in one passage through normal kidney, gauges not only glomerular filtrating power but also the efficiency of the tubular epithelium to secrete.

An acid is a substance which liberates H^+ ions (proton donor). A base is a substance which can accept a H^+ ion (proton acceptor). Acids are formed in the body during the breakdown of food, during cell metabolism and, especially, by the production of CO_2' and its combination with water. However the concentration of free H^+ in the body fluids is kept relatively constant at about pH 7.4 (pH 7.4 = 4×10^{-8} mol/litre).

This equilibrium is maintained by **buffer systems** binding free H^+; by the **lungs** eliminating CO_2 and finally by the kidneys excreting H^+ and conserving base (mainly HCO_3^-).

The main extracellular fluid buffer system is the **bicarbonate buffer** system
$$CO_2 + H_2O \rightleftharpoons H_2CO_3 \rightleftharpoons H^+ + HCO_3^-.$$

Thus if H^+ is liberated it combines with HCO_3^- forming carbonic acid which breaks down to $CO_2 + H_2O$. *NB*: HCO_3^- is 'used up' in this reaction. The Henderson-Hasselbalch equation shows the relationship between pH, CO_2 and HCO_3^-

$$pH \propto \frac{\text{Concentration of } HCO_3^-}{\text{Dissolved } CO_2}$$

The lungs decrease H^+ (increase pH) by eliminating CO_2. The kidneys decrease H^+ (increase pH) by **reabsorption** of HCO_3^- as well as by **excretion** of H^+.

REABSORPTION OF BICARBONATE

Bicarbonate is in a concentration of about 24 mmol/l in filtrate. Most is reabsorbed in the proximal tubule by this mechanism – – – – –

Secretion of free hydrogen ions in exchange for sodium ions (secondary active transport) – – –

Secreted hydrogen ions react with bicarbonate to form carbon acid.

This carbonic acid breaks down to form CO_2 and H_2O. Reaction is facilitated by carbonic anhydrase in cell membrane of proximal tubule.

This mechanism reabsorbs base (HCO_3^-)

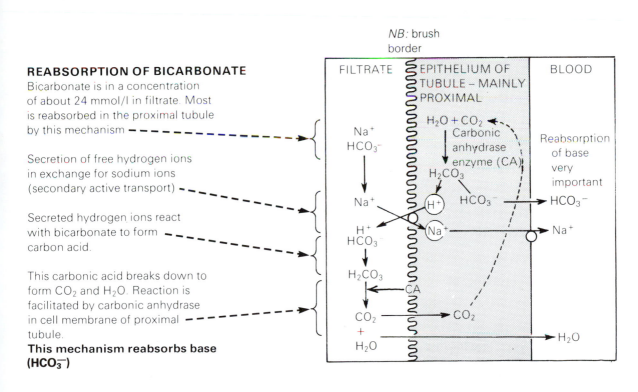

MAINTENANCE OF ACID-BASE BALANCE 2

A large increase in concentration of free H^+ in the tubular filtrate (to pH 4.5) would prevent the secretion of H^+ ions from the tubular cells. Two mechanisms bind free H^+ in filtrate and allow continued **secretion** of H^+.

PHOSPHATE MECHANISM

Dibasic phosphate
in filtrate

Hydrogen ion is **secreted**
in exchange for a sodium ion
(secondary active transport)
or by primary active transport
(proton pump).

Secreted hydrogen ion is
bound and **excreted** as
monobasic phosphate.

This mechanism excretes H^+
and **reabsorbs** some base (HCO_3^-).

AMMONIA MECHANISM
This is the more important mechanism
for buffering H^+ in tubule.

Sodium and chloride ions
in filtrate
Hydrogen ion is **secreted** in
exchange for a sodium ion
(secondary active transport)
or by a proton pump:
Secreted H^+ combines with
ammonia (NH_3) which diffuses
from the tubular cell.

Thus an ammonium ion (NH_4^+)
is formed.
Cell membrane is permeable to
NH_3 but not to NH_4^+ so NH_4^+
containing secreted H^+ is **excreted**.

This mechanism excretes H^+
and **reabsorbs** some base (HCO_3^-).

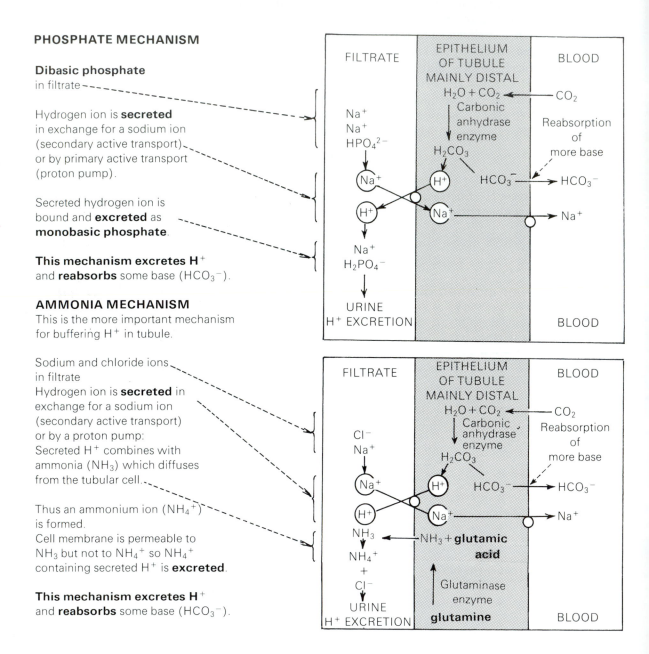

DEFENCE OF BODY FLUID TONICITY

The **tonicity** or **osmolarity** of body fluids is controlled by **thirst** (which alters water intake) and *VASOPRESSIN* (*antidiuretic hormone, ADH*) released from the posterior pituitary gland (page 186) (which alters water excretion by the kidney).

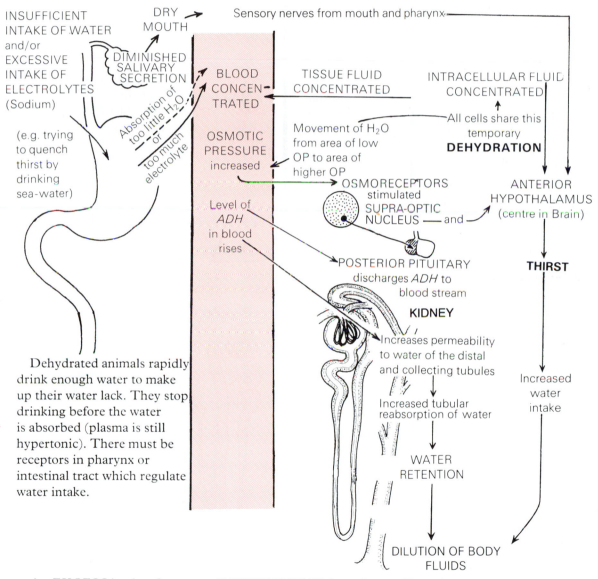

INSUFFICIENT INTAKE OF WATER and/or EXCESSIVE INTAKE OF ELECTROLYTES (Sodium)

(e.g. trying to quench thirst by drinking sea-water)

DRY MOUTH

DIMINISHED SALIVARY SECRETION

Absorption of too little H_2O or too much electrolyte

Sensory nerves from mouth and pharynx

BLOOD CONCENTRATED

OSMOTIC PRESSURE increased

Level of *ADH* in blood rises

TISSUE FLUID CONCENTRATED

Movement of H_2O from area of low OP to area of higher OP

INTRACELLULAR FLUID CONCENTRATED

All cells share this temporary **DEHYDRATION**

OSMORECEPTORS stimulated SUPRA-OPTIC NUCLEUS — and —

POSTERIOR PITUITARY discharges *ADH* to blood stream

ANTERIOR HYPOTHALAMUS (centre in Brain)

THIRST

KIDNEY Increases permeability to water of the distal and collecting tubules

Increased tubular reabsorption of water

WATER RETENTION

Increased water intake

Dehydrated animals rapidly drink enough water to make up their water lack. They stop drinking before the water is absorbed (plasma is still hypertonic). There must be receptors in pharynx or intestinal tract which regulate water intake.

DILUTION OF BODY FLUIDS

An *EXCESS* intake of water or *INSUFFICIENT* electrolytes will produce mainly decreased osmoreceptor stimulation and hence decreased ADH release leading to increased tonicity of body fluids.

DEFENCE OF BODY FLUID VOLUME

The **volume** of the extracellular fluid (ECF) is determined mainly by the amount of osmotically active solute it contains. Na^+ is the most important active solute in the body, hence mechanisms that control Na^+ balance will also control ECF volume.

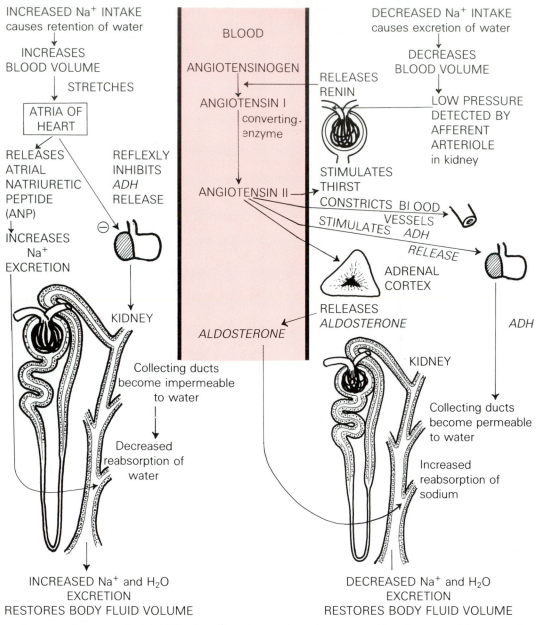

INCREASED Na^+ INTAKE causes retention of water

INCREASES BLOOD VOLUME

STRETCHES

ATRIA OF HEART

RELEASES ATRIAL NATRIURETIC PEPTIDE (ANP)

INCREASES Na^+ EXCRETION

REFLEXLY INHIBITS *ADH* RELEASE

KIDNEY

Collecting ducts become impermeable to water

Decreased reabsorption of water

INCREASED Na^+ and H_2O EXCRETION
RESTORES BODY FLUID VOLUME

BLOOD

ANGIOTENSINOGEN

ANGIOTENSIN I

converting enzyme

ANGIOTENSIN II

ALDOSTERONE

DECREASED Na^+ INTAKE causes excretion of water

DECREASES BLOOD VOLUME

LOW PRESSURE DETECTED BY AFFERENT ARTERIOLE in kidney

RELEASES RENIN

STIMULATES THIRST

CONSTRICTS BLOOD VESSELS

STIMULATES *ADH* RELEASE

ADRENAL CORTEX

RELEASES *ALDOSTERONE*

ADH

KIDNEY

Collecting ducts become permeable to water

Increased reabsorption of sodium

DECREASED Na^+ and H_2O EXCRETION
RESTORES BODY FLUID VOLUME

Control of *vasopressin (ADH)* release by changes in **volume** overrides the control by **osmotic** changes.

160

A resistant, distensible **transitional epithelium** lines all urinary passages.

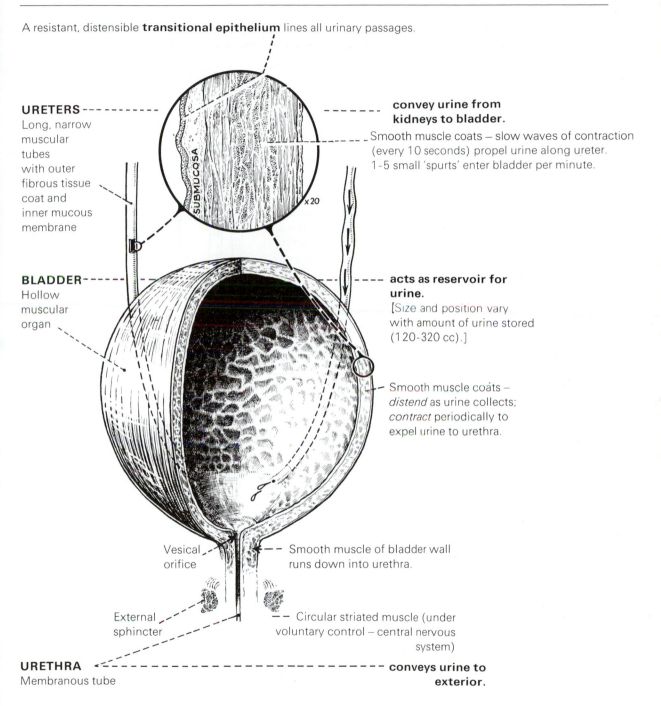

URETERS
Long, narrow
muscular
tubes
with outer
fibrous tissue
coat and
inner mucous
membrane

SUBMUCOSA

×20

**convey urine from
kidneys to bladder.**

Smooth muscle coats – slow waves of contraction
(every 10 seconds) propel urine along ureter.
1-5 small 'spurts' enter bladder per minute.

BLADDER
Hollow
muscular
organ

**acts as reservoir for
urine.**
[Size and position vary
with amount of urine stored
(120-320 cc).]

Smooth muscle coats –
distend as urine collects;
contract periodically to
expel urine to urethra.

Vesical
orifice

Smooth muscle of bladder wall
runs down into urethra.

External
sphincter

Circular striated muscle (under
voluntary control – central nervous
system)

URETHRA
Membranous tube

**conveys urine to
exterior.**

STORAGE AND EXPULSION OF URINE

Urine is formed continuously by the kidneys. It collects, drop by drop, in the urinary bladder which expands to hold about 300 ml. When the bladder is full the desire to void urine is experienced.

MICTURITION is essentially **reflex** – carried out through centres in spinal cord

STIMULUS
Distension of **receptors** in smooth muscle

AFFERENT PATHWAYS

MIDBRAIN PONS

In adult – the reflex can be controlled and inhibited **voluntarily** through HIGHER CENTRES

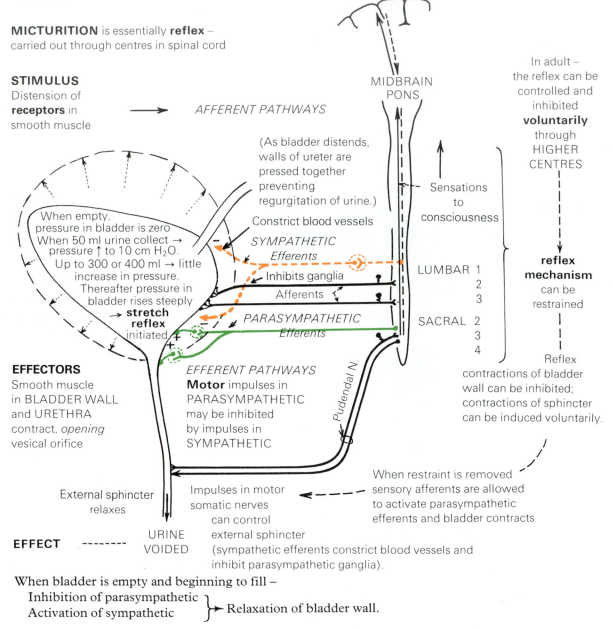

(As bladder distends, walls of ureter are pressed together preventing regurgitation of urine.)

Constrict blood vessels

When empty, pressure in bladder is zero
When 50 ml urine collect → pressure ↑ to 10 cm H$_2$O.
Up to 300 or 400 ml → little increase in pressure. Thereafter pressure in bladder rises steeply
→ **stretch reflex** initiated

SYMPATHETIC Efferents

Inhibits ganglia
Afferents

PARASYMPATHETIC Efferents

Sensations to consciousness

LUMBAR 1 2 3

SACRAL 2 3 4

reflex mechanism can be restrained

Reflex contractions of bladder wall can be inhibited; contractions of sphincter can be induced voluntarily.

EFFECTORS
Smooth muscle in BLADDER WALL and URETHRA contract, *opening* vesical orifice

EFFERENT PATHWAYS
Motor impulses in PARASYMPATHETIC may be inhibited by impulses in SYMPATHETIC

Pudendal N.

External sphincter relaxes

Impulses in motor somatic nerves can control external sphincter (sympathetic efferents constrict blood vessels and inhibit parasympathetic ganglia).

When restraint is removed sensory afferents are allowed to activate parasympathetic efferents and bladder contracts

EFFECT - - - - - - - - URINE VOIDED

When bladder is empty and beginning to fill –
Inhibition of parasympathetic ⎫
Activation of sympathetic ⎬→ Relaxation of bladder wall.

VOLUME: In adult
1000-1500 ml/24 hours

SPECIFIC GRAVITY: 1.010-1.035

} Vary with fluid intake and with fluid output from other routes – skin, lungs, gut.
[Volume reduced during sleep and muscular exercise: specific gravity greater on protein diet.]

REACTION: Usually slightly acid – (pH 4.5-8)

Varies with diet
[acid on ordinary mixed diet: alkaline on vegetarian diet].

COLOUR:

Yellow due to **urochrome** pigment – probably from destruction of tissue protein.

More concentrated and **darker** in early morning – less water excreted at night but unchanged amounts of urinary solids.

ODOUR
Aromatic when fresh → **ammoniacal** on standing due to bacterial decomposition of **urea** to **ammonia**.

COMPOSITION

Water – – – – – – – – – 1000-1500 ml/24 h

Inorganic substances	millimoles excreted in 24 h
Sodium – – – – – – – – –200	
Chloride – – – – – – – – 200	
Calcium – – – – – – – – – –5	
Potassium – – – – – – – 50	
Phosphates – – – – – – –25	
Sulphates – – – – – – – –50	

[These figures are approximate and vary widely in healthy individuals]

Organic substances

Urea – – – – – – – – 400 – – – – – – derived from breakdown of protein – therefore varies with protein in diet.

Uric Acid – – – – – – – –4 – – – – – comes from purine of food and body tissues.

Creatinine – – – – – – –10 – – – – – from breakdown of body tissues; uninfluenced by amount of dietary protein.

Ammonia – – – – – – – 40 – – – – – formed in kidney from glutamine brought to it by blood stream; varies with amounts of acid substances requiring neutralization in the kidney.

[In the **newborn**, volume and specific gravity are low and composition varies.]

163

CHAPTER 8

ENDOCRINE SYSTEM

ENDOCRINE SYSTEM

The **DUCTLESS GLANDS** produce **hormones** ('chemical messengers') which they pass into the **blood stream** for **general circulation** to excite or inhibit activity of other organs or tissues.

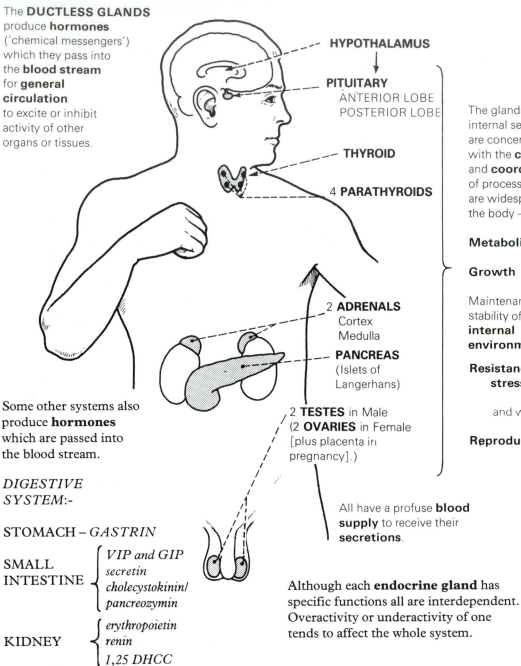

HYPOTHALAMUS

PITUITARY
ANTERIOR LOBE
POSTERIOR LOBE

THYROID

4 PARATHYROIDS

2 **ADRENALS**
Cortex
Medulla

PANCREAS
(Islets of Langerhans)

2 **TESTES** in Male
(2 **OVARIES** in Female
[plus placenta in pregnancy].)

The glands of internal secretion are concerned with the **control** and **coordination** of processes which are widespread in the body – such as:

Metabolism

Growth

Maintenance of stability of **internal environment**

Resistance to **stress**

and with

Reproduction.

Some other systems also produce **hormones** which are passed into the blood stream.

DIGESTIVE SYSTEM:-

STOMACH – *GASTRIN*

SMALL INTESTINE
{ *VIP and GIP secretin cholecystokinin/ pancreozymin*

KIDNEY
{ *erythropoietin renin 1,25 DHCC*

All have a profuse **blood supply** to receive their **secretions**.

Although each **endocrine gland** has specific functions all are interdependent. Overactivity or underactivity of one tends to affect the whole system.

166

STRUCTURE:

2 LOBES (joined by ISTHMUS) composed of

lie in front of TRACHEA

FOLLICLES lined by CUBICAL EPITHELIUM

(NO DUCTS)

(×150)

Blood vessels to and from THYROID

Gland weighs about 25 g in adult

Richly supplied with BLOOD CAPILLARIES

PARAFOLLICULAR or 'C'CELLS

FUNCTION:

Cubical epithelium extracts from the blood stream (and concentrates) *IODIDE* (**iodide trapping**)

IODINE links with *TYROSINE* } *MONOIODOTYROSINE DIIODO-Tyrosine*

TRIIODO-THYRONINE (T_3)

TETRAIODO-THYRONINE (Thyroxine) (T_4)

BLOOD

Stored in colloid linked with protein as *THYROGLOBULIN* | when required a protein-splitting enzyme releases T_3 and T_4

T_3 and T_4 are carried by the blood to all body tissues. T_4 is usually converted in the cell cytoplasm to T_3 which binds to receptors in the nuclei to increase mRNA and ribosomal RNA and hence protein synthesis. This brings about increased **oxygen consumption, heat production** and **metabolism**. Normal thyroid output is required for normal **growth**.

[**Parafollicular** cells secrete *calcitonin* – which **lowers blood calcium** by suppressing calcium mobilization from bone and increasing **calcium** excretion in the urine.]

REGULATION OF SECRETION:

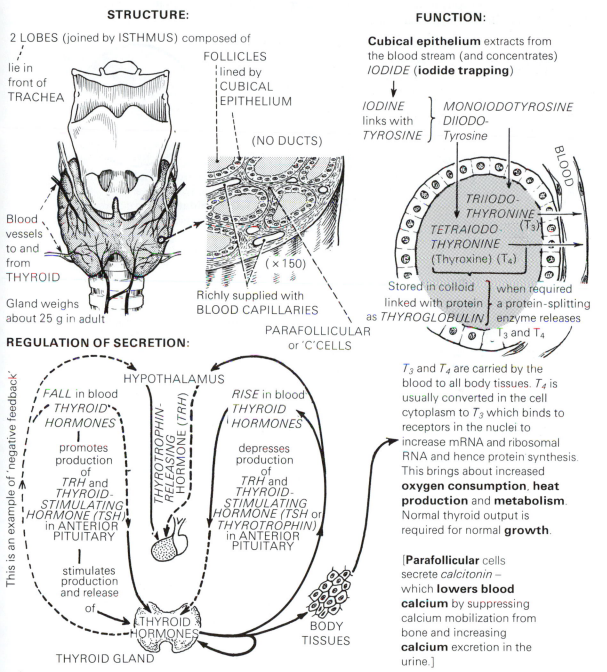

This is an example of 'negative feedback'

HYPOTHALAMUS

FALL in blood *THYROID HORMONES*

promotes production of *TRH* and *THYROID-STIMULATING HORMONE (TSH)* in ANTERIOR PITUITARY

stimulates production and release of

THYROTROPHIN-RELEASING HORMONE (TRH)

RISE in blood *THYROID HORMONES*

depresses production of *TRH* and *THYROID-STIMULATING HORMONE (TSH or THYROTROPHIN)* in ANTERIOR PITUITARY

THYROID HORMONES

THYROID GLAND

BODY TISSUES

Increased O_2 consumption is probably due to increased activity of the Na^+, K^+ – ATPase pumps which maintain intracellular K^+ high and Na^+ low.

UNDERACTIVITY OF THYROID

If the **thyroid** shows atrophy or destruction of its secretory cells or is inadequately stimulated by the anterior pituitary or hypothalamus:

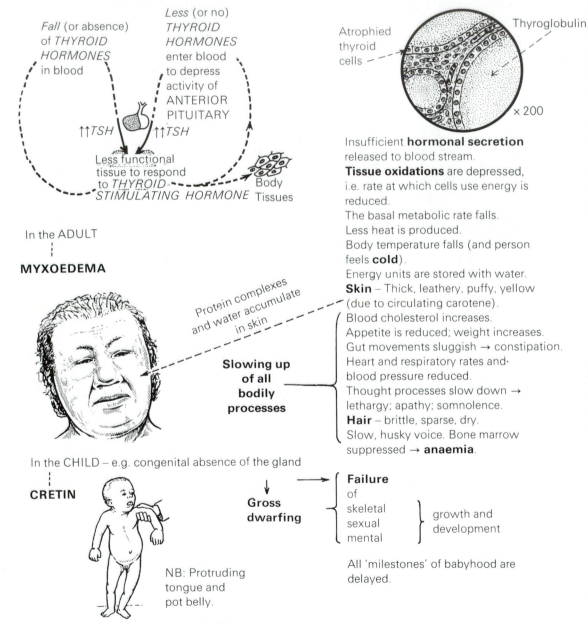

Fall (or absence) of *THYROID HORMONES* in blood

Less (or no) *THYROID HORMONES* enter blood to depress activity of ANTERIOR PITUITARY

↑↑*TSH*   ↑↑*TSH*

Less functional tissue to respond to *THYROID STIMULATING HORMONE*

Body Tissues

Atrophied thyroid cells

Thyroglobulin

× 200

Insufficient **hormonal secretion** released to blood stream.
Tissue oxidations are depressed, i.e. rate at which cells use energy is reduced.
The basal metabolic rate falls.
Less heat is produced.
Body temperature falls (and person feels **cold**).
Energy units are stored with water.
Skin – Thick, leathery, puffy, yellow (due to circulating carotene).
Blood cholesterol increases.
Appetite is reduced; weight increases.
Gut movements sluggish → constipation.
Heart and respiratory rates and blood pressure reduced.
Thought processes slow down → lethargy; apathy; somnolence.
Hair – brittle, sparse, dry.
Slow, husky voice. Bone marrow suppressed → **anaemia**.

In the ADULT

MYXOEDEMA

Protein complexes and water accumulate in skin

Slowing up of all bodily processes

In the CHILD – e.g. congenital absence of the gland

CRETIN

↓
Gross dwarfing

→ **Failure** of skeletal sexual mental } growth and development

All 'milestones' of babyhood are delayed.

NB: Protruding tongue and pot belly.

THYROXINE (taken by mouth) restores individuals to normal.

Commonest form is **Graves' disease**. Produces increased thyroid hormone secretion, enlarged thyroid (**goitre**) and protrusion of eyeballs (**exophthalmos**). Caused by production of antibodies against person's own thyroid cells. These antibodies, *Thyroid-Stimulating Immunoglobulins (TSI)*, act like *Thyroid-Stimulating Hormone (TSH)* and release thyroid hormones (T_3 and T_4).

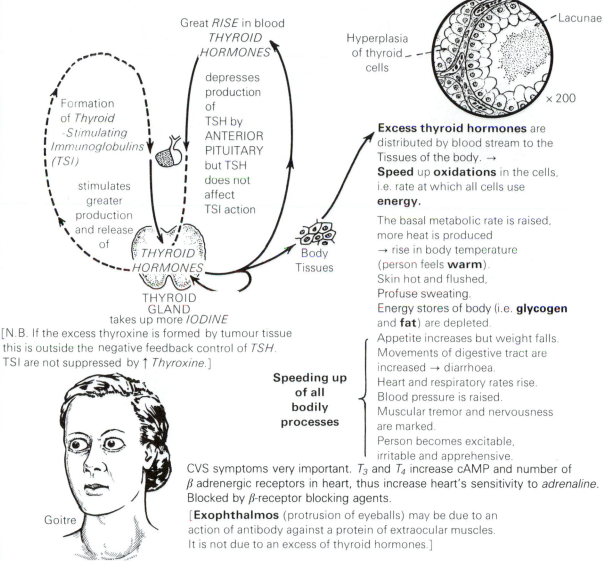

Great *RISE* in blood *THYROID HORMONES*

depresses production of TSH by ANTERIOR PITUITARY but TSH does not affect TSI action

Formation of *Thyroid -Stimulating Immunoglobulins (TSI)*

stimulates greater production and release of

THYROID HORMONES

THYROID GLAND takes up more *IODINE*

Body Tissues

Hyperplasia of thyroid cells

Lacunae

× 200

Excess thyroid hormones are distributed by blood stream to the Tissues of the body. →
Speed up **oxidations** in the cells, i.e. rate at which all cells use **energy.**

The basal metabolic rate is raised, more heat is produced
→ rise in body temperature (person feels **warm**).
Skin hot and flushed,
Profuse sweating.
Energy stores of body (i.e. **glycogen** and **fat**) are depleted.

[N.B. If the excess thyroxine is formed by tumour tissue this is outside the negative feedback control of *TSH*. TSI are not suppressed by ↑ *Thyroxine*.]

Speeding up of all bodily processes

Appetite increases but weight falls.
Movements of digestive tract are increased → diarrhoea.
Heart and respiratory rates rise.
Blood pressure is raised.
Muscular tremor and nervousness are marked.
Person becomes excitable, irritable and apprehensive.

CVS symptoms very important. T_3 and T_4 increase cAMP and number of β adrenergic receptors in heart, thus increase heart's sensitivity to *adrenaline*. Blocked by β-receptor blocking agents.

[**Exophthalmos** (protrusion of eyeballs) may be due to an action of antibody against a protein of extraocular muscles. It is not due to an excess of thyroid hormones.]

Goitre

Surgical removal of part of the overactive gland or destruction by radioactive iodine reduces the thyroid activity.

PARATHYROIDS

Four small glands composed of cords of cells which secrete parathyroid hormone –

parathormone or PTH

↓

CAPILLARIES

↓

General circulation

↓

to **all tissues** of the body

But not all tissues are sensitive to it.

It plays an important role in **calcium** and **phosphate metabolism**.

PARATHYROID GLANDS

Situated behind THYROID

OESOPHAGUS

TRACHEA

Each weighs from 20-50 mg in Adult

× 200

Function of EOSINOPHIL cells is unknown

There hormones, *parathormone, 1, 25-dihydroxycholecalciferol* (1,25-DHCC) and *calcitonin* act on **kidney** and **gut** to keep blood ionized **calcium** constant (necessary for normal nerve and muscle excitability and blood coagulation).

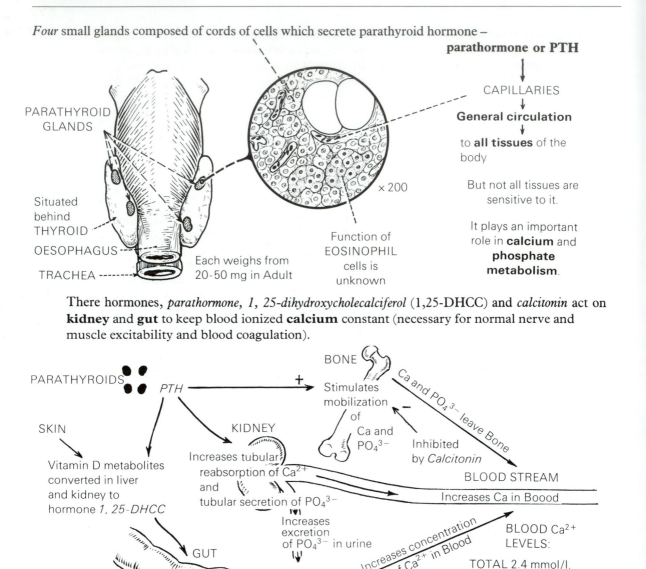

PARATHYROIDS

PTH

SKIN

Vitamin D metabolites converted in liver and kidney to hormone *1, 25-DHCC*

KIDNEY

Increases tubular reabsorption of Ca^{2+} and tubular secretion of PO_4^{3-}

Increases excretion of PO_4^{3-} in urine

GUT

Increases absorption of dietary Ca^{2+} and reduces loss of Ca^{2+} in faeces

BONE

+

Stimulates mobilization of Ca and PO_4^{3-}

Ca and PO_4^{3-} leave Bone

−

Inhibited by *Calcitonin*

BLOOD STREAM

Increases Ca in Boood

Increases concentration of Ca^{2+} in Blood

BLOOD Ca^{2+} LEVELS:

TOTAL 2.4 mmol/l.
PROTEIN BOUND 1.2 mmol/l.
ACTIVE IONIZED Ca^{2+} 1.2 mmol/l.

In bones and kidneys PTH activates adenylate cyclase thus increasing cAMP. In bones, **osteoblasts** pump Ca^{2+} into extracellular fluid. This pump is stimulated by 1, 25-DHCC.
   Calcium ions in extracellular fluid control parathyroid activity. ↑ Ca^{2+} depresses PTH secretion. ↓ Ca^{2+} increases PTH secretion.

UNDERACTIVITY OF PARATHYROIDS

Atrophy or removal of parathyroid tissue causes a fall in **blood calcium** level and increased excitability of neuromuscular tissue. This leads to severe convulsive disorder – **tetany**.

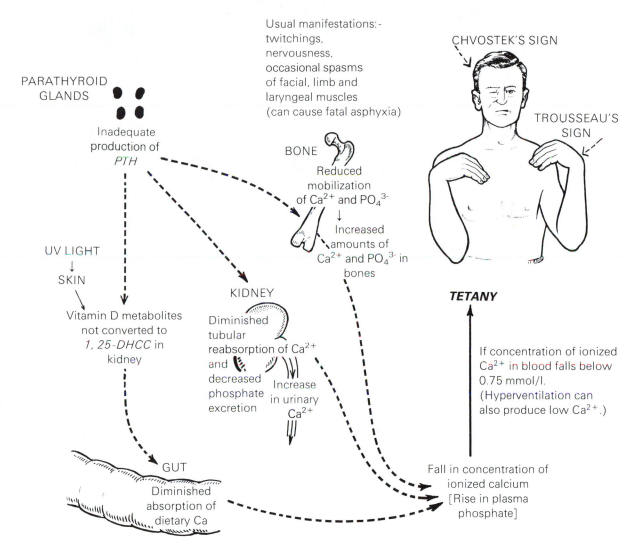

Usual manifestations:-
twitchings,
nervousness,
occasional spasms
of facial, limb and
laryngeal muscles
(can cause fatal asphyxia)

CHVOSTEK'S SIGN

TROUSSEAU'S SIGN

PARATHYROID GLANDS

Inadequate production of *PTH*

BONE
Reduced mobilization of Ca^{2+} and PO_4^{3-}
↓
Increased amounts of Ca^{2+} and PO_4^{3-} in bones

UV LIGHT
↓
SKIN
↓
Vitamin D metabolites not converted to *1, 25-DHCC* in kidney

KIDNEY
Diminished tubular reabsorption of Ca^{2+} and decreased phosphate excretion
Increase in urinary Ca^{2+}

TETANY

If concentration of ionized Ca^{2+} in blood falls below 0.75 mmol/l. (Hyperventilation can also produce low Ca^{2+}.)

GUT
Diminished absorption of dietary Ca

Fall in concentration of ionized calcium [Rise in plasma phosphate]

[Note the inverse relationship between plasma calcium and inorganic phosphate]

Symptoms are relieved by injection of calcium, large doses of a Vit. D compound and parathormone.

OVERACTIVITY OF PARATHYROIDS

Overactivity of the parathyroids (due often to tumour) leads to rise in **blood calcium** level and eventually to **osteitis fibrosa cystica**.

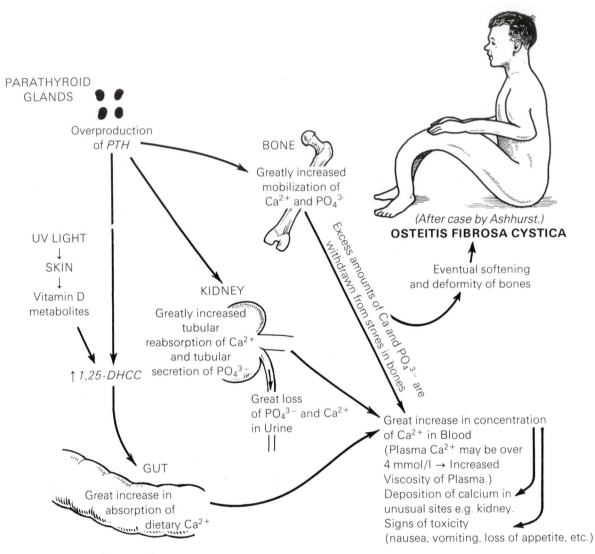

PARATHYROID
GLANDS

Overproduction
of *PTH*

BONE

Greatly increased
mobilization of
Ca^{2+} and PO_4^{3-}

(After case by Ashhurst.)
OSTEITIS FIBROSA CYSTICA

UV LIGHT
↓
SKIN
↓
Vitamin D
metabolites

KIDNEY

Greatly increased
tubular
reabsorption of Ca^{2+}
and tubular
secretion of PO_4^{3-}

↑ *1,25-DHCC*

Excess amounts of Ca and PO_4^{3-} are withdrawn from stores in bones

Eventual softening
and deformity of bones

Great loss
of PO_4^{3-} and Ca^{2+}
in Urine

Great increase in concentration
of Ca^{2+} in Blood
(Plasma Ca^{2+} may be over
4 mmol/l → Increased
Viscosity of Plasma.)
Deposition of calcium in
unusual sites e.g. kidney.
Signs of toxicity
(nausea, vomiting, loss of appetite, etc.)

GUT

Great increase in
absorption of
dietary Ca^{2+}

The increased level of blood calcium eventually leads to excessive loss of **calcium** in **urine** (in spite of ↑ reabsorption) and also of **water** since the salts are excreted in solution. **Polyuria** and **thirst** result.

Excision of the overactive parathyroid tissue abolishes syndrome.

The adrenal cortex is essential to life and plays an important rôle in states of stress.

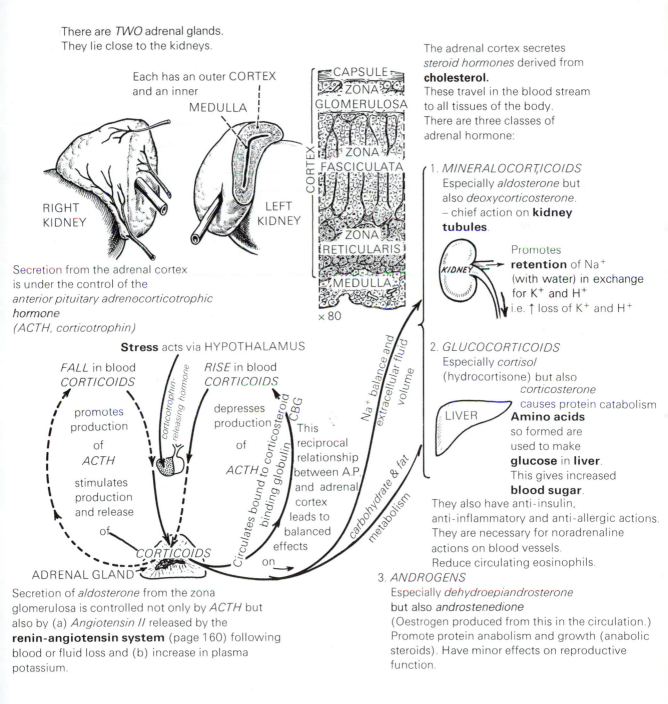

There are *TWO* adrenal glands.
They lie close to the kidneys.

Each has an outer CORTEX and an inner

MEDULLA

RIGHT KIDNEY

LEFT KIDNEY

CAPSULE
ZONA GLOMERULOSA
ZONA FASCICULATA
ZONA RETICULARIS
MEDULLA
CORTEX
× 80

Secretion from the adrenal cortex is under the control of the *anterior pituitary adrenocorticotrophic hormone (ACTH, corticotrophin)*

Stress acts via HYPOTHALAMUS

FALL in blood *CORTICOIDS*

RISE in blood *CORTICOIDS*

corticotrophin-releasing hormone

promotes production of *ACTH* stimulates production and release of

depresses production of *ACTH*

Circulates bound to corticosteroid binding globulin CBG

This reciprocal relationship between A.P. and adrenal cortex leads to balanced effects on

CORTICOIDS

ADRENAL GLAND

Secretion of *aldosterone* from the zona glomerulosa is controlled not only by *ACTH* but also by (a) *Angiotensin II* released by the **renin-angiotensin system** (page 160) following blood or fluid loss and (b) increase in plasma potassium.

The adrenal cortex secretes *steroid hormones* derived from **cholesterol.**
These travel in the blood stream to all tissues of the body.
There are three classes of adrenal hormone:

1. *MINERALOCORTICOIDS*
Especially *aldosterone* but also *deoxycorticosterone*.
– chief action on **kidney tubules.**

KIDNEY

Promotes **retention** of Na^+ (with water) in exchange for K^+ and H^+ i.e. ↑ loss of K^+ and H^+

Na^+ balance and extracellular fluid volume

2. *GLUCOCORTICOIDS*
Especially *cortisol* (hydrocortisone) but also *corticosterone* causes protein catabolism

LIVER **Amino acids** so formed are used to make **glucose** in **liver**. This gives increased **blood sugar**.

carbohydrate & fat metabolism

They also have anti-insulin, anti-inflammatory and anti-allergic actions. They are necessary for noradrenaline actions on blood vessels. Reduce circulating eosinophils.

3. *ANDROGENS*
Especially *dehydroepiandrosterone* but also *androstenedione*
(Oestrogen produced from this in the circulation.) Promote protein anabolism and growth (anabolic steroids). Have minor effects on reproductive function.

UNDERACTIVITY OF ADRENAL CORTEX

Atrophy of adrenal cortex (occasionally occurs with destructive disease of the gland, e.g. tuberculosis, cancer)

gives
inadequate production of all *corticoids*:-

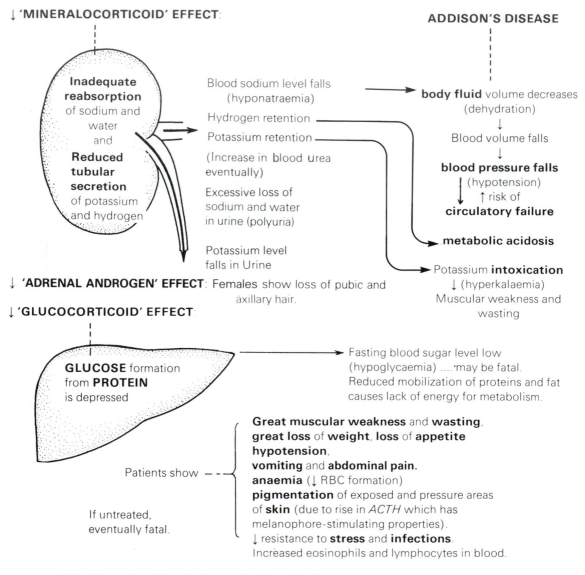

↓ 'MINERALOCORTICOID' EFFECT:

ADDISON'S DISEASE

Inadequate reabsorption of sodium and water and **Reduced tubular secretion** of potassium and hydrogen

Blood sodium level falls (hyponatraemia)

Hydrogen retention

Potassium retention

(Increase in blood urea eventually)

Excessive loss of sodium and water in urine (polyuria)

Potassium level falls in Urine

body fluid volume decreases (dehydration)
↓
Blood volume falls
↓
blood pressure falls
(hypotension)
↑ risk of
circulatory failure

metabolic acidosis

Potassium **intoxication**
↓ (hyperkalaemia)
Muscular weakness and wasting

↓ 'ADRENAL ANDROGEN' EFFECT: Females show loss of pubic and axillary hair.

↓ 'GLUCOCORTICOID' EFFECT:

GLUCOSE formation from **PROTEIN** is depressed

Fasting blood sugar level low (hypoglycaemia) ....may be fatal. Reduced mobilization of proteins and fat causes lack of energy for metabolism.

Patients show - - -

Great muscular weakness and **wasting**, **great loss** of **weight**, **loss** of **appetite** hypotension, **vomiting** and **abdominal pain**. **anaemia** (↓ RBC formation) **pigmentation** of exposed and pressure areas of **skin** (due to rise in *ACTH* which has melanophore-stimulating properties). ↓ resistance to **stress** and **infections**. Increased eosinophils and lymphocytes in blood.

If untreated, eventually fatal.

Administration of *cortisol*, a synthetic mineralocorticoid, and sodium chloride restores individual to normal.

OVERACTIVITY OF ADRENAL CORTEX

Overactivity or tumour of adrenal cortex may give **overproduction of any or all of the corticoids:**

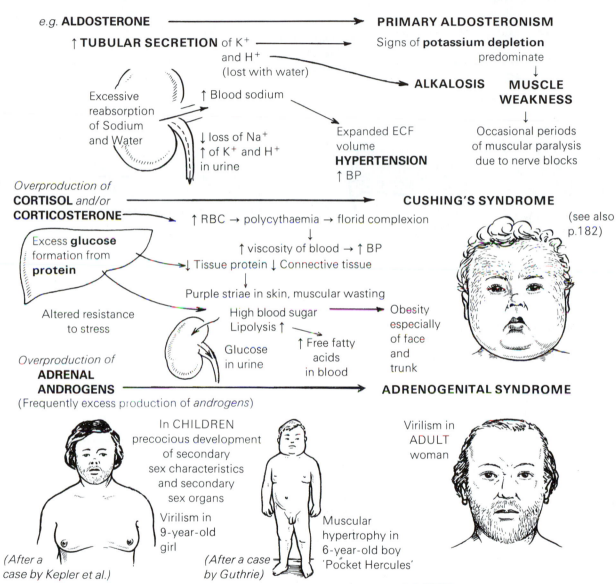

e.g. **ALDOSTERONE** ⟶ **PRIMARY ALDOSTERONISM**

↑ **TUBULAR SECRETION** of K^+ ⟶ Signs of **potassium depletion**
and H^+ predominate
(lost with water)
↓
ALKALOSIS   **MUSCLE
WEAKNESS**

Excessive reabsorption of Sodium and Water

↑ Blood sodium

↓ loss of Na^+
↑ of K^+ and H^+
in urine

Expanded ECF volume
HYPERTENSION
↑ BP

↓
Occasional periods of muscular paralysis due to nerve blocks

Overproduction of
CORTISOL *and/or*
CORTICOSTERONE ⟶ **CUSHING'S SYNDROME**

(see also p.182)

↑ RBC → polycythaemia → florid complexion
↓
↑ viscosity of blood → ↑ BP
↓ Tissue protein ↓ Connective tissue
↓
Purple striae in skin, muscular wasting

Excess **glucose** formation from **protein**

Altered resistance to stress

High blood sugar
Lipolysis ↑
Glucose in urine
↑ Free fatty acids in blood

Obesity especially of face and trunk

Overproduction of
**ADRENAL
ANDROGENS** ⟶ **ADRENOGENITAL SYNDROME**
(Frequently excess production of *androgens*)

In CHILDREN precocious development of secondary sex characteristics and secondary sex organs

Virilism in 9-year-old girl

Virilism in ADULT woman

Muscular hypertrophy in 6-year-old boy 'Pocket Hercules'

(After a case by Kepler et al.)

(After a case by Guthrie)

Administration of *cortisone* depresses pituitary secretion of *ACTH* → inhibits production of the abnormal steroids.

Removal of the over-secreting tissue or tumour restores individual. In **secondary** hyperaldosteronism *excess aldosterone* is the result of *increased* **renin** secretion.

175

ADRENAL MEDULLA

The adrenal medulla arises from the same primitive tissue as the postganglionic cells of the sympathetic nervous system.

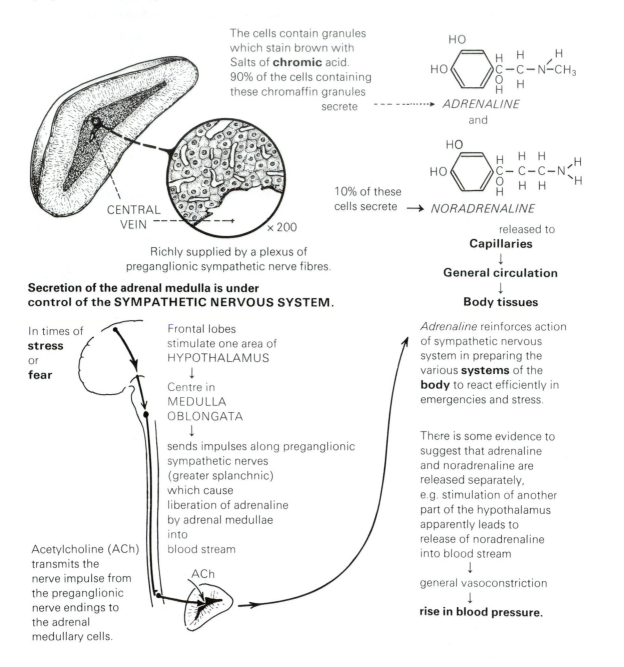

The cells contain granules which stain brown with Salts of **chromic** acid. 90% of the cells containing these chromaffin granules secrete ⟶ *ADRENALINE*

and

10% of these cells secrete ⟶ *NORADRENALINE*

released to
Capillaries
↓
General circulation
↓
Body tissues

CENTRAL VEIN

× 200

Richly supplied by a plexus of preganglionic sympathetic nerve fibres.

Secretion of the adrenal medulla is under control of the SYMPATHETIC NERVOUS SYSTEM.

In times of
stress
or
fear

Frontal lobes stimulate one area of HYPOTHALAMUS
↓
Centre in MEDULLA OBLONGATA
↓
sends impulses along preganglionic sympathetic nerves (greater splanchnic) which cause liberation of adrenaline by adrenal medullae into blood stream

Acetylcholine (ACh) transmits the nerve impulse from the preganglionic nerve endings to the adrenal medullary cells.

ACh

Adrenaline reinforces action of sympathetic nervous system in preparing the various **systems** of the **body** to react efficiently in emergencies and stress.

There is some evidence to suggest that adrenaline and noradrenaline are released separately, e.g. stimulation of another part of the hypothalamus apparently leads to release of noradrenaline into blood stream
↓
general vasoconstriction
↓
rise in blood pressure.

Under quiet resting conditions the blood contains very little *adrenaline*. During excitement or circumstances which demand special efforts *adrenaline* is released into the blood stream, and is responsible for the following actions summed up as the '**fight or flight' function** of the adrenal medullae.

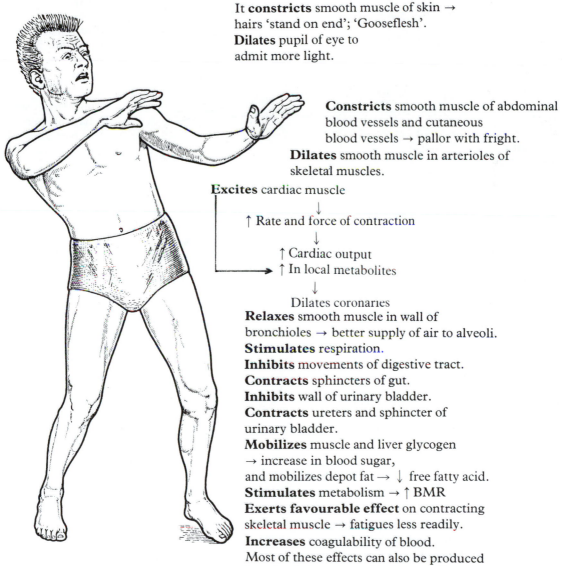

It **constricts** smooth muscle of skin → hairs 'stand on end'; 'Gooseflesh'.
Dilates pupil of eye to admit more light.

Constricts smooth muscle of abdominal blood vessels and cutaneous blood vessels → pallor with fright.
Dilates smooth muscle in arterioles of skeletal muscles.

Excites cardiac muscle
↓
↑ Rate and force of contraction
↓
↑ Cardiac output
↑ In local metabolites
↓
Dilates coronaries

Relaxes smooth muscle in wall of bronchioles → better supply of air to alveoli.
Stimulates respiration.
Inhibits movements of digestive tract.
Contracts sphincters of gut.
Inhibits wall of urinary bladder.
Contracts ureters and sphincter of urinary bladder.
Mobilizes muscle and liver glycogen → increase in blood sugar, and mobilizes depot fat → ↓ free fatty acid.
Stimulates metabolism → ↑ BMR
Exerts favourable effect on contracting skeletal muscle → fatigues less readily.
Increases coagulability of blood.
Most of these effects can also be produced by stimulating sympathetic nerve fibres.

The adrenal medullae are not essential to life – but without them the body is less able to face emergencies and conditions of stress.

DEVELOPMENT OF PITUITARY

The pituitary gland consists of **anterior, intermediate** and **posterior** parts which differ in **origin, structure** and **function**.

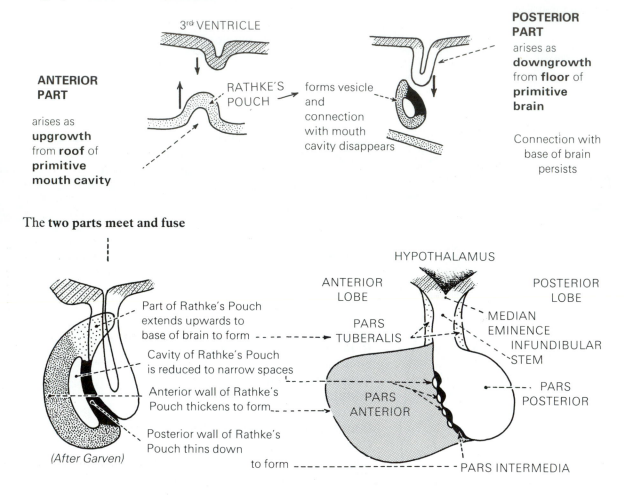

ANTERIOR PART

arises as **upgrowth** from **roof** of **primitive mouth cavity**

3rd VENTRICLE

RATHKE'S POUCH → forms vesicle and connection with mouth cavity disappears

POSTERIOR PART

arises as **downgrowth** from **floor** of **primitive brain**

Connection with base of brain persists

The **two parts meet and fuse**

Part of Rathke's Pouch extends upwards to base of brain to form

Cavity of Rathke's Pouch is reduced to narrow spaces

Anterior wall of Rathke's Pouch thickens to form

Posterior wall of Rathke's Pouch thins down

(After Garven)

HYPOTHALAMUS

ANTERIOR LOBE

POSTERIOR LOBE

PARS TUBERALIS

MEDIAN EMINENCE

INFUNDIBULAR STEM

PARS ANTERIOR

PARS POSTERIOR

to form — PARS INTERMEDIA

Pars tuberalis + infundibular stem = **infundibulum** or **pituitary stalk**.
Neurohypophysis = pars posterior (posterior or neural lobe) + infundibular stem + median eminence.
Adenohypophysis = pars anterior (pars distalis or glandularis) + pars tuberalis (pars intermedia is also sometimes included).
The **adult pituitary (hypophysis)** is a small (8×12 mm) oval gland which lies in the **Sella Turcica** – a small cavity in the bone at the base of the skull. It weighs only 500 mg but, along with the adjacent hypothalamus, it exerts a major control over endocrine function.

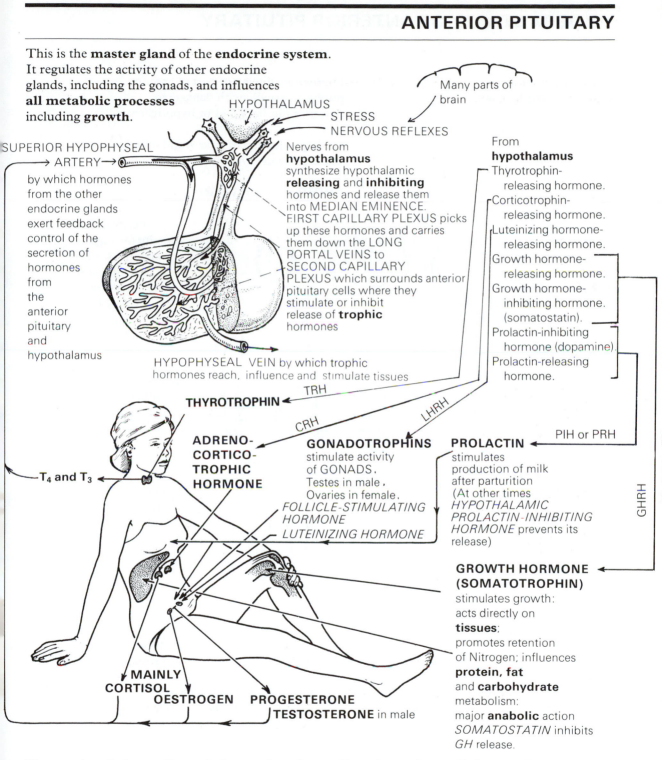

This is the **master gland** of the **endocrine system**. It regulates the activity of other endocrine glands, including the gonads, and influences **all metabolic processes** including **growth**.

Many parts of brain

HYPOTHALAMUS
STRESS
NERVOUS REFLEXES

SUPERIOR HYPOPHYSEAL ARTERY by which hormones from the other endocrine glands exert feedback control of the secretion of hormones from the anterior pituitary and hypothalamus

Nerves from **hypothalamus** synthesize hypothalamic **releasing** and **inhibiting** hormones and release them into MEDIAN EMINENCE. FIRST CAPILLARY PLEXUS picks up these hormones and carries them down the LONG PORTAL VEINS to SECOND CAPILLARY PLEXUS which surrounds anterior pituitary cells where they stimulate or inhibit release of **trophic** hormones

From **hypothalamus**
Thyrotrophin-releasing hormone.
Corticotrophin-releasing hormone.
Luteinizing hormone-releasing hormone.
Growth hormone-releasing hormone.
Growth hormone-inhibiting hormone. (somatostatin).
Prolactin-inhibiting hormone (dopamine).
Prolactin-releasing hormone.

HYPOPHYSEAL VEIN by which trophic hormones reach, influence and stimulate tissues

TRH
THYROTROPHIN
T_4 and T_3

CRH
LHRH

ADRENO-CORTICO-TROPHIC HORMONE

GONADOTROPHINS stimulate activity of GONADS. Testes in male. Ovaries in female. *FOLLICLE-STIMULATING HORMONE* *LUTEINIZING HORMONE*

PROLACTIN stimulates production of milk after parturition (At other times *HYPOTHALAMIC PROLACTIN-INHIBITING HORMONE* prevents its release)

PIH or PRH
GHRH

GROWTH HORMONE (SOMATOTROPHIN) stimulates growth: acts directly on **tissues**; promotes retention of Nitrogen; influences **protein, fat** and **carbohydrate** metabolism: major **anabolic** action *SOMATOSTATIN* inhibits *GH* release.

MAINLY CORTISOL
OESTROGEN **PROGESTERONE** **TESTOSTERONE** in male

The anterior pituitary cell population consists of 15-20% corticotrophs, 3-5% thyrotrophs, 10-15% gonadotrophs, 40-50% somatotrophs, 10-25% mammotrophs; identified by immunohistochemistry.

UNDERACTIVITY OF ANTERIOR PITUITARY

Deficiency or absence
of **somatotroph** cells
↓
Underproduction of
*growth hormone
(somatotrophin)*
↓

LORAIN DWARF
Delayed skeletal
growth and
retarded sexual
development
but alert, intelligent,
well proportioned
child.

Destructive disease of part of anterior
pituitary (usually with damage to
posterior pituitary and/or hypothalamus)
↓
Underproduction of *growth* and other
endocrine-trophic hormones
↓

FRÖHLICH'S DWARF

Stunting of growth,
obesity (large
appetite for sugar);
arrested sexual
development;
lethargic;
somnolent;
mentally
subnormal.

If atrophy of
other
endocrine
glands
↓
Signs of
deficiency
of their
hormones.

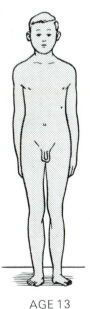

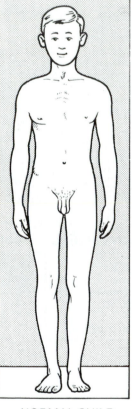

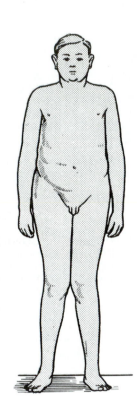

AGE 13   NORMAL CHILD   AGE 13
                    AGE 13

A similar condition occurs in adults
without dwarfing but with suppression
of sex functions and regression of
secondary sex characteristics.
   Growth and *gonadotrophic hormones*
aid in restoring patient to normal.

Growth hormone restores
growth and development
pattern to normal.

OVERACTIVITY OF PITUITARY SOMATOTROPH CELLS

Functional overactivity (or tumour) chiefly of the **SOMATOTROPH** cells of the anterior pituitary leads to ⟶ **GIANTISM** in the CHILD: **ACROMEGALY** in the ADULT

↓

Overproduction of *growth Hormone*

↓

General Circulation

↓

Increases **NITROGEN** retention. Influences Protein, Carbohydrate and Fat metabolism of **ALL CELLS** of the body.

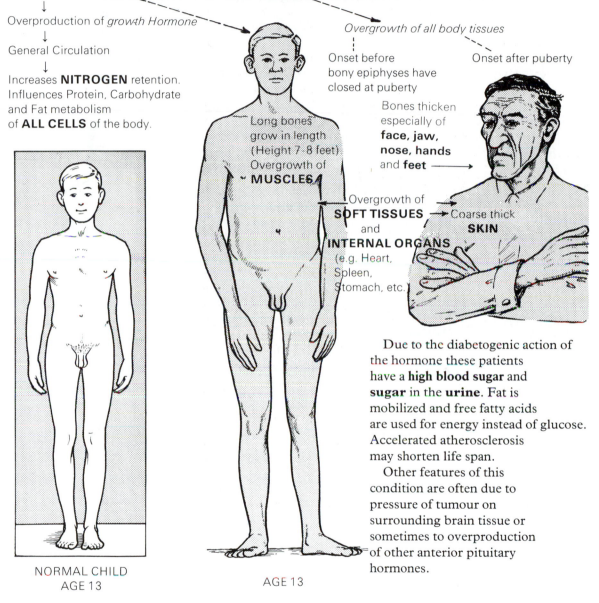

Overgrowth of all body tissues

Onset before bony epiphyses have closed at puberty

Onset after puberty

Bones thicken especially of **face, jaw, nose, hands** and **feet** ⟶

Long bones grow in length (Height 7-8 feet) Overgrowth of **MUSCLES**

Overgrowth of **SOFT TISSUES** and **INTERNAL ORGANS** (e.g. Heart, Spleen, Stomach, etc.

⟶ Coarse thick **SKIN**

NORMAL CHILD AGE 13

AGE 13

Due to the diabetogenic action of the hormone these patients have a **high blood sugar** and **sugar** in the **urine**. Fat is mobilized and free fatty acids are used for energy instead of glucose. Accelerated atherosclerosis may shorten life span.

Other features of this condition are often due to pressure of tumour on surrounding brain tissue or sometimes to overproduction of other anterior pituitary hormones.

Destruction of the overactive tissue – usually by surgery or radiation therapy – prevents progress of the condition.

OVERACTIVITY OF PITUITARY CORTICOTROPH CELLS

Overactivity (often due to tumour) of the **corticotroph** cells of the anterior pituitary

gives

Cushing's syndrome

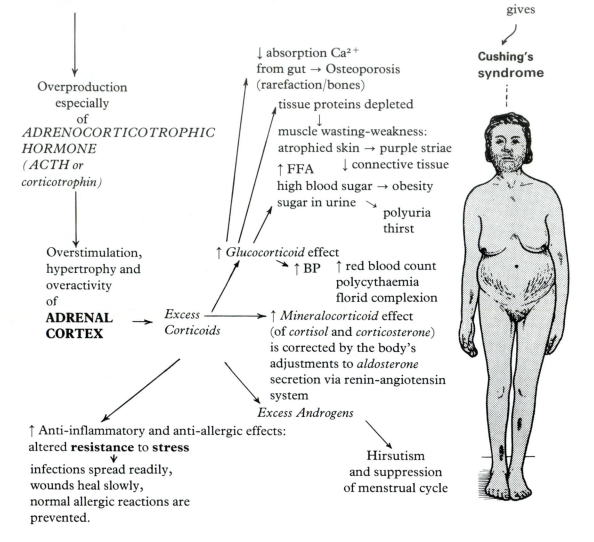

Overproduction especially of *ADRENOCORTICOTROPHIC HORMONE* (*ACTH or corticotrophin*)

↓ absorption Ca^{2+} from gut → Osteoporosis (rarefaction/bones)

tissue proteins depleted
↓
muscle wasting-weakness: atrophied skin → purple striae

↑ FFA   ↓ connective tissue

high blood sugar → obesity

sugar in urine ↘ polyuria thirst

Overstimulation, hypertrophy and overactivity of **ADRENAL CORTEX** →

↑ *Glucocorticoid* effect
↘ ↑ BP   ↑ red blood count polycythaemia florid complexion

Excess Corticoids → ↑ *Mineralocorticoid* effect (of *cortisol* and *corticosterone*) is corrected by the body's adjustments to *aldosterone* secretion via renin-angiotensin system

Excess Androgens

↑ Anti-inflammatory and anti-allergic effects: altered **resistance** to **stress**
↓
infections spread readily, wounds heal slowly, normal allergic reactions are prevented.

Hirsutism and suppression of menstrual cycle

This condition is usually indistinguishable clinically from that seen in primary overactivity or tumour of the adrenal cortex itself.

The syndrome is here shown in the adult woman.

Overproduction of *thyroid stimulating hormone* → Overactivity of **thyroid** gland.

Complete atrophy (or insufficiency) of all secreting cells of anterior pituitary in adult –

**SIMMOND'S
DISEASE**

Appearance of premature senility

Failure to
**produce
any hormones** ⟶ Features usually
associated with
very **old age**

Lack of
growth hormone
   Grave upset in
   tissue metabolism⟶ {
   Hair grey, sparse:
     loss of body hair.
   Skin dry, sallow,
     wrinkled.
   Body emaciated
     (great loss of weight)
   Bones frail

Lack of
gonadotrophins ⟶ **Sex organs** atrophy.
Menstruation ceases.
Reproductive cycle stops.
Secondary sex
characteristics
gradually regress.

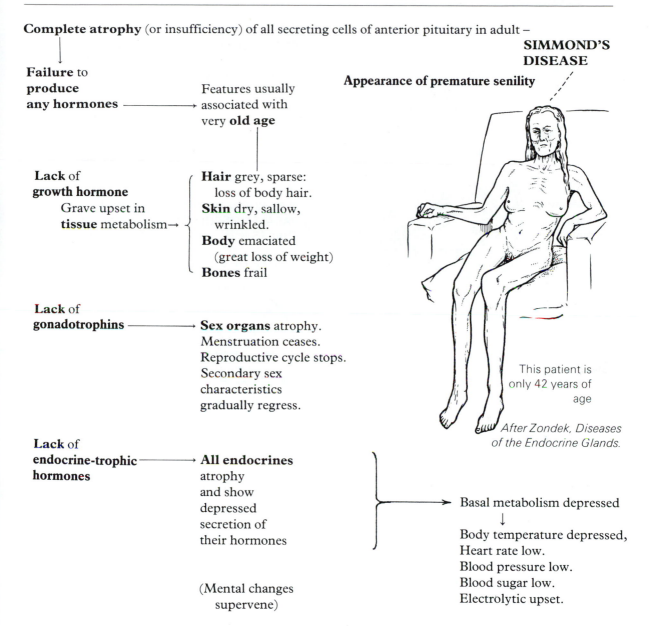

This patient is
only 42 years of
age

*After Zondek, Diseases
of the Endocrine Glands.*

Lack of
endocrine-trophic ⟶ **All endocrines**
hormones
atrophy
and show
depressed
secretion of
their hormones

(Mental changes
supervene)

⟶ Basal metabolism depressed
↓
Body temperature depressed,
Heart rate low.
Blood pressure low.
Blood sugar low.
Electrolytic upset.

Anterior pituitary hormones may relieve the condition but rarely succeed in completely
restoring the patient to normal.

POSTERIOR PITUITARY

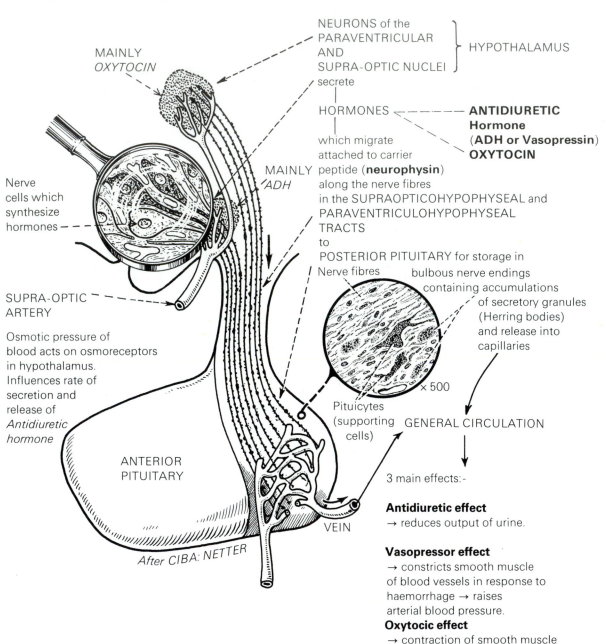

NEURONS of the
PARAVENTRICULAR
AND
SUPRA-OPTIC NUCLEI } HYPOTHALAMUS

MAINLY
OXYTOCIN

secrete

HORMONES ←————— **ANTIDIURETIC
Hormone
(ADH or Vasopressin)
OXYTOCIN**

which migrate
attached to carrier
MAINLY peptide (**neurophysin**)
ADH along the nerve fibres
in the SUPRAOPTICOHYPOPHYSEAL and
PARAVENTRICULOHYPOPHYSEAL
TRACTS
to
POSTERIOR PITUITARY for storage in
Nerve fibres   bulbous nerve endings
containing accumulations
of secretory granules
(Herring bodies)
and release into
capillaries

Nerve
cells which
synthesize
hormones

SUPRA-OPTIC
ARTERY

Osmotic pressure of
blood acts on osmoreceptors
in hypothalamus.
Influences rate of
secretion and
release of
*Antidiuretic
hormone*

Pituicytes   × 500
(supporting
cells)   GENERAL CIRCULATION

ANTERIOR
PITUITARY

3 main effects:-

VEIN

Antidiuretic effect
→ reduces output of urine.

Vasopressor effect
→ constricts smooth muscle
of blood vessels in response to
haemorrhage → raises
arterial blood pressure.
Oxytocic effect
→ contraction of smooth muscle
of uterus after childbirth and of
myoepithelial cells in lactating mammary
glands.

After CIBA: NETTER

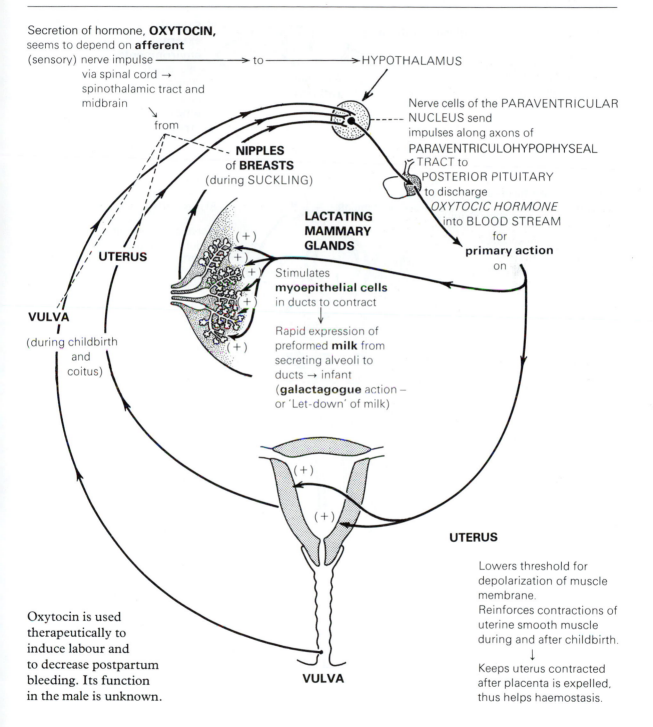

Secretion of hormone, **OXYTOCIN,** seems to depend on **afferent** (sensory) nerve impulse ⟶ to ⟶ HYPOTHALAMUS
via spinal cord →
spinothalamic tract and
midbrain

from

NIPPLES of **BREASTS**
(during SUCKLING)

Nerve cells of the PARAVENTRICULAR NUCLEUS send
impulses along axons of
PARAVENTRICULOHYPOPHYSEAL TRACT to
POSTERIOR PITUITARY
to discharge
OXYTOCIC HORMONE
into BLOOD STREAM
for
primary action
on

LACTATING MAMMARY GLANDS

(+)
(+)
(+)
(+)
(+)

Stimulates
myoepithelial cells
in ducts to contract
↓
Rapid expression of
preformed **milk** from
secreting alveoli to
ducts → infant
(**galactagogue** action –
or 'Let-down' of milk)

UTERUS

VULVA

(during childbirth
and
coitus)

(+)

(+)

UTERUS

Lowers threshold for
depolarization of muscle
membrane.
Reinforces contractions of
uterine smooth muscle
during and after childbirth.
↓
Keeps uterus contracted
after placenta is expelled,
thus helps haemostasis.

VULVA

Oxytocin is used
therapeutically to
induce labour and
to decrease postpartum
bleeding. Its function
in the male is unknown.

ANTIDIURETIC HORMONE

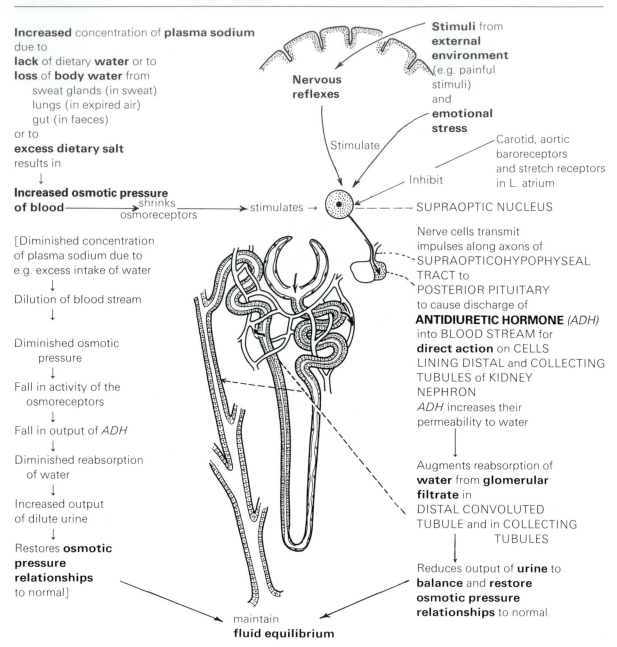

Increased concentration of **plasma sodium** due to **lack** of dietary **water** or to **loss** of **body water** from
    sweat glands (in sweat)
    lungs (in expired air)
    gut (in faeces)
or to
excess dietary salt
results in
↓
Increased osmotic pressure of blood —→ shrinks osmoreceptors —→ stimulates →

Nervous reflexes

Stimuli from **external environment** (e.g. painful stimuli) and **emotional stress**

Carotid, aortic baroreceptors and stretch receptors in L. atrium

Stimulate

Inhibit

SUPRAOPTIC NUCLEUS

[Diminished concentration of plasma sodium due to e.g. excess intake of water
↓
Dilution of blood stream
↓
Diminished osmotic pressure
↓
Fall in activity of the osmoreceptors
↓
Fall in output of *ADH*
↓
Diminished reabsorption of water
↓
Increased output of dilute urine
↓
Restores **osmotic pressure relationships** to normal]

Nerve cells transmit impulses along axons of SUPRAOPTICOHYPOPHYSEAL TRACT to POSTERIOR PITUITARY to cause discharge of **ANTIDIURETIC HORMONE** *(ADH)* into BLOOD STREAM for **direct action** on CELLS LINING DISTAL and COLLECTING TUBULES of KIDNEY NEPHRON *ADH* increases their permeability to water

Augments reabsorption of **water** from **glomerular filtrate** in DISTAL CONVOLUTED TUBULE and in COLLECTING TUBULES

Reduces output of **urine** to **balance** and **restore osmotic pressure relationships** to normal.

maintain
fluid equilibrium

ADH binds to V_2 receptors on capillary side of duct cells → activates adenylate cyclase → increases cyclic AMP → activates a protein kinase on luminal side of cell → phosphorylates a membrane protein → increases permeability of cell membrane to water.

Damage, by **injury** or **disease**, to

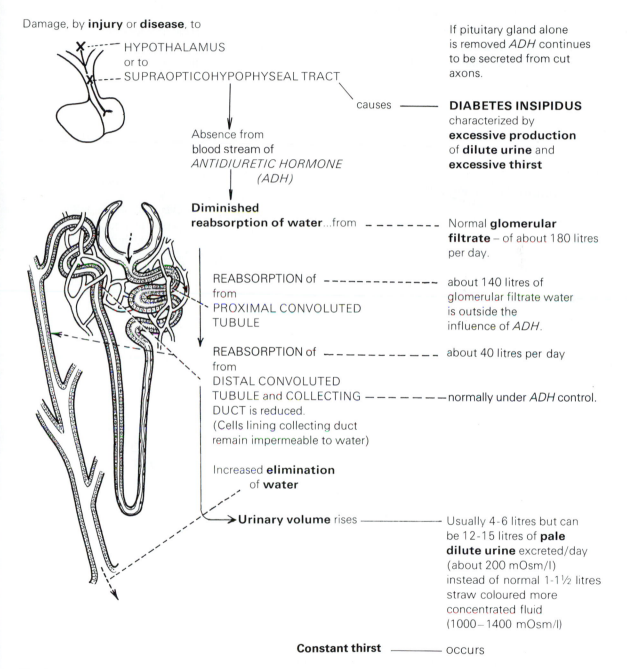

HYPOTHALAMUS
or to
SUPRAOPTICOHYPOPHYSEAL TRACT

If pituitary gland alone
is removed *ADH* continues
to be secreted from cut
axons.

causes —— **DIABETES INSIPIDUS**
characterized by
excessive production
of **dilute urine** and
excessive thirst

Absence from
blood stream of
ANTIDIURETIC HORMONE
(ADH)

**Diminished
reabsorption of water**...from – – – – – – – – Normal **glomerular
filtrate** – of about 180 litres
per day.

REABSORPTION of – – – – – – – – – – – about 140 litres of
from   glomerular filtrate water
PROXIMAL CONVOLUTED   is outside the
TUBULE   influence of *ADH*.

REABSORPTION of – – – – – – – – – – about 40 litres per day
from
DISTAL CONVOLUTED
TUBULE and COLLECTING – – – – – – –normally under *ADH* control.
DUCT is reduced.
(Cells lining collecting duct
remain impermeable to water)

Increased **elimination**
of **water**

Urinary volume rises ——————— Usually 4-6 litres but can
be 12-15 litres of **pale
dilute urine** excreted/day
(about 200 mOsm/l)
instead of normal 1-1½ litres
straw coloured more
concentrated fluid
(1000–1400 mOsm/l)

Constant thirst ———— occurs

Replacement of *ADH* restores to normal elimination of water and symptoms of thirst.

ALDOSTERONE AND ANTIDIURETIC HORMONE (ADH) IN THE MAINTENANCE OF BLOOD VOLUME

A reduction in the total volume of **extracellular fluid** (e.g. after haemorrhage or loss of isotonic secretions from the gut in vomiting or diarrhoea) leads to chain of **compensatory** mechanisms in which *aldosterone* plays an important role. See pages 159 and 160.

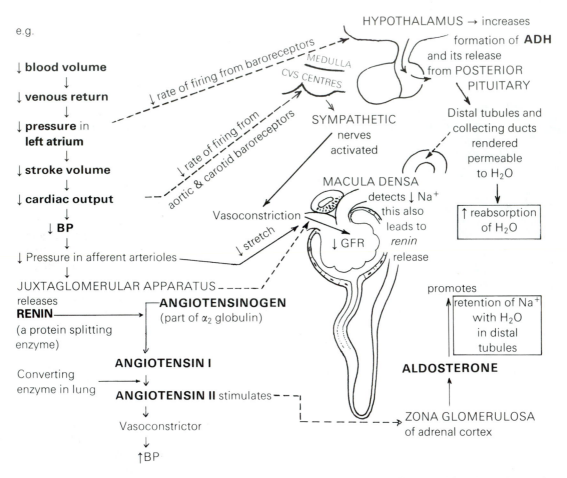

e.g.

↓ **blood volume**
↓
↓ **venous return**
↓
↓ **pressure** in **left atrium**
↓
↓ **stroke volume**
↓
↓ **cardiac output**
↓
↓ **BP**
↓
↓ Pressure in afferent arterioles
↓
JUXTAGLOMERULAR APPARATUS
releases
RENIN
(a protein splitting enzyme)

↓ rate of firing from baroreceptors

↓ rate of firing from aortic & carotid baroreceptors

MEDULLA
CVS CENTRES

HYPOTHALAMUS → increases formation of **ADH** and its release from POSTERIOR PITUITARY

SYMPATHETIC nerves activated

MACULA DENSA detects ↓ Na^+ this also leads to *renin* release

Vasoconstriction

↓ stretch

↓ GFR

Distal tubules and collecting ducts rendered permeable to H_2O

↑ reabsorption of H_2O

ANGIOTENSINOGEN
(part of α_2 globulin)

ANGIOTENSIN I
↓
Converting enzyme in lung →
ANGIOTENSIN II stimulates
↓
Vasoconstrictor
↓
↑BP

promotes
retention of Na^+ with H_2O in distal tubules

ALDOSTERONE
↑
ZONA GLOMERULOSA of adrenal cortex

Decreased pressure in atria decreases circulating **atrial natriuretic peptide**, thus decreasing excretion of Na^+

These measures serve to maintain **blood volume** till the long term replacement of the lost RBC, plasma proteins and electrolytes can be achieved.

ISLETS OF LANGERHANS make up 1-2% of pancreatic tissue. Consist of four cell types: A(α) secrete *glucagon*, B(β) secrete *insulin*, D(δ) secrete *somatostatin*, F secrete *pancreatic polypeptide* (unknown function).

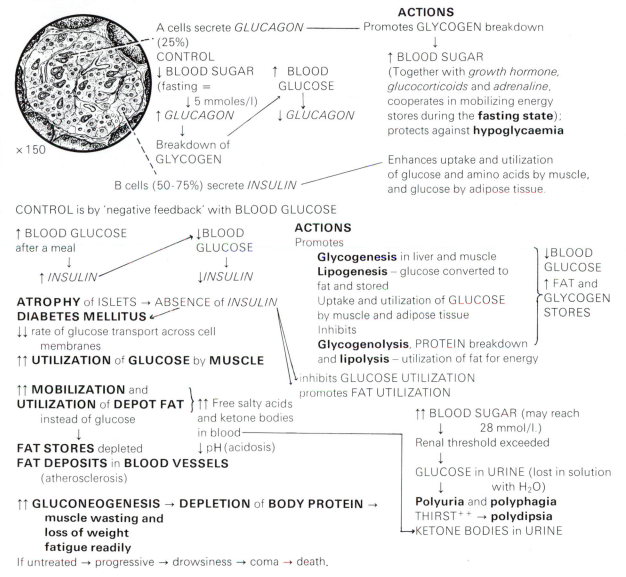

×150

A cells secrete *GLUCAGON* ——————

(25%)

CONTROL

↓ BLOOD SUGAR   ↑ BLOOD
(fasting =   GLUCOSE
   ↓ 5 mmoles/l)   ↓
↑ *GLUCAGON*   ↓ *GLUCAGON*
↓
Breakdown of
GLYCOGEN

B cells (50-75%) secrete *INSULIN* ——————

ACTIONS

Promotes GLYCOGEN breakdown
↓
↑ BLOOD SUGAR
(Together with *growth hormone*, *glucocorticoids* and *adrenaline*, cooperates in mobilizing energy stores during the **fasting state**); protects against **hypoglycaemia**

Enhances uptake and utilization of glucose and amino acids by muscle, and glucose by adipose tissue.

CONTROL is by 'negative feedback' with BLOOD GLUCOSE

↑ BLOOD GLUCOSE   ↓ BLOOD
after a meal   GLUCOSE
↓   ↓
↑ *INSULIN*   ↓ *INSULIN*

ACTIONS
Promotes
   Glycogenesis in liver and muscle
   Lipogenesis – glucose converted to
   fat and stored
   Uptake and utilization of GLUCOSE
   by muscle and adipose tissue
Inhibits
   Glycogenolysis, PROTEIN breakdown
   and **lipolysis** – utilization of fat for energy

↓ BLOOD GLUCOSE ↑ FAT and GLYCOGEN STORES

ATROPHY of ISLETS → ABSENCE of *INSULIN*
DIABETES MELLITUS
↓↓ rate of glucose transport across cell
    membranes
↑↑ **UTILIZATION** of **GLUCOSE** by **MUSCLE**

inhibits GLUCOSE UTILIZATION
promotes FAT UTILIZATION

↑↑ **MOBILIZATION** and
UTILIZATION of **DEPOT FAT**
    instead of glucose
↓
FAT STORES depleted
FAT DEPOSITS in **BLOOD VESSELS**
    (atherosclerosis)

↑↑ Free salty acids
and ketone bodies
in blood
↓ pH (acidosis)

↑↑ BLOOD SUGAR (may reach
↓   28 mmol/l.)
Renal threshold exceeded
↓
GLUCOSE in URINE (lost in solution
↓   with H_2O)
Polyuria and **polyphagia**
THIRST++ → **polydipsia**
KETONE BODIES in URINE

↑↑ **GLUCONEOGENESIS** → **DEPLETION** of **BODY PROTEIN** →
    muscle wasting and
    loss of weight
    fatigue readily

If untreated → progressive → drowsiness → coma → death.

Excess insulin (**hyperinsulinism**) → low blood sugar (hypoglycaemia) → irritability; sweating; hunger. If untreated → reduction of metabolism of nervous tissues → giddiness → coma → death. *Somatostatin* prevents excessive levels of nutrients in plasma by reducing rate of food digestion and absorption – Inhibits insulin and glucagon secretion.

CHAPTER 9

REPRODUCTIVE SYSTEM

MALE REPRODUCTIVE SYSTEM

PRIMARY SEX ORGANS ⟶ produce the MALE GERM CELLS – **SPERMATOZOA**
TESTES (Two)   and the MALE SEX HORMONE – **TESTOSTERONE**

Testosterone $\xrightarrow{5\alpha\text{-reductase}}$ dihydrotestosterone

These are responsible for
maturation at puberty of:

ACCESSORY SEX ORGANS

EPIDIDYMIS (Two) ⎫
VAS DEFERENS (Two) ⎬ transfer spermatozoa from the testes.

SEMINAL VESICLES (Two) ⎱ secrete fluid medium for transport of spermatozoa.
PROSTATE GLAND ⎰

PENIS ——————— transfers spermatozoa from male to female.

and appearance of

SECONDARY SEX CHARACTERISTICS

Laryngeal changes → Deep voice,
growth of pubic, axillary and facial
hair.
Receding hair at temples.
Increase in muscle and
skeletal mass (protein anabolism)
giving growth spurt and
characteristic male shape
of body.
Sex drive. ? Aggression.
Increased sebaceous
gland secretion –
oversecretion causes
acne.

VAS DEFERENS

SEMINAL VESICLE

EPIDIDYMIS

TESTIS

SCROTUM

PENIS

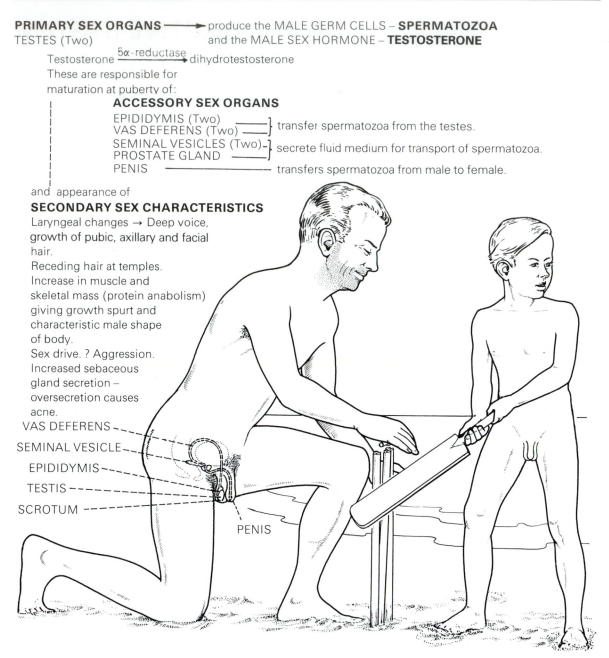

In the male the process of spermatogenesis starts just after
puberty and is normally continuous until old age.

There are *two* TESTES.
These produce the MALE GERM CELLS:-

SPERMATOGENESIS

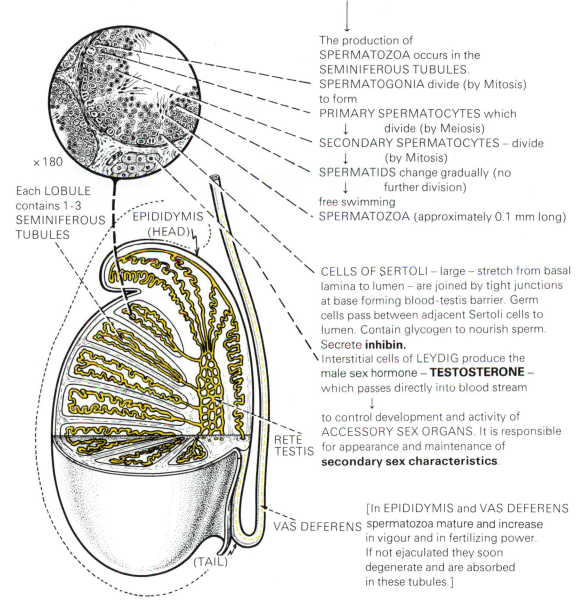

× 180

Each LOBULE contains 1-3 SEMINIFEROUS TUBULES

EPIDIDYMIS (HEAD)

RETE TESTIS

VAS DEFERENS

(TAIL)

The production of SPERMATOZOA occurs in the SEMINIFEROUS TUBULES.
SPERMATOGONIA divide (by Mitosis) to form
PRIMARY SPERMATOCYTES which
     ↓   divide (by Meiosis)
SECONDARY SPERMATOCYTES – divide
     ↓   (by Mitosis)
SPERMATIDS change gradually (no
     ↓   further division)
free swimming
SPERMATOZOA (approximately 0.1 mm long)

CELLS OF SERTOLI – large – stretch from basal lamina to lumen – are joined by tight junctions at base forming blood-testis barrier. Germ cells pass between adjacent Sertoli cells to lumen. Contain glycogen to nourish sperm.
Secrete **inhibin.**
Interstitial cells of LEYDIG produce the male sex hormone – **TESTOSTERONE** – which passes directly into blood stream
     ↓
to control development and activity of ACCESSORY SEX ORGANS. It is responsible for appearance and maintenance of **secondary sex characteristics**.

[In EPIDIDYMIS and VAS DEFERENS spermatozoa mature and increase in vigour and in fertilizing power. If not ejaculated they soon degenerate and are absorbed in these tubules.]

Events occurring in the testes are under **control** of **hormones,** chiefly those of ANTERIOR PITUITARY and the HYPOTHALAMUS.

MALE ACCESSORY SEX ORGANS

These are the organs adapted for **transfer** of live **spermatozoa** from male to female.

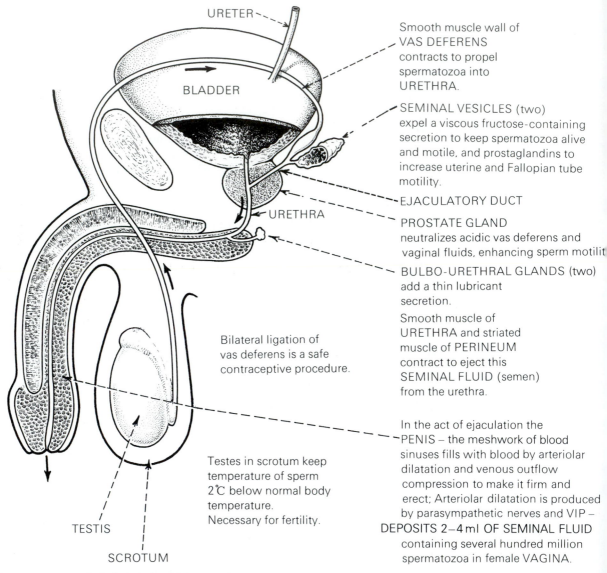

URETER

BLADDER

Smooth muscle wall of
VAS DEFERENS
contracts to propel
spermatozoa into
URETHRA.

SEMINAL VESICLES (two)
expel a viscous fructose-containing
secretion to keep spermatozoa alive
and motile, and prostaglandins to
increase uterine and Fallopian tube
motility.

URETHRA

EJACULATORY DUCT

PROSTATE GLAND
neutralizes acidic vas deferens and
vaginal fluids, enhancing sperm motilit

BULBO-URETHRAL GLANDS (two)
add a thin lubricant
secretion.

Smooth muscle of
URETHRA and striated
muscle of PERINEUM
contract to eject this
SEMINAL FLUID (semen)
from the urethra.

Bilateral ligation of
vas deferens is a safe
contraceptive procedure.

In the act of ejaculation the
PENIS – the meshwork of blood
sinuses fills with blood by arteriolar
dilatation and venous outflow
compression to make it firm and
erect; Arteriolar dilatation is produced
by parasympathetic nerves and VIP –

Testes in scrotum keep
temperature of sperm
2℃ below normal body
temperature.
Necessary for fertility.

DEPOSITS 2–4 ml OF SEMINAL FLUID
containing several hundred million
spermatozoa in female VAGINA.

TESTIS

SCROTUM

Sperm must remain in female tract for several
hours to acquire ability to penetrate ovum –
capacitation.

CONTROL OF EVENTS IN THE TESTIS

Between the ages of 13 and 16 years the hypothalamus begins to secrete *luteinizing hormone-releasing hormone (LHRH or Gonadotrophin-releasing hormone)*.

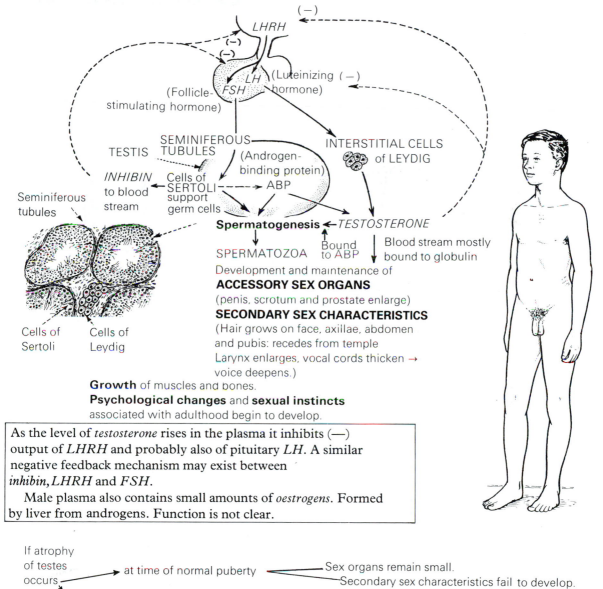

As the level of *testosterone* rises in the plasma it inhibits (—) output of *LHRH* and probably also of pituitary *LH*. A similar negative feedback mechanism may exist between *inhibin, LHRH* and *FSH*.

Male plasma also contains small amounts of *oestrogens*. Formed by liver from androgens. Function is not clear.

If atrophy of testes occurs —→ at time of normal puberty ———— Sex organs remain small.
Secondary sex characteristics fail to develop.

after puberty → Spermatogenesis stops → Sterility.
Testosterone production falls → Atrophy of secondary sex organs.

Injections of *Testosterone* in cases of delayed puberty → Changes associated with puberty.

Use of *inhibin* as a male contraceptive hormone is a possibility.

FEMALE REPRODUCTIVE SYSTEM

PRIMARY SEX ORGANS ———————————————→ produce the FEMALE GERM CELLS – **OVA**
OVARIES (two) and the FEMALE SEX HORMONES. –
OESTROGENS and **PROGESTERONE**

Oestrogen and *progesterone* are responsible
for maturation at Puberty of:
ACCESSORY SEX ORGANS
FALLOPIAN TUBES (two)– – – – – for the transfer of the ova from ovaries
VAGINA — – — — — — — — — — for the reception of the male germ cells.
UTERUS — — — — — — — — —- for the nutrition and development of the
fertilized egg cell → developing embrvo.
MAMMARY GLANDS (two) for the nutrition of the new individual after birth.

and
appearance and
maintenance of
SECONDARY SEX
CHARACTERISTICS

Development of
breasts
Typical feminine
proportions of body
Narrow shoulders
Broad hips

Androgens from
adrenal cortex
are responsible
for sex drive
and growth of
pubic and axillary
hair.

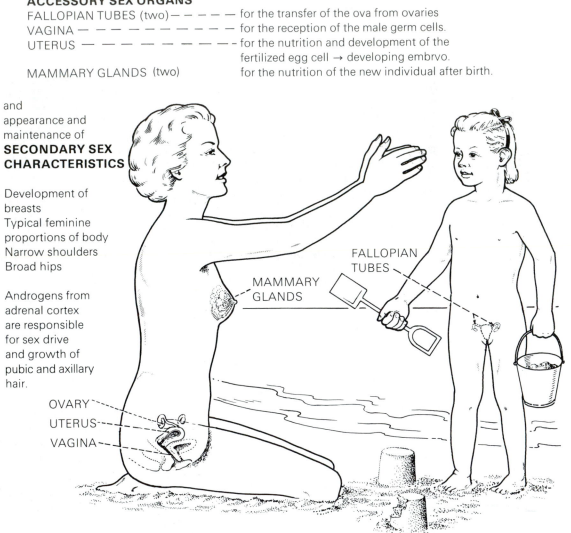

MAMMARY
GLANDS

FALLOPIAN
TUBES

OVARY
UTERUS
VAGINA

In the female the cyclical production of ova starts just after puberty and continues (unless
interrupted by pregnancy or disease) until the menopause.

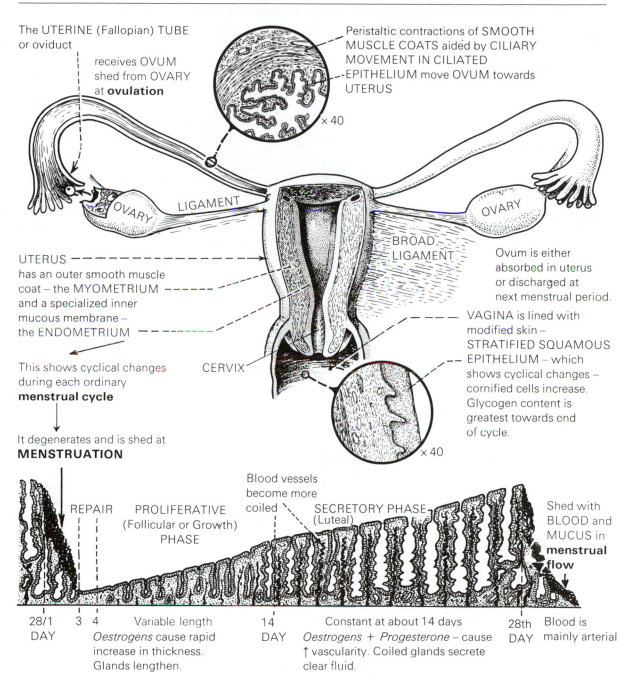

The UTERINE (Fallopian) TUBE or oviduct

receives OVUM shed from OVARY at **ovulation**

Peristaltic contractions of SMOOTH MUSCLE COATS aided by CILIARY MOVEMENT IN CILIATED EPITHELIUM move OVUM towards UTERUS

× 40

LIGAMENT

OVARY

OVARY

BROAD LIGAMENT

UTERUS – – – – –
has an outer smooth muscle coat – the MYOMETRIUM – – – – –
and a specialized inner mucous membrane –
the ENDOMETRIUM – – – –

Ovum is either absorbed in uterus or discharged at next menstrual period.

This shows cyclical changes during each ordinary **menstrual cycle**

CERVIX

VAGINA is lined with modified skin – STRATIFIED SQUAMOUS EPITHELIUM – which shows cyclical changes – cornified cells increase. Glycogen content is greatest towards end of cycle.

It degenerates and is shed at **MENSTRUATION**

× 40

REPAIR

PROLIFERATIVE (Follicular or Growth) PHASE

Blood vessels become more coiled

SECRETORY PHASE (Luteal)

Shed with BLOOD and MUCUS in **menstrual flow**

| 28/1 DAY | 3 | 4 | Variable length | 14 DAY | Constant at about 14 days | 28th DAY | Blood is mainly arterial |

Oestrogens cause rapid increase in thickness. Glands lengthen.

Oestrogens + *Progesterone* – cause ↑ vascularity. Coiled glands secrete clear fluid.

Rhythmical changes occur in uterus, uterine tubes and vagina under the action of ovarian hormones.

OVARY IN ORDINARY ADULT CYCLE

There are *two* OVARIES. These produce the Female GERM CELLS. The production of OVA is a cyclical process – **OOGENESIS.**

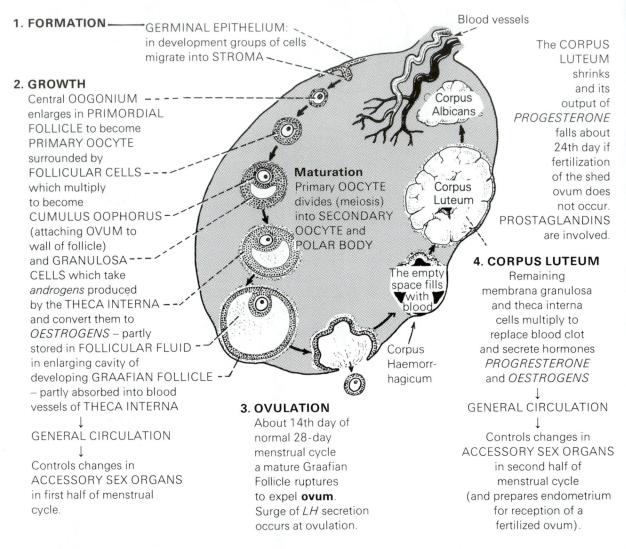

1. FORMATION — GERMINAL EPITHELIUM: in development groups of cells migrate into STROMA

Blood vessels

2. GROWTH
Central OOGONIUM enlarges in PRIMORDIAL FOLLICLE to become PRIMARY OOCYTE surrounded by FOLLICULAR CELLS which multiply to become CUMULUS OOPHORUS (attaching OVUM to wall of follicle) and GRANULOSA CELLS which take *androgens* produced by the THECA INTERNA and convert them to *OESTROGENS* – partly stored in FOLLICULAR FLUID in enlarging cavity of developing GRAAFIAN FOLLICLE – partly absorbed into blood vessels of THECA INTERNA
↓
GENERAL CIRCULATION
↓
Controls changes in ACCESSORY SEX ORGANS in first half of menstrual cycle.

Maturation
Primary OOCYTE divides (meiosis) into SECONDARY OOCYTE and POLAR BODY

Corpus Albicans

Corpus Luteum

The empty space fills with blood

Corpus Haemorr-hagicum

The CORPUS LUTEUM shrinks and its output of *PROGESTERONE* falls about 24th day if fertilization of the shed ovum does not occur. PROSTAGLANDINS are involved.

4. CORPUS LUTEUM
Remaining membrana granulosa and theca interna cells multiply to replace blood clot and secrete hormones *PROGRESTERONE* and *OESTROGENS*
↓
GENERAL CIRCULATION
↓
Controls changes in ACCESSORY SEX ORGANS in second half of menstrual cycle (and prepares endometrium for reception of a fertilized ovum).

3. OVULATION
About 14th day of normal 28-day menstrual cycle a mature Graafian Follicle ruptures to expel **ovum.** Surge of *LH* secretion occurs at ovulation.

For simplicity the development of only one Graafian follicle is shown here. Several grow in each cycle but in the human subject usually only one follicle ruptures. The others atrophy: i.e. *one mature ovum* is shed each month.

Events in the ovary are under control of anterior pituitary hormones *FSH* and *LH*, and hypothalamic *Luteinizing Hormone-Releasing Hormone (LHRH)*.

When pregnancy occurs the ordinary ovarian cycle is suspended.

After the first 14 days the developing placenta secretes *HUMAN CHORIONIC GONADOTROPHIC (HCG)* hormone.

Under its influence the CORPUS LUTEUM —— continues to grow and secrete *oestrogens, progesterone* and *relaxin* until it may come to occupy 30-50% of the total volume of the ovary.

HCG can be detected in urine 14 days after conception. Basis of pregnancy test.

The large amount of *PROGESTERONE*

reaches its peak at about 6 weeks after conception

helps to maintain the PREGNANCY in its early stages and is essential for development of the PLACENTA – the special structure through which the child receives its nourishment from the mother. The placenta also produces progesterone which gradually takes over from the

falls off about 2nd month

CORPUS LUTEUM ——— ceases to contribute significantly after 4th month.

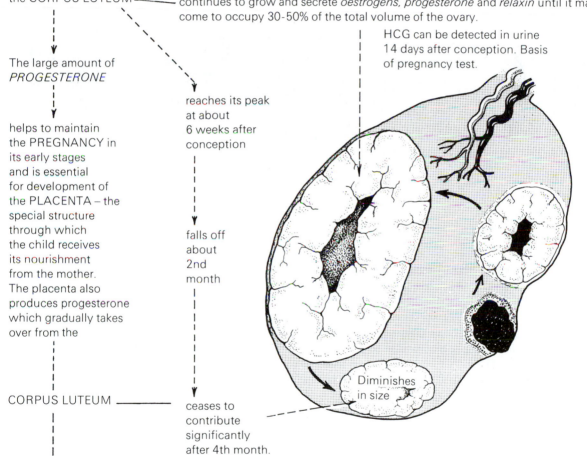

Diminishes in size

PLACENTAL PROGESTERONE takes over to maintain the pregnancy. Increases resting membrane potential of uterine muscle; hence decreases its excitability. Also decreases its sensitivity to *oxytocin* and its number of oestrogen receptors. Helps prepare mammary glands for lactation.

RELAXIN – secreted by corpus luteum and placenta. Relaxes pelvic bones and ligaments. Softens cervix. Ensures uterine quiescence and prevents early abortion of the pregnancy.

CONTROL OF EVENTS IN THE OVARY

Between the ages of 10 and 14 years the HYPOTHALAMUS begins to secrete
LUTEINIZING HORMONE-RELEASING HORMONE (*LHRH*). The girl enters
puberty. Thereafter the cycle is controlled thus:

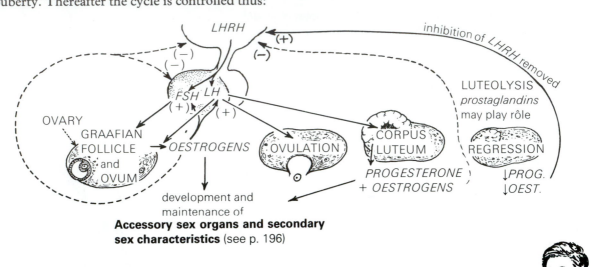

development and
maintenance of
**Accessory sex organs and secondary
sex characteristics** (see p. 196)

The cycle begins with a rise in LHRH. This increases LH and FSH secretion. LH acts on
the theca interna cells which produce androgens. Androgens are converted to *oestrogens* by
granulosa cells under the influence of FSH. The oestrogens enter the follicular fluid.

As the level of oestrogen rises it first inhibits (−) output of LHRH, FSH and LH.

About the 12th or 13th day the prolonged high level of oestrogen, by enhancing the
sensitivity of LH-releasing mechanism to LHRH, causes a positive feedback (+) effect. A
sudden surge of LH (and of FSH) secretion leads to **ovulation** nine hours later and the
formation of the CORPUS LUTEUM. As the level of progesterone rises (along with oestrogen)
it inhibits (−) LHRH, LH and FSH.

A few days before menstruation the corpus luteum involutes. As the levels of progesterone
and oestrogen fall, LHRH is freed from inhibition. The cycle starts again.

The ovarian cycle is repeated monthly from **puberty** to the **menopause** unless interrupted by **pregnancy** or **disease**.

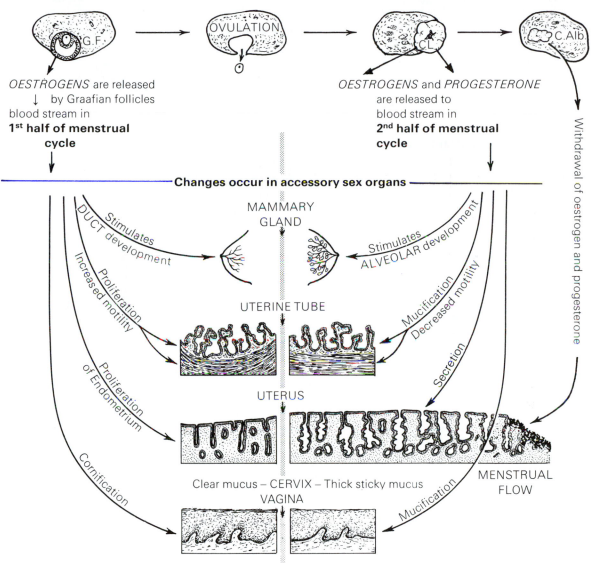

OESTROGENS are released
↓ by Graafian follicles
blood stream in
**1st half of menstrual
cycle**

OESTROGENS and PROGESTERONE
are released to
blood stream in
**2nd half of menstrual
cycle**

Withdrawal of oestrogen and progesterone

— **Changes occur in accessory sex organs** —

MAMMARY GLAND

Stimulates DUCT development

Stimulates ALVEOLAR development

Increased motility
Proliferation

Mucification
Decreased motility

UTERINE TUBE

Proliferation of Endometrium

Secretion

UTERUS

Clear mucus – CERVIX – Thick sticky mucus

MENSTRUAL FLOW

Cornification

VAGINA

Mucification

Ovarian hormones are therefore directly responsible for regular cycle of events in accessory sex organs.

There are three natural oestrogens: *oestradiol* is the major and most potent. It is formed from *androgen* precursors, as is a second called *oestrone*. Oestrone is metabolized to *oestriol* mainly in the liver.

UTERUS AND UTERINE TUBES

The **uterus** (or womb) is the organ which bears the developing child till birth.

IN CHILDHOOD –
it is a small undeveloped organ
situated deep in the pelvis.

AT PUBERTY

LHRH

A P

FSH and *LH*

Outer walls of
smooth muscle –
the MYOMETRIUM

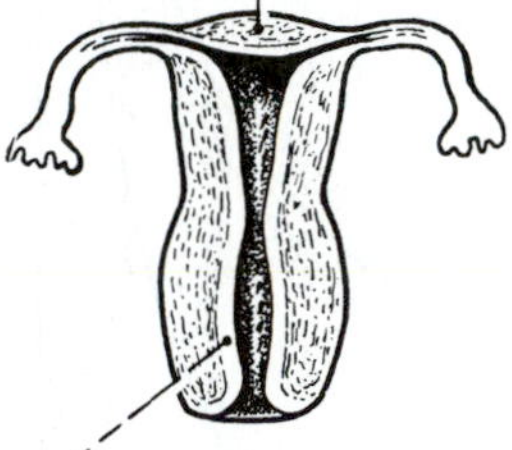

OESTROGEN
from developing GRAAFIAN
FOLLICLES

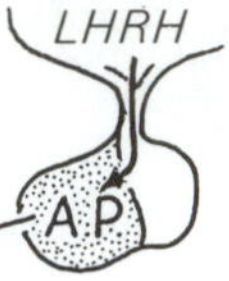

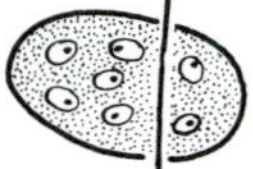

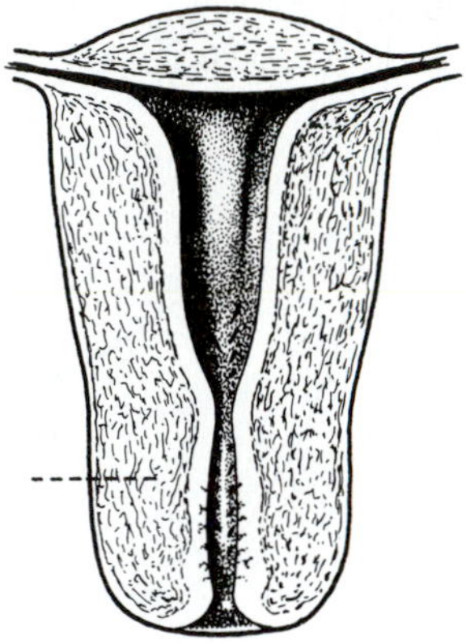

Stimulates
growth of
MYOMETRIUM

and

Inner lining or
Mucous membrane –
the ENDOMETRIUM

growth of
GLANDS and
FIBROUS TISSUE
STROMA of
ENDOMETRIUM

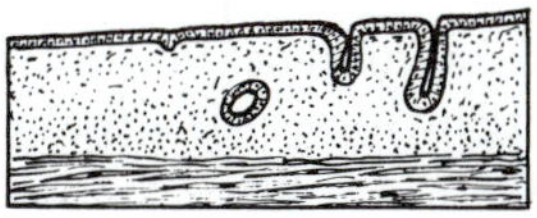

IN MATURITY

[From puberty to menopause
(unless interrupted by
pregnancy or disease) –
when ovarian cycle is
fully established]

MENSTRUATION

If fertilization does not take place
the OVUM, which is about 100 μm in
diameter (cf RBC = 7μm), is either
absorbed in uterus or discharged
at next menstrual period.

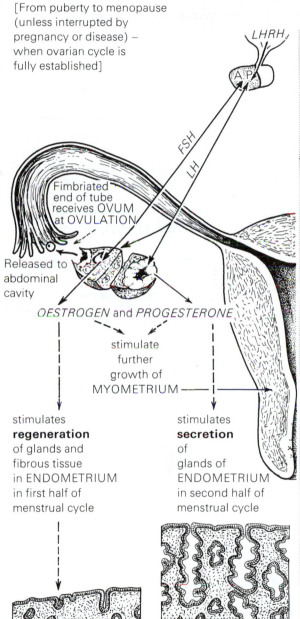

LHRH

A.P.

FSH

LH

Fimbriated
end of tube
receives OVUM
at OVULATION

Released to
abdominal
cavity

OESTROGEN and *PROGESTERONE*

stimulate
further
growth of
MYOMETRIUM

stimulates
regeneration
of glands and
fibrous tissue
in ENDOMETRIUM
in first half of
menstrual cycle

stimulates
secretion
of
glands of
ENDOMETRIUM
in second half of
menstrual cycle

FALL in BLOOD
LEVELS of
OESTROGEN and
PROGESTERONE
towards end of menstrual
cycle

↓

brings about
constriction
of spiral arteries

↓

Degeneration and
breakdown of glands and other
tissues of ENDOMETRIUM to give
Menstrual flow.

UTERINE TUBES IN CYCLE ENDING IN PREGNANCY

The fimbriated end of the **uterine tube** receives the **ovum** at **ovulation**. **Peristaltic** contractions of the muscular tube aided by ciliary movements of its lining cells transfer the **ovum** towards the **uterus**. The uterine tube also transmits **spermatozoa** towards the **ova**.

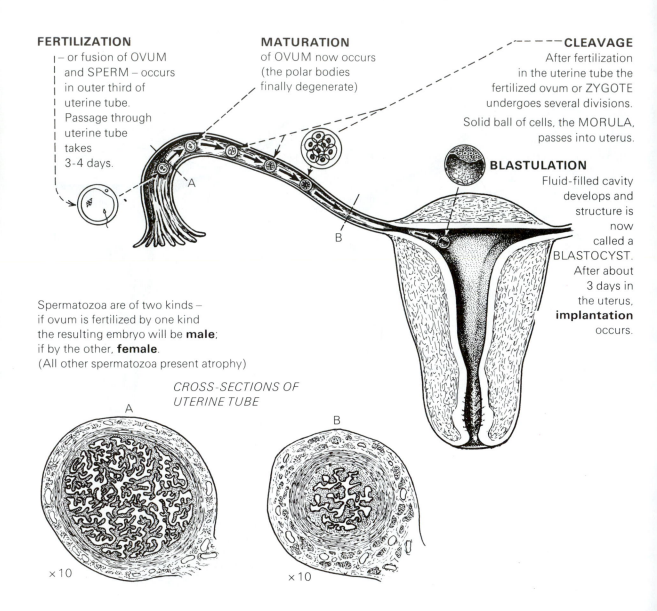

FERTILIZATION
– or fusion of OVUM and SPERM – occurs in outer third of uterine tube. Passage through uterine tube takes 3-4 days.

MATURATION
of OVUM now occurs (the polar bodies finally degenerate)

CLEAVAGE
After fertilization in the uterine tube the fertilized ovum or ZYGOTE undergoes several divisions.

Solid ball of cells, the MORULA, passes into uterus.

BLASTULATION
Fluid-filled cavity develops and structure is now called a BLASTOCYST. After about 3 days in the uterus, **implantation** occurs.

Spermatozoa are of two kinds – if ovum is fertilized by one kind the resulting embryo will be **male**; if by the other, **female**.
(All other spermatozoa present atrophy)

CROSS-SECTIONS OF UTERINE TUBE

A

× 10

B

× 10

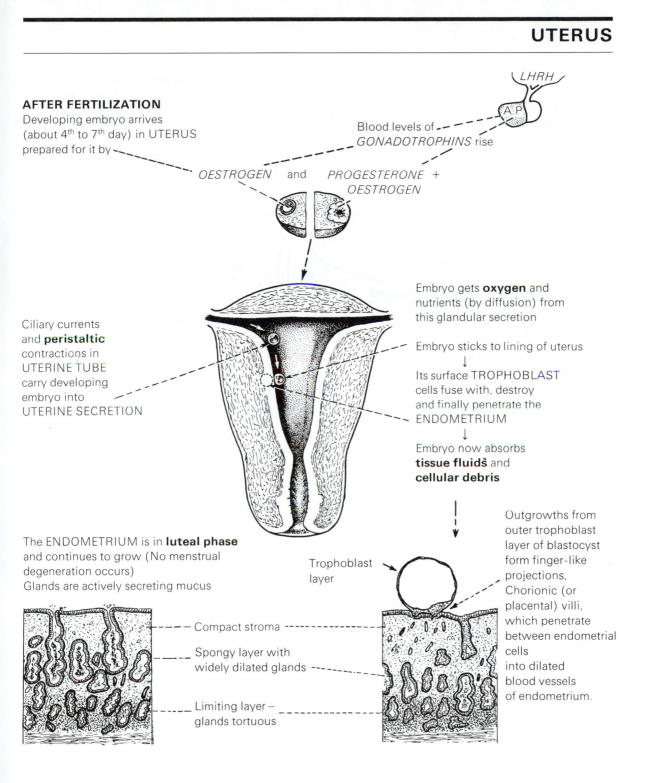

LHRH

A P

AFTER FERTILIZATION
Developing embryo arrives
(about 4th to 7th day) in UTERUS
prepared for it by

Blood levels of
GONADOTROPHINS rise

OESTROGEN   and   *PROGESTERONE +*
OESTROGEN

Embryo gets **oxygen** and
nutrients (by diffusion) from
this glandular secretion

Ciliary currents
and **peristaltic**
contractions in
UTERINE TUBE
carry developing
embryo into
UTERINE SECRETION

Embryo sticks to lining of uterus
↓
Its surface TROPHOBLAST
cells fuse with, destroy
and finally penetrate the
ENDOMETRIUM
↓
Embryo now absorbs
tissue fluids and
cellular debris

Outgrowths from
outer trophoblast
layer of blastocyst
form finger-like
projections,
Chorionic (or
placental) villi,
which penetrate
between endometrial
cells
into dilated
blood vessels
of endometrium.

The ENDOMETRIUM is in **luteal phase**
and continues to grow (No menstrual
degeneration occurs)
Glands are actively secreting mucus

Trophoblast
layer

Compact stroma

Spongy layer with
widely dilated glands

Limiting layer –
glands tortuous

UTERUS

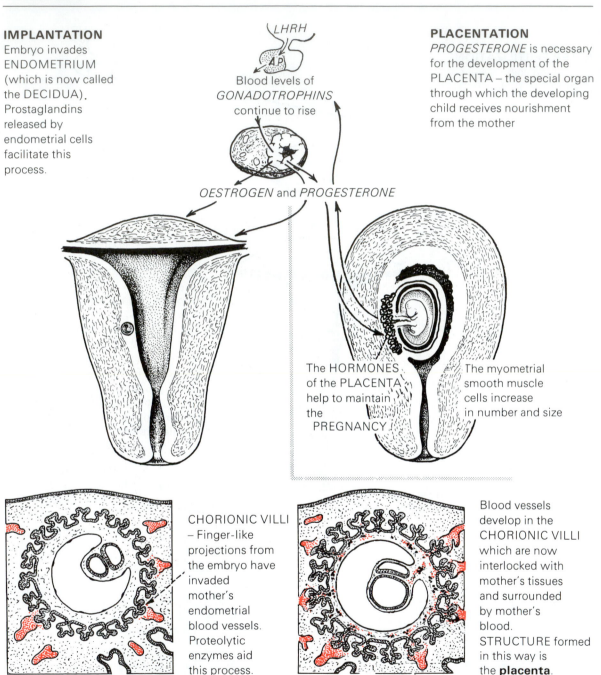

IMPLANTATION

Embryo invades ENDOMETRIUM (which is now called the DECIDUA). Prostaglandins released by endometrial cells facilitate this process.

LHRH

A.P.

Blood levels of *GONADOTROPHINS* continue to rise

OESTROGEN and *PROGESTERONE*

PLACENTATION

PROGESTERONE is necessary for the development of the PLACENTA – the special organ through which the developing child receives nourishment from the mother

The HORMONES of the PLACENTA help to maintain the PREGNANCY

The myometrial smooth muscle cells increase in number and size

CHORIONIC VILLI – Finger-like projections from the embryo have invaded mother's endometrial blood vessels. Proteolytic enzymes aid this process.

Blood vessels develop in the CHORIONIC VILLI which are now interlocked with mother's tissues and surrounded by mother's blood. STRUCTURE formed in this way is the **placenta**.

After 2 months the developing embryo is called a **fetus**.

The **placenta** functions for the **fetus** as alimentary tract, kidneys and lungs. It increases in weight throughout pregnancy.

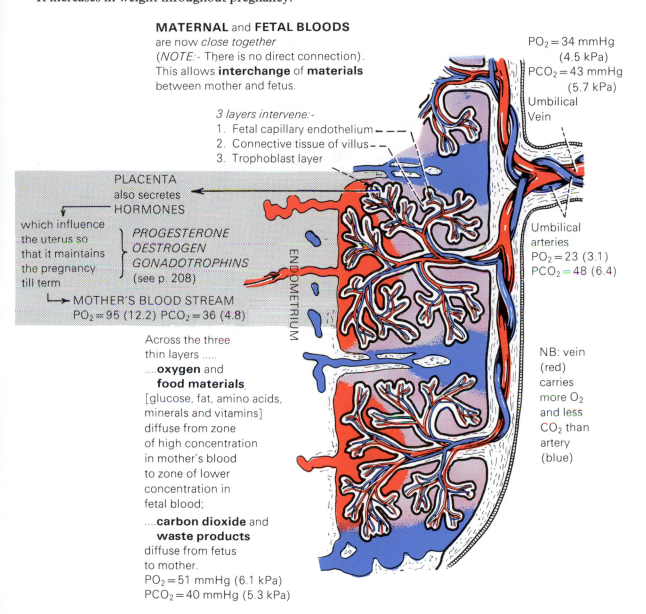

MATERNAL and **FETAL BLOODS**
are now *close together*
(*NOTE:*- There is no direct connection).
This allows **interchange** of **materials**
between mother and fetus.

3 layers intervene:-
1. Fetal capillary endothelium
2. Connective tissue of villus
3. Trophoblast layer

$PO_2 = 34$ mmHg
(4.5 kPa)
$PCO_2 = 43$ mmHg
(5.7 kPa)
Umbilical
Vein

PLACENTA
also secretes
HORMONES
which influence
the uterus so
that it maintains
the pregnancy
till term

PROGESTERONE
OESTROGEN
GONADOTROPHINS
(see p. 208)

MOTHER'S BLOOD STREAM
$PO_2 = 95$ (12.2) $PCO_2 = 36$ (4.8)

ENDOMETRIUM

Umbilical
arteries
$PO_2 = 23$ (3.1)
$PCO_2 = 48$ (6.4)

Across the three
thin layers .....
....**oxygen** and
food materials
[glucose, fat, amino acids,
minerals and vitamins]
diffuse from zone
of high concentration
in mother's blood
to zone of lower
concentration in
fetal blood;
....**carbon dioxide** and
waste products
diffuse from fetus
to mother.
$PO_2 = 51$ mmHg (6.1 kPa)
$PCO_2 = 40$ mmHg (5.3 kPa)

NB: vein
(red)
carries
more O_2
and less
CO_2 than
artery
(blue)

Substances of small molecular weight usually pass in either direction by diffusion. Larger molecules are probably transported by special carrier systems.

UTERUS

AS PREGNANCY ADVANCES

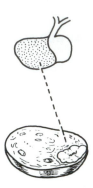

Fetus grows larger and comes to fill UTERINE CAVITY

CORPUS LUTEUM remains, but after the 4th month its contribution to *oestrogen* and *progesterone* supply is dwarfed by that of the placenta.

Fetus is attached by UMBILICAL CORD to PLACENTA

Fetus is bathed in AMNIOTIC FLUID which is derived from amniotic epithelium, fetal urine and lung fluid and contained within amniotic and chorionic membranes. Maintains fetus in shock-proof, constant temperature environment.

Growth of MYOMETRIUM – increase in number and size of smooth muscle cells and of the blood vessels.

Stretching of MYOMETRIUM

AMNIOTIC FLUID

PLACENTAL HORMONES
(1) *HUMAN CHORIONIC GONADOTROPHIN (HCG)* maintains corpus luteum. Placenta takes over main secretion of (2) *OESTROGENS* and (3) *PROGESTERONE* after the sixth week.
(4) *HUMAN CHORIONIC SOMATOMAMMOTROPHIN (HCS)* or *human placental lactogen (HPL)* has anabolic and lactogenic activity and maintains supply of nutrients, especially glucose, to fetus.
(5) *RELAXIN* relaxes pelvic joints, softens cervix, decreases uterine activity.

Major *oestrogen* of pregnancy is *oestriol*. Synthesized by the placenta from precursors synthesized in the adrenal gland of the fetus.

Amniotic fluid can be sampled – **amniocentesis** – to detect fetal abnormalities.

FETAL CIRCULATION

For the fetus the **placenta** acts as the organ of transfer for oxygen, nutritives and waste products. Only a small volume of blood passes through the fetal lungs.

BLOOD RETURNING TO HEART

...To RIGHT ATRIUM
Small amount from heart, head, neck and arms → S.V.C.
LARGE AMOUNT via UMBILICAL VEINS through LIVER – short circuits to I.V.C. via DUCTUS VENOSUS.
Some of this passes to right atrium.
Small amount from abdominal cavity and legs.

FORAMEN OVALE
(Opening between right and left atria)

DUCTUS VENOSUS

...To LEFT ATRIUM
Small amount from 2 lungs
LARGE AMOUNT from INFERIOR VENA CAVA through

FORAMEN OVALE.
(thus by-passing pulmonary circulation)

BLOOD LEAVING HEART

...From RIGHT VENTRICLE
Small amount to 2 lungs
LARGE AMOUNT to AORTA through

DUCTUS ARTERIOSUS
(thus by-passing pulmonary circulation) joins
OUTPUT from LEFT VENTRICLE
↓
Small amount to heart, head, neck and arms.
LARGE AMOUNT to PLACENTA through UMBILICAL ARTERIES
Small amount to abdominal cavity and legs.

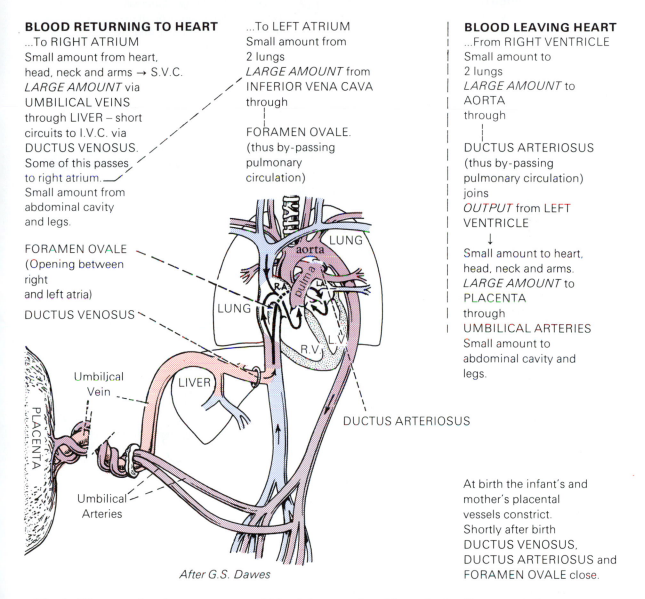

After G.S. Dawes

Umbilical Vein

LIVER

PLACENTA

Umbilical Arteries

LUNG

aorta

LUNG

R.A. pulm'y L.A.

R.V. L.V.

DUCTUS ARTERIOSUS

At birth the infant's and mother's placental vessels constrict. Shortly after birth DUCTUS VENOSUS, DUCTUS ARTERIOSUS and FORAMEN OVALE close.

Head of fetus receives better oxygenated blood than trunk and lower body. **Oxygenated blood** → umbilical vein → ductus venosus → IVC → R. atrium → foramen ovale → L. atrium → L. ventricle → aorta → **head.**

UTERUS

PARTURITION

About 40 weeks after conception the process of **childbirth** begins. When uterine contractions are strong, coordinated and occur at 10-15 min intervals, **labour** has started.

After the 32nd week of pregnancy *relaxin* and *oestrogens* increase OXYTOCIN receptors on uterus and uterine *PROSTAGLANDIN* synthesis. Both factors increase uterine contractions.

1ST stage usually lasts up to 14 hours with a first birth

MYOMETRIUM
Uterine muscle is now very greatly stretched.
↓
Rhythmic contractions which begin to increase in strength and frequency
↓
Press on amniotic fluid

2ND stage LABOUR usually lasts up to 2 hours with a first birth

Uterine contractions increase in strength and frequency (aided by voluntary contractions of abdominal muscles)
↓
Child is slowly forced through CERVIX and is delivered from VAGINA

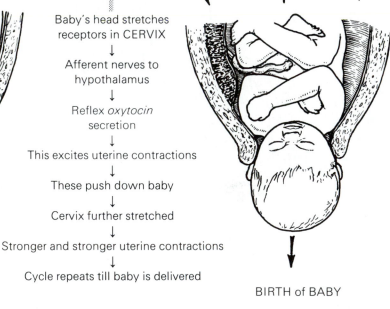

Baby's head stretches receptors in CERVIX
↓
Afferent nerves to hypothalamus
↓
Reflex *oxytocin* secretion
↓
This excites uterine contractions
↓
These push down baby
↓
Cervix further stretched
↓
Stronger and stronger uterine contractions
↓
Cycle repeats till baby is delivered

RELAXIN
softens cervix and it dilates to about 10 cm.

Membranes rupture and AMNIOTIC FLUID escapes

BIRTH of BABY

AFTER PARTURITION

3RD stage labour
5-15 minutes
after birth of child

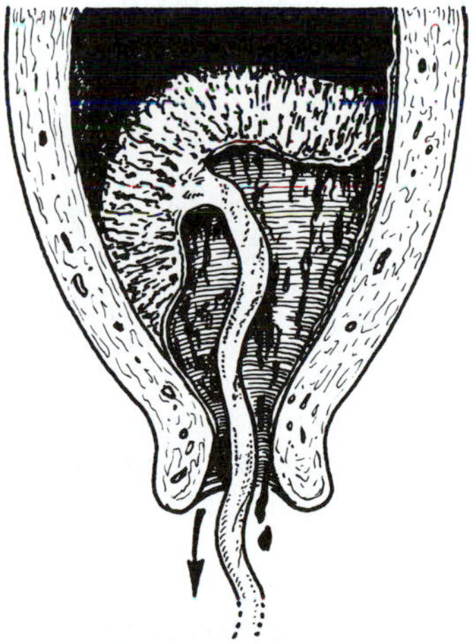

IN PUERPERIUM (Immediately following childbirth.)

HYPOTHALAMUS

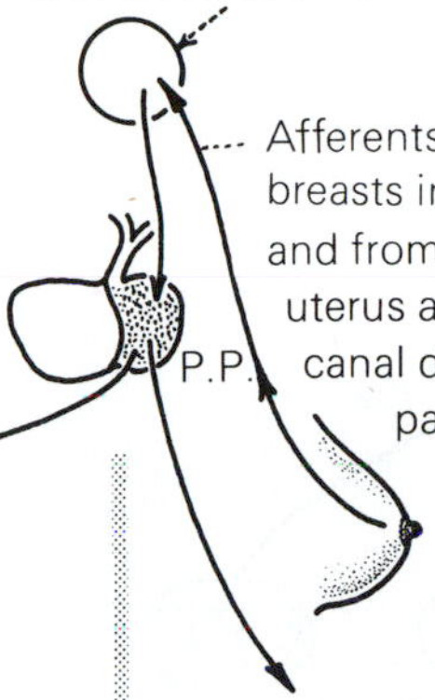

Afferents from breasts in suckling and from the uterus and birth canal during parturition

P.P.

OXYTOCIN
released to blood stream
stimulates
contractions of UTERINE MUSCLE
↓
Detach and **deliver** PLACENTA and the membranes as the AFTERBIRTH.

Fall in
blood levels
of
*OESTROGEN,
PROGESTERONE*
and
*OTHER PLACENTAL
HORMONES*
after loss of
placenta

OXYTOCIN
released to
blood stream

MYOMETRIUM
Uterine muscle contracts down to close off blood vessels torn and bleeding after separation of placenta

A large part of endometrium – decidua – is shed with the placenta.
Only the limiting layer is left.

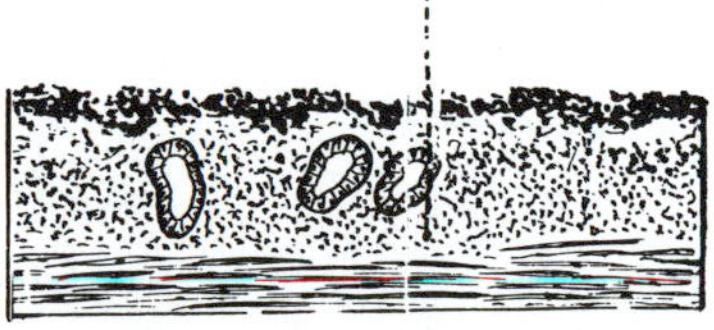

UTERUS

INVOLUTION

HYPOTHALAMUS

LHRH

Afferents
from breast
in suckling

A.P. P.P.

LH

FSH

OXYTOCIN
released to blood
stream

MYOMETRIUM
Muscle cells and
blood vessels
regress.
Uterus shrinks
till it is only a little
larger than it was
before conception.

ENDOMETRIUM
is reformed from
limiting layer.

Under action
of
*ovarian
hormones*

Gradual restoration
of **menstrual
cycle**

AFTER MENOPAUSE

When ovarian tissue ceases to be
responsive to *anterior pituitary
gonadotrophins*

↓

. Fall in production of
oestrogen and *progesterone*

↓

Shrinkage of
MYOMETRIUM

Atrophy and shrinkage of
glands and stroma of
ENDOMETRIUM

↓

Cessation of
menstrual cycle

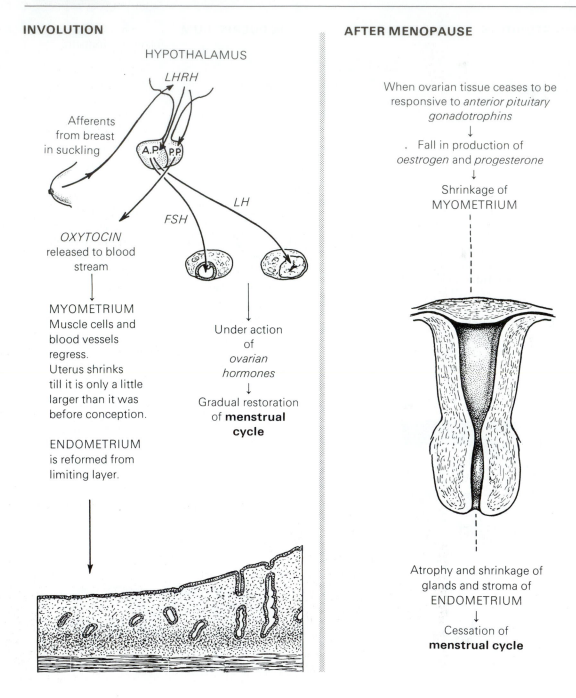

There are two mammary glands.

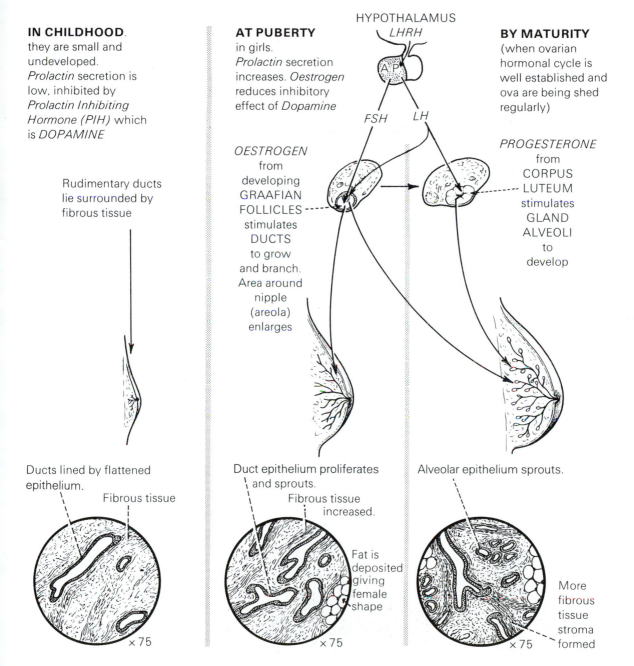

IN CHILDHOOD.
they are small and
undeveloped.
Prolactin secretion is
low, inhibited by
*Prolactin Inhibiting
Hormone (PIH)* which
is *DOPAMINE*

Rudimentary ducts
lie surrounded by
fibrous tissue

AT PUBERTY
in girls.
Prolactin secretion
increases. *Oestrogen*
reduces inhibitory
effect of *Dopamine*

HYPOTHALAMUS
LHRH
A P

FSH   LH

OESTROGEN
from
developing
GRAAFIAN
FOLLICLES
stimulates
DUCTS
to grow
and branch.
Area around
nipple
(areola)
enlarges

BY MATURITY
(when ovarian
hormonal cycle is
well established and
ova are being shed
regularly)

PROGESTERONE
from
CORPUS
LUTEUM
stimulates
GLAND
ALVEOLI
to
develop

Ducts lined by flattened
epithelium.
Fibrous tissue

× 75

Duct epithelium proliferates
and sprouts.
Fibrous tissue
increased.

Fat is
deposited
giving
female
shape

× 75

Alveolar epithelium sprouts.

More
fibrous
tissue
stroma
formed

× 75

MAMMARY GLANDS

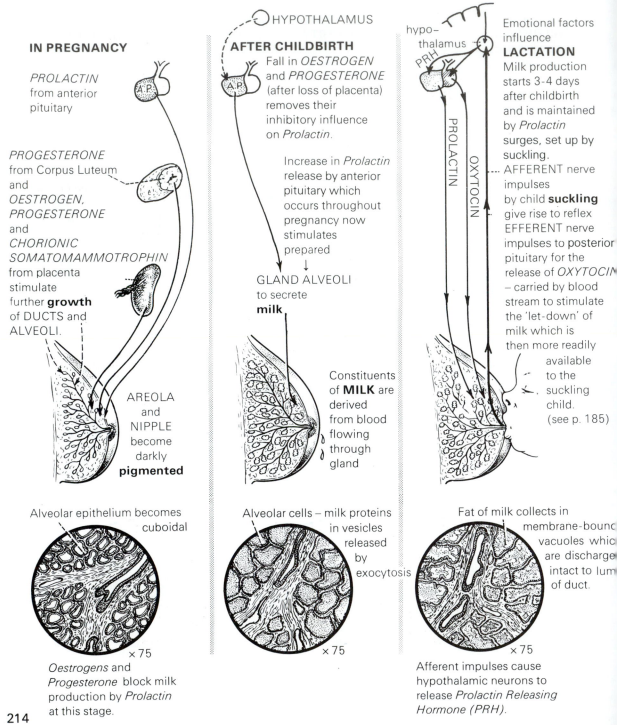

IN PREGNANCY

PROLACTIN from anterior pituitary

PROGESTERONE from Corpus Luteum and *OESTROGEN, PROGESTERONE* and *CHORIONIC SOMATOMAMMOTROPHIN* from placenta stimulate further **growth** of DUCTS and ALVEOLI.

A.P.

AREOLA and NIPPLE become darkly **pigmented**

Alveolar epithelium becomes cuboidal

×75

Oestrogens and *Progesterone* block milk production by *Prolactin* at this stage.

HYPOTHALAMUS

AFTER CHILDBIRTH

Fall in *OESTROGEN* and *PROGESTERONE* (after loss of placenta) removes their inhibitory influence on *Prolactin*.

Increase in *Prolactin* release by anterior pituitary which occurs throughout pregnancy now stimulates prepared

↓

GLAND ALVEOLI to secrete **milk**

A.P.

Constituents of **MILK** are derived from blood flowing through gland

Alveolar cells – milk proteins in vesicles released by exocytosis

×75

hypo– thalamus

PRH

Emotional factors influence **LACTATION** Milk production starts 3-4 days after childbirth and is maintained by *Prolactin* surges, set up by suckling. AFFERENT nerve impulses by child **suckling** give rise to reflex EFFERENT nerve impulses to posterior pituitary for the release of *OXYTOCIN* – carried by blood stream to stimulate the 'let-down' of milk which is then more readily available to the suckling child. (see p. 185)

PROLACTIN

OXYTOCIN

Fat of milk collects in membrane-bound vacuoles which are discharged intact to lumen of duct.

×75

Afferent impulses cause hypothalamic neurons to release *Prolactin Releasing Hormone (PRH)*.

214

POST LACTATION

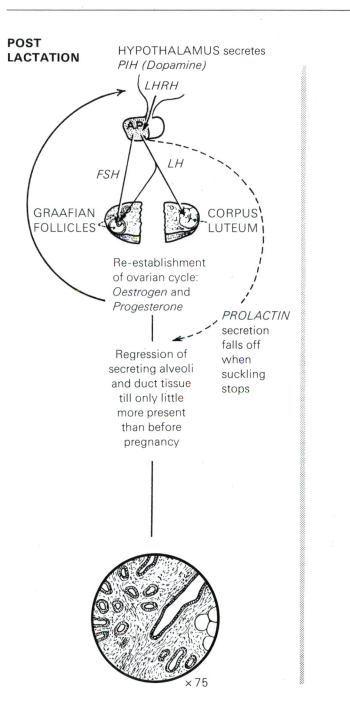

HYPOTHALAMUS secretes *PIH (Dopamine)*

LHRH

AP

FSH

LH

GRAAFIAN FOLLICLES

CORPUS LUTEUM

Re-establishment of ovarian cycle: *Oestrogen* and *Progesterone*

PROLACTIN secretion falls off when suckling stops

Regression of secreting alveoli and duct tissue till only little more present than before pregnancy

×75

AT MENOPAUSE

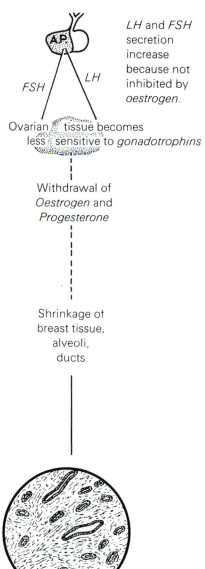

LH and *FSH* secretion increase because not inhibited by *oestrogen*.

AP

FSH

LH

Ovarian tissue becomes less sensitive to *gonadotrophins*

Withdrawal of *Oestrogen* and *Progesterone*

Shrinkage of breast tissue, alveoli, ducts

×75

215

MENOPAUSE

Between the ages of 45 and 55 years **ovarian** tissue gradually ceases to respond to stimulation by *anterior pituitary gonadotrophic hormones*.

FSH
LH
A.P.

Ovarian cycle———becomes irregular and finally ceases → Ovary becomes small and fibrosed and no longer produces ripe ova.

OESTROGEN and *PROGESTERONE* levels in blood stream fall.

TISSUES of the body ———— begin to show changes which mark the end of **reproductive life**.

Sometimes final redistribution of fat → less typically feminine distribution.

Regression of secondary sex characteristics.

Ducts ⎫
Alveoil ⎬ Atrophy

Breasts shrink.
Hair becomes sparse in axillae and pubis.

Accessory sex organs atrophy.

Fallopian tubes shrink.
Uterine cycle and menstruation cease.
(Muscle and lining shrink).
Vaginal epithelium becomes thin and loses its secretions.
External genitalia shrink.

Uterus

Vagina

Psychological and personality changes
Sexual drive is frequently not diminished – may be increased. Irritability and anxiety attacks may occur accompanied by 'hot flushes' (vasodilatation of arterioles), feeling of warmth and excessive sweating.

Incidence of high blood pressure and atherosclerosis rises to that of men. Marked bone demineralization (**osteoporosis**) occurs, because of oestrogen deficiency.

Oestrogen **supplements** reduce many of the symptoms of the menopause but they may facilitate breast or cervical cancer. Some oestrogen secretion continues. Androgen precursors from ovarian stromal and adrenal cells are converted to *oestrone* by liver and adipose tissue. This diminishes menopausal symptoms.

PITUITARY, OVARIAN AND ENDOMETRIAL CYCLES

HYPOTHALAMUS secretes *luteinizing hormone-releasing hormone* into hypothalamic-hypophyseal portal circulation (page 179).

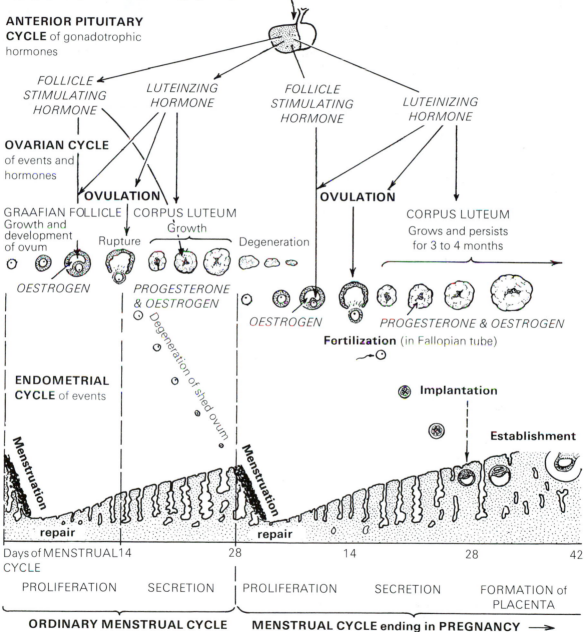

Secretion of LHRH, and thus LH and FSH, is powerfully inhibited by progesterone and oestrogens. Since these hormones are present in high concentration throughout pregnancy, follicle development, ovulation and menstrual cycles stop for the duration of pregnancy.

CHAPTER 10

CENTRAL NERVOUS SYSTEM LOCOMOTOR SYSTEM

NERVOUS SYSTEM

Most functions of the body are controlled by either the **nervous** or **endocrine** systems. Usually rapid activities, e.g. muscular contraction, are controlled by the nervous system and slower activities, e.g. metabolic functions, are controlled by the endocrine system.

The **NERVOUS** system is specialized in:
(a) **Irritability** – the ability to receive and respond to stimuli from the external and internal environments.
(b) **Conduction** – the ability to transmit signals to and from **central integrating** centres.
(c) **Integration** – the ability to analyse information from the environment in order to generate **behaviour** appropriate to that information.

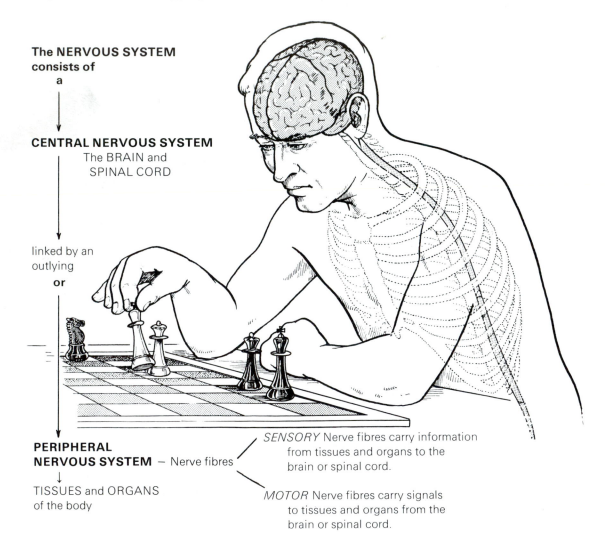

The NERVOUS SYSTEM consists of a

↓

CENTRAL NERVOUS SYSTEM
The BRAIN and
SPINAL CORD

↓

linked by an outlying

or

PERIPHERAL NERVOUS SYSTEM – Nerve fibres

↓

TISSUES and ORGANS
of the body

SENSORY Nerve fibres carry information from tissues and organs to the brain or spinal cord.

MOTOR Nerve fibres carry signals to tissues and organs from the brain or spinal cord.

DEVELOPMENT OF THE NERVOUS SYSTEM

The nervous system develops in the embryo from a simple tube of **ectoderm**:- The **primitive neural tube**.

The **cells** lining it become the nervous tissue of the **brain** and **spinal cord**. The **canal** becomes distended to form the **ventricles** of the **brain** and **central canal** of the **spinal cord**:-

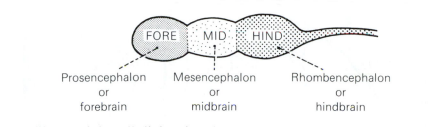

FORE · MID · HIND

| Prosencephalon
or
forebrain | Mesencephalon
or
midbrain | Rhombencephalon
or
hindbrain |

Each of these swellings and the cells lining them become more complicated:-

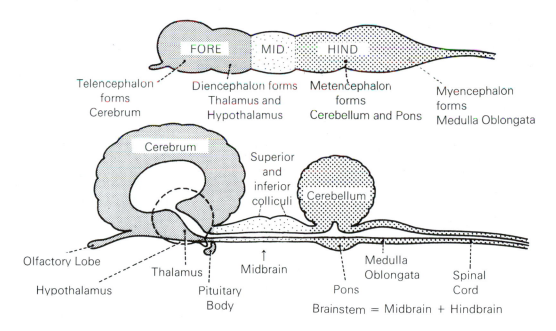

FORE · MID · HIND

Telencephalon forms Cerebrum

Diencephalon forms Thalamus and Hypothalamus

Metencephalon forms Cerebellum and Pons

Myencephalon forms Medulla Oblongata

Cerebrum

Superior and inferior colliculi

Cerebellum

Olfactory Lobe

Hypothalamus

Thalamus

Midbrain

Pituitary Body

Medulla Oblongata

Pons

Spinal Cord

Brainstem = Midbrain + Hindbrain

There are now many layers of cells forming the brain and spinal cord. The ventricles of the brain and the central canal of the spinal cord are filled with **cerebrospinal fluid**.

CEREBRUM

The largest part of the human brain is the **cerebrum** – made up of **two cerebral hemispheres.**
Each of these is divided into **lobes**.

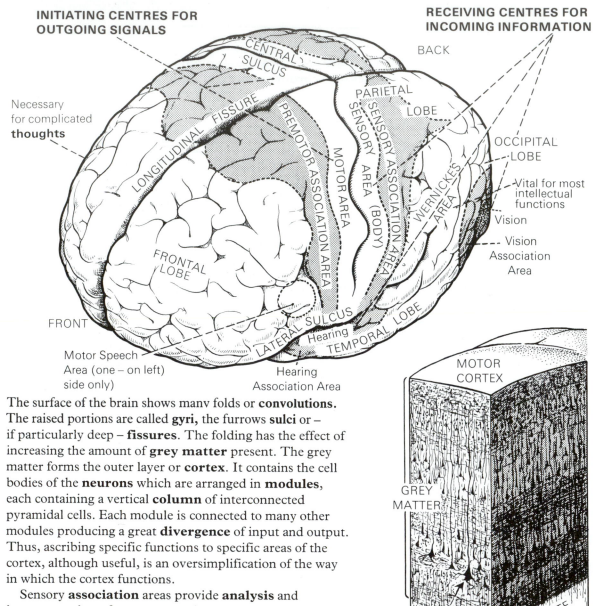

INITIATING CENTRES FOR OUTGOING SIGNALS

RECEIVING CENTRES FOR INCOMING INFORMATION

Necessary for complicated **thoughts**

CENTRAL SULCUS

BACK

LONGITUDINAL FISSURE

PARIETAL LOBE

PREMOTOR ASSOCIATION AREA

SENSORY ASSOCIATION AREA

OCCIPITAL LOBE

MOTOR AREA

SENSORY AREA (BODY)

WERNICKES AREA

Vital for most intellectual functions

Vision

Vision Association Area

FRONTAL LOBE

FRONT

Motor Speech Area (one – on left) side only)

LATERAL SULCUS

Hearing

Hearing Association Area

TEMPORAL LOBE

MOTOR CORTEX

GREY MATTER

WHITE MATTER

The surface of the brain shows many folds or **convolutions.**
The raised portions are called **gyri,** the furrows **sulci** or –
if particularly deep – **fissures**. The folding has the effect of
increasing the amount of **grey matter** present. The grey
matter forms the outer layer or **cortex**. It contains the cell
bodies of the **neurons** which are arranged in **modules,**
each containing a vertical **column** of interconnected
pyramidal cells. Each module is connected to many other
modules producing a great **divergence** of input and output.
Thus, ascribing specific functions to specific areas of the
cortex, although useful, is an oversimplification of the way
in which the cortex functions.

Sensory **association** areas provide **analysis** and
interpretation of sensory experiences.

90% of all nerve cells in the body are in the cerebral cortex.

GIANT PYRAMIDAL CELL (BETZ CELL)

This view shows surface **grey matter** containing nerve cells and inner **white matter** made up of nerve fibres.

Deep in the substance of the cerebal hemispheres there are additional masses of **grey matter**:-

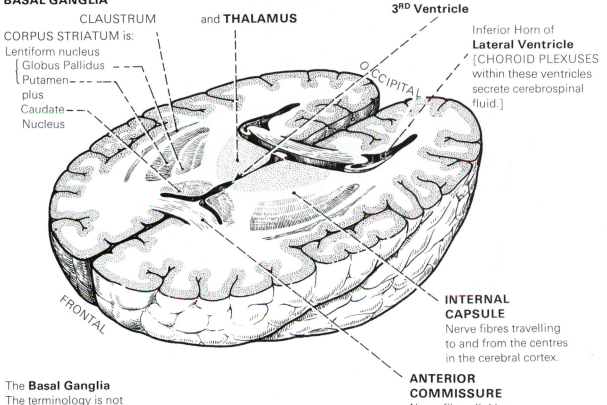

BASAL GANGLIA

CLAUSTRUM   and **THALAMUS**

3RD Ventricle

CORPUS STRIATUM is:

Lentiform nucleus
{ Globus Pallidus – – –
{ Putamen– – – – –
plus
Caudate– – –
Nucleus

OCCIPITAL

Inferior Horn of
Lateral Ventricle
[CHOROID PLEXUSES
within these ventricles
secrete cerebrospinal
fluid.]

FRONTAL

INTERNAL CAPSULE
Nerve fibres travelling
to and from the centres
in the cerebral cortex.

ANTERIOR COMMISSURE
Nerve fibres linking
the two hemispheres.

The **Basal Ganglia**
The terminology is not
completely agreed but
usually now
Basal Ganglia = Globus Pallidus
+ Putamen + Caudate nucleus.
Claustrum is often excluded.
Structures associated with
basal ganglia functionally
are subthalamic nucleus
and substantia nigra.
This complex is concerned
with **planning** and
**programming voluntary
muscle movement**.

The **Thalamus**
is an important
relay centre for
sensory fibres on
their way to the
cerebral cortex.
'Crude' sensation and the perception of
pain may occur here. Also relays part of
the reticular activating system which
controls the level of alertness and state
of consciousness.

223

VERTICAL SECTION THROUGH BRAIN

This is a vertical section through the **longitudinal fissure** which separates the two cerebral hemispheres. At the bottom of the cleft are tracts of nerve fibres which link the two hemispheres – the **corpus callosum**.

FOREBRAIN

Cerebral Hemisphere

Thalamus
– relay centres for sensation: pain perceived here.

Hypothalamus
Contains nuclei which control thirst, appetite, temperature, autonomic nervous system and pituitary. Integrates responses to stress.

Opening of lateral ventricle

PARIETAL LOBE

OCCIPITAL LOBE

CORPUS CALLOSUM
FORNIX

FRONTAL LOBE

PONS

Pituitary gland

The **grey matter** in the brainstem is formed by groups of nerve cell bodies called nuclei. These are distributed irregularly through the white matter.

3rd Ventricle

Colliculi

Cerebellum
Coordinates signals from muscle, joint, visual, auditory and equilibrium receptors with instructions from cortex.

MIDBRAIN
Superior and inferior colliculi are centres for visual and auditory reflexes. Contains nuclei of III, IV Cranial nerves, also the Red Nucleus and Substantia Nigra which help to control skilled muscular movements.
The **white matter** carries nerve fibres linking Red Nucleus with Cerebral Cortex, Thalamus, Cerebellum, Corpus Striatum and Spinal Cord. It also carries Ascending Sensory fibres in Lateral and Medial Lemnisci, and Descending Motor fibres on their way to Pons and Spinal Cord.

HINDBRAIN:
[PONS, CEREBELLUM, MEDULLA OBLONGATA]

Pons: Groups of Neurons form sensory nucleus of V and also nuclei of VI and VII Cranial nerves. Other nerve cells here relay impuses along their axons to Cerebellum and Cerebrum. Rubrospinal tract. Lateral and Medial Lemnisci pass through Pons as do nerve fibres linking Cerebral Cortex with Medulla Oblongata and Spinal Cord.

Medulla Oblongata
Contains centres controlling heart rate, blood vessels, respiration and nuclei of VIII, IX, X, XI, XII Cranial nerves, Gracile and Cuneate nuclei – second sensory neurons in cutaneous pathways. Tracts of Sensory fibres decussate and ascend to other side of Cerebral Cortex. Some fibres remain uncrossed. The larger part of each Motor corticospinal tract crosses and descends in other side of Spinal Cord.

CORONAL SECTION THROUGH BRAIN

This is a section through the **central (transverse) sulcus**. It shows each of the major developments of the brain –

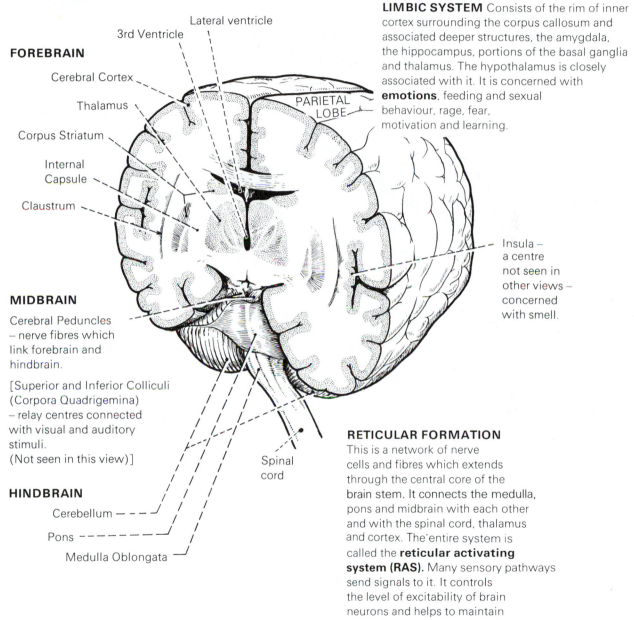

FOREBRAIN

Cerebral Cortex

Thalamus

Corpus Striatum

Internal Capsule

Claustrum

3rd Ventricle

Lateral ventricle

PARIETAL LOBE

LIMBIC SYSTEM Consists of the rim of inner cortex surrounding the corpus callosum and associated deeper structures, the amygdala, the hippocampus, portions of the basal ganglia and thalamus. The hypothalamus is closely associated with it. It is concerned with **emotions**, feeding and sexual behaviour, rage, fear, motivation and learning.

Insula – a centre not seen in other views – concerned with smell.

MIDBRAIN

Cerebral Peduncles – nerve fibres which link forebrain and hindbrain.

[Superior and Inferior Colliculi (Corpora Quadrigemina) – relay centres connected with visual and auditory stimuli. (Not seen in this view)]

Spinal cord

HINDBRAIN

Cerebellum — – – –

Pons – – – – – –

Medulla Oblongata

RETICULAR FORMATION
This is a network of nerve cells and fibres which extends through the central core of the brain stem. It connects the medulla, pons and midbrain with each other and with the spinal cord, thalamus and cortex. The entire system is called the **reticular activating system (RAS).** Many sensory pathways send signals to it. It controls the level of excitability of brain neurons and helps to maintain **consciousness** and the **waking state**. Inhibition of RAS leads to sleep or coma.

CRANIAL NERVES

Twelve pairs of nerves arise directly from the undersurface of the brain to supply head and neck and most of the viscera.

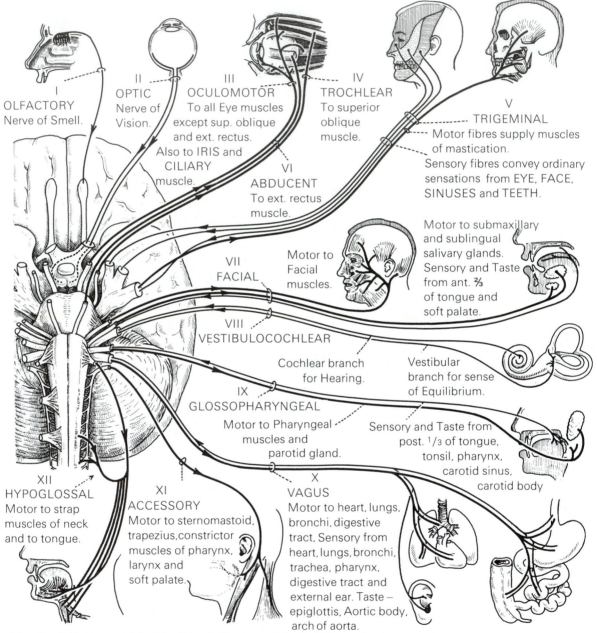

I
OLFACTORY
Nerve of Smell.

II
OPTIC
Nerve of Vision.

III
OCULOMOTOR
To all Eye muscles except sup. oblique and ext. rectus. Also to IRIS and CILIARY muscle.

IV
TROCHLEAR
To superior oblique muscle.

VI
ABDUCENT
To ext. rectus muscle.

V
TRIGEMINAL
Motor fibres supply muscles of mastication. Sensory fibres convey ordinary sensations from EYE, FACE, SINUSES and TEETH.

Motor to submaxillary and sublingual salivary glands. Sensory and Taste from ant. ⅔ of tongue and soft palate.

VII
FACIAL

Motor to Facial muscles.

VIII
VESTIBULOCOCHLEAR

Cochlear branch for Hearing.

Vestibular branch for sense of Equilibrium.

IX
GLOSSOPHARYNGEAL
Motor to Pharyngeal muscles and parotid gland.

Sensory and Taste from post. ⅓ of tongue, tonsil, pharynx, carotid sinus, carotid body

XII
HYPOGLOSSAL
Motor to strap muscles of neck and to tongue.

XI
ACCESSORY
Motor to sternomastoid, trapezius, constrictor muscles of pharynx, larynx and soft palate.

X
VAGUS
Motor to heart, lungs, bronchi, digestive tract, Sensory from heart, lungs, bronchi, trachea, pharynx, digestive tract and external ear. Taste – epiglottis, Aortic body, arch of aorta.

(After Frank H. Netter, M.D., The Ciba Collection of Medical Illustrations)

The **spinal cord** lies within the vertebral canal. It is continuous above with the medulla oblongata.

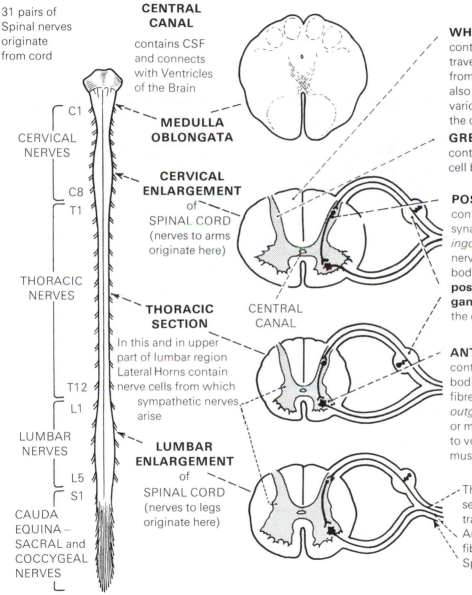

31 pairs of Spinal nerves originate from cord

CERVICAL NERVES

C1

C8

T1

THORACIC NERVES

T12

L1

LUMBAR NERVES

L5

S1

CAUDA EQUINA – SACRAL and COCCYGEAL NERVES

CENTRAL CANAL

contains CSF and connects with Ventricles of the Brain

MEDULLA OBLONGATA

CERVICAL ENLARGEMENT
of
SPINAL CORD
(nerves to arms originate here)

CENTRAL CANAL

THORACIC SECTION

In this and in upper part of lumbar region Lateral Horns contain nerve cells from which sympathetic nerves arise

LUMBAR ENLARGEMENT
of
SPINAL CORD
(nerves to legs originate here)

WHITE MATTER
contains nerve fibres travelling to and from brain and also linking various parts of the cord itself.

GREY MATTER
contains nerve cell bodies.

POSTERIOR HORNS
contain cells which synapse with *ingoing (afferent)* nerves whose cell bodies lie in the **posterior root ganglia** outside the cord.

ANTERIOR HORNS
contain cell bodies whose fibres carry *outgoing (efferent)* or motor signals to voluntary muscles.

The Posterior sensory fibres travel with the Anterior or motor fibres in the same Spinal nerve.

The spinal nerves travel to all parts of the trunk and limbs.

SYNAPSE

The structural unit of the nervous system is the **neuron**.
Neurons are linked together in the nervous system.....

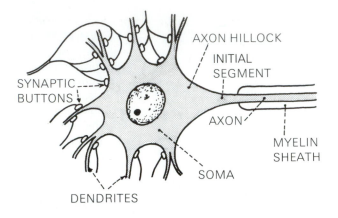

AXON HILLOCK

INITIAL
SEGMENT

SYNAPTIC
BUTTONS

AXON

MYELIN
SHEATH

SOMA

DENDRITES

The AXON of a neuron ends in small swellings – SYNAPTIC BUTTONS or END FEET. These terminate very close to the DENDRITES, SOMA or AXON of the next cell. In most cases there is no direct protoplasmic union between neurons at the **synapse** though connection of neurons by **gap junctions** (page 60) sometimes occurs.

One neuron usually connects with a great many others, often widely scattered in different parts of the brain and spinal cord. In this way intricate chains of nerve cells forming complex pathways for *incoming* and *outgoing* information can be built up within the central nervous system.

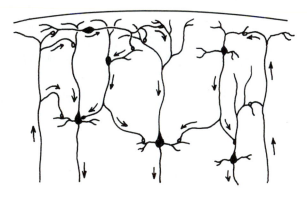

When the **nerve impulse** – a small brief change in membrane potential – reaches a synapse it causes the release from the nerve endings of a **chemical** substance which diffuses across the gap and alters the membrane potential of the next neuron. This alteration of potential spreads across the **soma** of the next neuron and, if large enough, generates more nerve impulses at its **axon hillock-initial segment**. These impulses then travel along the next axon.

A synapse permits transmission of the impulse in one direction only.

The **neuron** is the **anatomical** or **structural unit** of the nervous system: the **nervous reflex** is the **physiological** or **functional unit**.

A nervous reflex is an involuntary action caused by the stimulation of an *afferent (sensory)* nerve ending or **receptor.**

The structural basis of reflex action is the reflex arc. In its simplest form this consists of:-

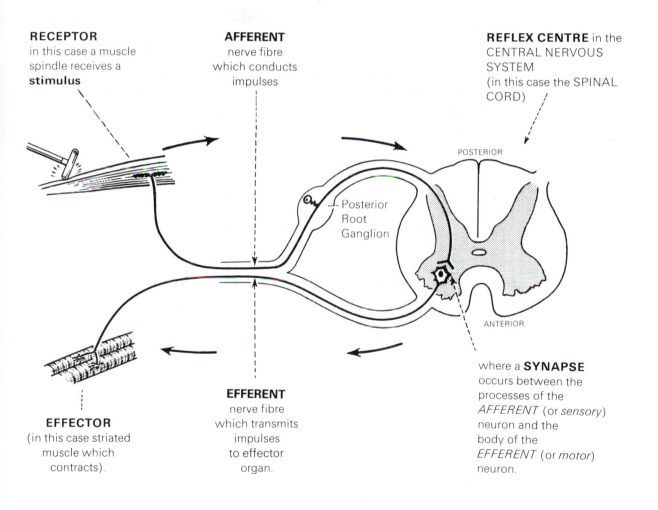

RECEPTOR
in this case a muscle
spindle receives a
stimulus

AFFERENT
nerve fibre
which conducts
impulses

REFLEX CENTRE in the
CENTRAL NERVOUS
SYSTEM
(in this case the SPINAL
CORD)

POSTERIOR

Posterior
Root
Ganglion

ANTERIOR

where a **SYNAPSE**
occurs between the
processes of the
AFFERENT (or *sensory*)
neuron and the
body of the
EFFERENT (or *motor*)
neuron.

EFFECTOR
(in this case striated
muscle which
contracts).

EFFERENT
nerve fibre
which transmits
impulses
to effector
organ.

Reflexes form the basis of all central nervous system (CNS) activity. They occur at all levels of the brain and spinal cord. Important bodily functions such as movements of respiration, digestion, etc., are all controlled through reflexes. We are made aware of some reflex acts; others occur without our knowledge.

STRETCH REFLEXES

In man a very few **reflex arcs** involve *two neurons only*. Two examples elicited by doctors when testing the nervous system are:-

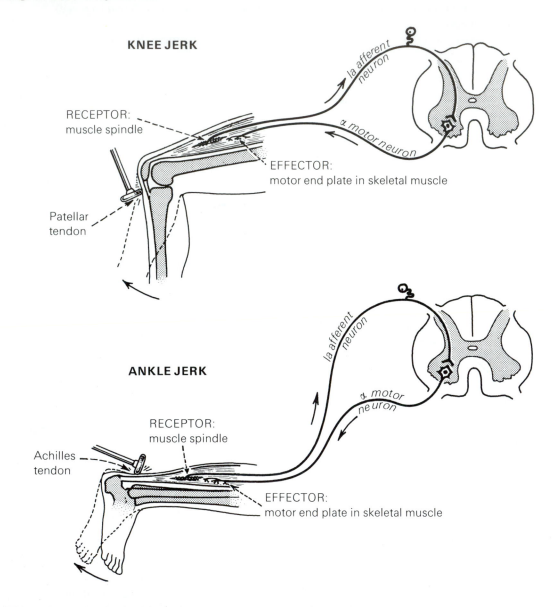

KNEE JERK

la afferent neuron

α motor neuron

RECEPTOR:
muscle spindle

EFFECTOR:
motor end plate in skeletal muscle

Patellar
tendon

ANKLE JERK

la afferent neuron

α motor neuron

RECEPTOR:
muscle spindle

Achilles
tendon

EFFECTOR:
motor end plate in skeletal muscle

When the tendon is sharply tapped the muscle is stretched (NB: the **stimulus** is by **stretch** of the **muscle spindle.**) Nerve impulses pass into the spinal cord – and out to the muscle which then contracts. This is a **monosynaptic** reflex.

In most **reflex arcs** in man *afferent* and *efferent neurons* are linked by at least one **interneuron**.

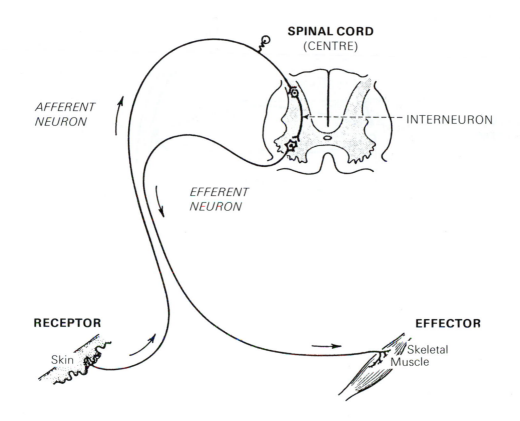

A chain of many interneurons is frequently found.

'EDIFICE' OF THE CNS *(After R.C. Garry)*

In the majority of reflex arcs in man a chain of many connector neurons is found. There may be link-ups with various levels of the brain and spinal cord.

This diagram gives a highly simplified concept of the type of **link-up** which can occur between different levels of the central nervous system.

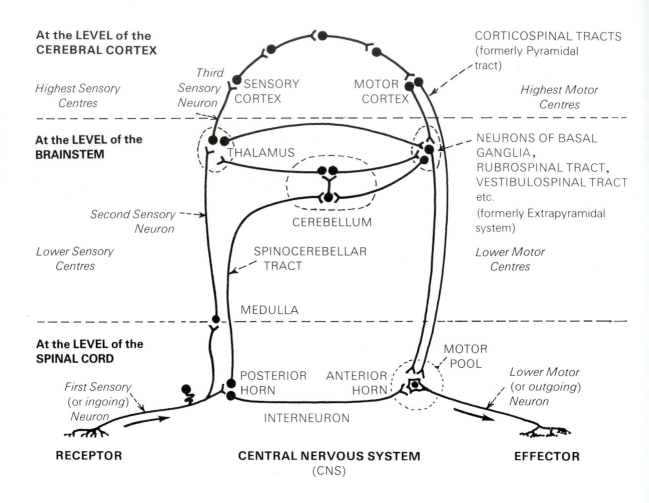

At the LEVEL of the CEREBRAL CORTEX

Highest Sensory Centres

Third Sensory Neuron SENSORY CORTEX   MOTOR CORTEX

CORTICOSPINAL TRACTS (formerly Pyramidal tract)

Highest Motor Centres

At the LEVEL of the BRAINSTEM

THALAMUS

NEURONS OF BASAL GANGLIA, RUBROSPINAL TRACT, VESTIBULOSPINAL TRACT etc. (formerly Extrapyramidal system)

Second Sensory Neuron

CEREBELLUM

Lower Sensory Centres

SPINOCEREBELLAR TRACT

Lower Motor Centres

MEDULLA

At the LEVEL of the SPINAL CORD

MOTOR POOL

First Sensory (or ingoing) Neuron

POSTERIOR HORN   ANTERIOR HORN

Lower Motor (or outgoing) Neuron

INTERNEURON

RECEPTOR   **CENTRAL NERVOUS SYSTEM** (CNS)   **EFFECTOR**

Every receptor neuron is thus potentially linked in the CNS with a large number of effector organs all over the body, and every effector neuron is similarly in communication with receptors all over the body.

Centres in the brain and brain stem can thus modify reflex acts which occur through the spinal cord. These centres can send 'suppressing' or 'facilitating' impulses along their pathways to the cells in the spinal cord.

Most **reflex actions** in man involve several **reflex arcs**.

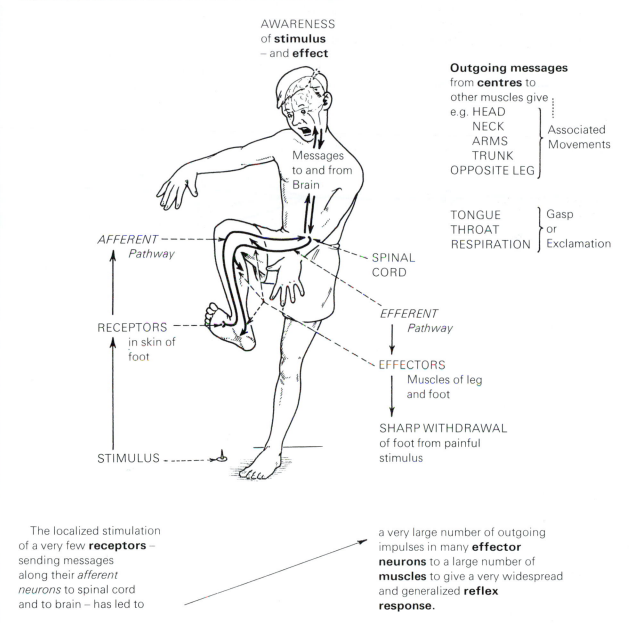

AWARENESS
of **stimulus**
– and **effect**

Messages
to and from
Brain

Outgoing messages
from **centres** to
other muscles give
e.g. HEAD
NECK
ARMS
TRUNK
OPPOSITE LEG } Associated Movements

TONGUE
THROAT
RESPIRATION } Gasp or Exclamation

AFFERENT
Pathway

SPINAL
CORD

RECEPTORS
in skin of
foot

EFFERENT
Pathway

EFFECTORS
Muscles of leg
and foot

STIMULUS

SHARP WITHDRAWAL
of foot from painful
stimulus

The localized stimulation of a very few **receptors** – sending messages along their *afferent* *neurons* to spinal cord and to brain – has led to a very large number of outgoing impulses in many **effector neurons** to a large number of **muscles** to give a very widespread and generalized **reflex response**.

This is possible because each receptor neuron is potentially connected within the central nervous system with many effector neurons.

ARRANGEMENT OF NEURONS

Some of the ways in which neurons can be linked are indicated here:-

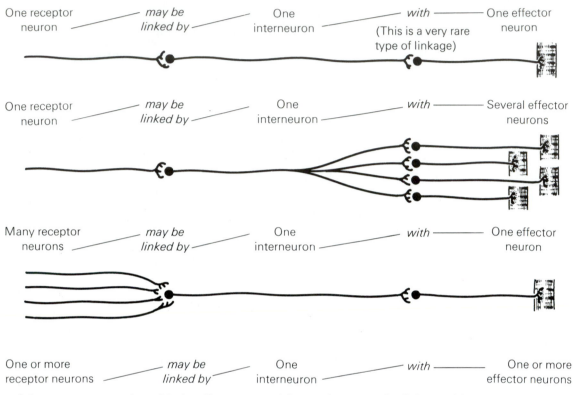

One receptor neuron — *may be linked by* — One interneuron — *with* — One effector neuron
(This is a very rare type of linkage)

One receptor neuron — *may be linked by* — One interneuron — *with* — Several effector neurons

Many receptor neurons — *may be linked by* — One interneuron — *with* — One effector neuron

One or more receptor neurons — *may be linked by* — One interneuron — *with* — One or more effector neurons

Other neurons synapsing with the effector neuron(s) may give a complex link-up with centres at higher and lower levels of the brain and spinal cord.

HIGHER LEVELS

LOWER LEVELS

Through such 'functional' link-ups, neurons in different parts of the central nervous system, when active, can influence each other. This makes it possible for **'conditioned' reflexes** to become established (for simple example see p. 67).

Such reflexes probably form the basis of all training so that it becomes difficult to say where **reflex** (or **involuntary**) behaviour ends and purely **voluntary** behaviour begins.

Man's awareness of the world is limited to those forms of energy, physical or chemical, to which he has receptors designed to respond. (Many 'events' in the Universe go undetected by man because he has no sense organ which can respond to them.)

Each sense organ is designed to respond to one type of stimulation.

EXTEROCEPTORS are stimulated by events in the **external environment.**

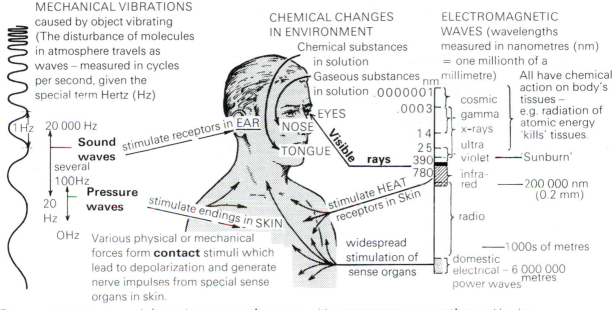

MECHANICAL VIBRATIONS caused by object vibrating (The disturbance of molecules in atmosphere travels as waves – measured in cycles per second, given the special term Hertz (Hz)

1 Hz   20 000 Hz
Sound waves stimulate receptors in EAR
several
100 Hz
Pressure waves stimulate endings in SKIN
20
Hz
OHz   Various physical or mechanical forces form **contact** stimuli which lead to depolarization and generate nerve impulses from special sense organs in skin.

CHEMICAL CHANGES IN ENVIRONMENT
Chemical substances in solution
Gaseous substances in solution
EYES
NOSE
TONGUE   Visible **rays**
stimulate HEAT receptors in Skin
widespread stimulation of sense organs

ELECTROMAGNETIC WAVES (wavelengths measured in nanometres (nm) = one millionth of a millimetre)
nm
.0000001
.0003
1 4
2 5
390
780
cosmic
gamma
x-rays
ultra violet
infra-red
radio
domestic electrical – 6 000 000 metres power waves

All have chemical action on body's tissues – e.g. radiation of atomic energy 'kills' tissues.
'Sunburn'
200 000 nm (0.2 mm)
1000s of metres

Exteroceptors may convey information to **consciousness** with **awareness** or **sensation** and lead to suitable **responses** planned in **cerebral cortex** or they may serve as *afferent* pathways for **reflex** (or **involuntary**) **action** with or without rising to consciousness.

PROPRIOCEPTORS are stimulated by changes in **locomotor system** of body

Labyrinth ....movements and position of head ⎤   Sense of **equilibrium** or
Muscles ....stretch   ⎮   **balance** and **awareness**
Tendons..............tension and stretch   ⎬   of **position** and **movement**
Joints ....stretch and pressure   ⎦   of body in space.

INTEROCEPTORS in **viscera** are stimulated by changes in **internal environment** (e.g. by distension in hollow organs).

Much of the proprio-and interoceptor information never rises to consciousness.
Overstimulation of any receptor can give rise to sensation of *pain*.
Most receptors show **adaptation** – if continously stimulated they send reduced numbers of impulses to the brain.

SMELL

Smell is a **chemical** sense, i.e. the receptors respond to **chemical stimuli**. To arouse the sensation a substance must first be in a **gaseous state** then go into **solution**.

The ORGAN OF SMELL is the NOSE ——— Also serves as the main air passage to respiratory system.

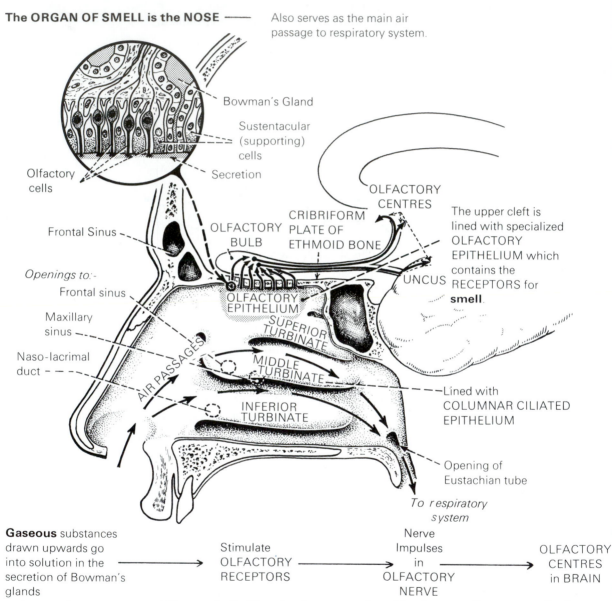

Bowman's Gland

Sustentacular (supporting) cells

Olfactory cells

Secretion

OLFACTORY CENTRES

CRIBRIFORM PLATE OF ETHMOID BONE

OLFACTORY BULB

Frontal Sinus

The upper cleft is lined with specialized OLFACTORY EPITHELIUM which contains the RECEPTORS for **smell**.

UNCUS

Openings to:-
Frontal sinus

OLFACTORY EPITHELIUM

SUPERIOR TURBINATE

Maxillary sinus

MIDDLE TURBINATE

Naso-lacrimal duct

AIR PASSAGES

INFERIOR TURBINATE

Lined with COLUMNAR CILIATED EPITHELIUM

Opening of Eustachian tube

To respiratory system

Gaseous substances drawn upwards go into solution in the secretion of Bowman's glands ——————→ Stimulate OLFACTORY RECEPTORS ——————→ Nerve Impulses in OLFACTORY NERVE ——————→ OLFACTORY CENTRES in BRAIN

Axons of receptors enter **olfactory bulb**. Terminations are gathered in clusters called **glomeruli** where they meet dendrites of **mitral** cells whose axons run back in olfactory nerve to terminate in **primary olfactory area** (uncus and adjacent parts of amygdaloid nucleus). These areas are linked to olfactory association areas, hypothalamus, autonomic nuclei and limbic system.

Taste is a **chemical** sense, i.e. receptors respond to **chemical stimuli**. To arouse the sensation a substance must be in **solution**.

The essential ORGAN OF TASTE is the TONGUE ——

The **voluntary** muscular organ concerned also in **mastication, swallowing** and **speech**.

Covered with STRATIFIED SQUAMOUS EPITHELIUM.

Projections on its upper surface are called

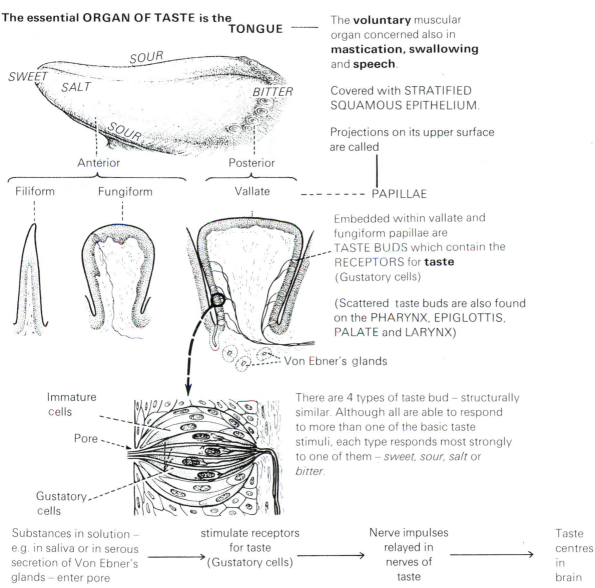

SOUR

SWEET

SALT

BITTER

SOUR

Anterior — Posterior

Filiform — Fungiform — Vallate - - - - - - - PAPILLAE

Embedded within vallate and fungiform papillae are TASTE BUDS which contain the RECEPTORS for **taste** (Gustatory cells)

(Scattered taste buds are also found on the PHARYNX, EPIGLOTTIS, PALATE and LARYNX)

Von Ebner's glands

Immature cells

Pore

Gustatory cells

There are 4 types of taste bud – structurally similar. Although all are able to respond to more than one of the basic taste stimuli, each type responds most strongly to one of them – *sweet, sour, salt* or *bitter*.

Substances in solution – e.g. in saliva or in serous secretion of Von Ebner's glands – enter pore	stimulate receptors for taste (Gustatory cells)	Nerve impulses relayed in nerves of taste	Taste centres in brain

Other tastes are probably due to combinations of these with smell or with ordinary skin sensations.

PATHWAYS AND CENTRES FOR TASTE

The **receptors** for **taste** are linked by a chain of three neurons with the **receiving centres** for **taste** in the **cerebral cortex**.

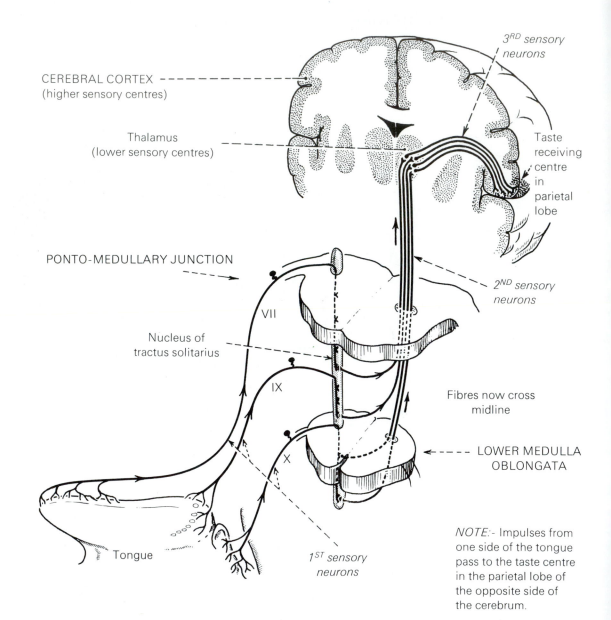

CEREBRAL CORTEX
(higher sensory centres)

Thalamus
(lower sensory centres)

PONTO-MEDULLARY JUNCTION

VII

Nucleus of
tractus solitarius

IX

X

Tongue

3RD sensory
neurons

Taste
receiving
centre
in
parietal
lobe

2ND sensory
neurons

Fibres now cross
midline

LOWER MEDULLA
OBLONGATA

1ST sensory
neurons

NOTE:- Impulses from
one side of the tongue
pass to the taste centre
in the parietal lobe of
the opposite side of
the cerebrum.

STRUCTURE
The eyeball has three coats:-

FUNCTION OF PARTS

1. OUTER COAT – SCLERA

PROTECTIVE LAYER

Tough fibrous tissue - - - - - of 'white of eye'

Preserves shape of eyeball and protects delicate inner layers.

Transparent CORNEA - - - - in front.
[Extrinsic muscles are - - - - attached to sclera.

Allows passage of **light rays**.
Permit and limit movements of eyeball within ORBIT.]

2. MIDDLE COAT – CHOROID

LAYER OF SUPPLY

Contains rich blood supply and melanin.
Circular opening at front – PUPIL.
Coloured muscular ring – IRIS – surrounds pupil.
CILIARY BODY.
CILIARY MUSCLE.
SUSPENSORY LIGAMENT suspends

Controls size of pupil; depth of focus; amount of light entering eye.

Produces AQUEOUS HUMOUR.
Circular – has sphincter-like action.
Relaxes to allow curvature of lens to alter for accommodation for **near vision**.

CRYSTALLINE LENS.

CHOROID – Posterior 5/6 of vascular coat.

Brings light rays to focus on light-sensitive RETINA

3. INNER COAT – the RETINA

LIGHT-SENSITIVE LAYER

Lines back of eye.
Contains RECEPTORS for **vision** - - - -

Highly specialized to respond to stimulation by light. Convert light energy into nerve impulses.

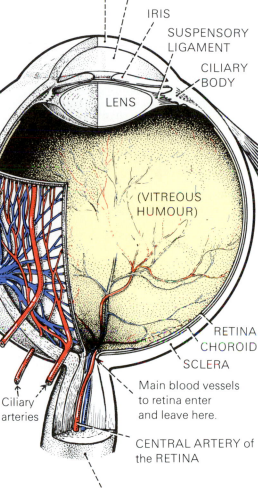

CORNEA

ANTERIOR CHAMBER contains aqueous humour

IRIS

SUSPENSORY LIGAMENT

CILIARY BODY

LENS

(VITREOUS HUMOUR)

RETINA
CHOROID
SCLERA

Main blood vessels to retina enter and leave here.

Ciliary arteries

CENTRAL ARTERY of the RETINA

OPTIC NERVE
Conveys these impulses to VISUAL CENTRES in OCCIPITAL (posterior) part of BRAIN.

PROTECTION OF THE EYE

The hidden posterior 4/5 of the eyeball is encased in a bony socket – the **orbital cavity.** A thick layer of areolar and adipose tissue forms a cushion between bone and eyeball. The exposed anterior 1/5 of the eyeball is protected from injury by:-

The EYELIDS - - - - - - - - - - - - - - - - - - close reflexly to protect eye from dust
    Fringed with EYELASHES   and other foreign particles.

CONJUNCTIVA   - - - - - - - - - - - - - - smooth surfaces which glide over each
   A delicate membrane lining eyelids   other when lids open and close.
and covering exposed surface of eye.

LACRIMAL GLANDS - - - - - - - - - - - - - continuously secrete TEARS. These
                          flow over, wash and lubricate surface
                          of eye. They contain an **enzyme –
                          lysozyme** – which destroys bacteria.
                          Secretion is controlled by parasympathetic
                          fibres of the facial (VII Cranial) nerve.

TARSAL GLANDS - secrete a fluid to prevent lids
                                  from sticking together

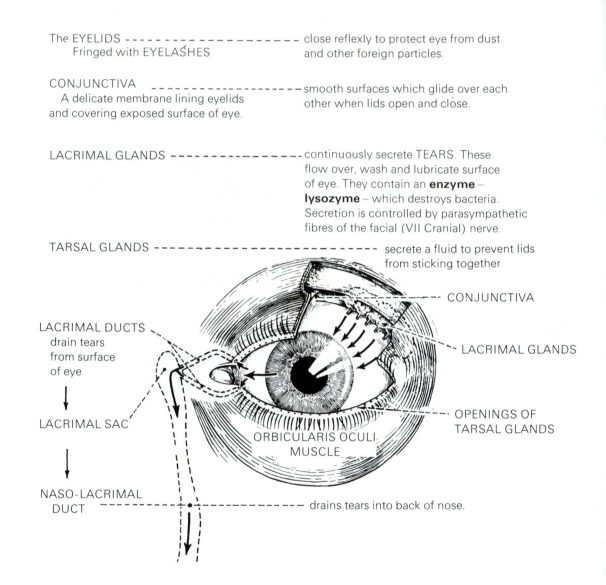

CONJUNCTIVA

LACRIMAL GLANDS

LACRIMAL DUCTS
drain tears
from surface
of eye

LACRIMAL SAC

OPENINGS OF
TARSAL GLANDS

ORBICULARIS OCULI
MUSCLE

NASO-LACRIMAL
DUCT - drains tears into back of nose.

The **eyeballs** are moved by **small muscles** which link the **selerotic coat** to the **bony socket.**

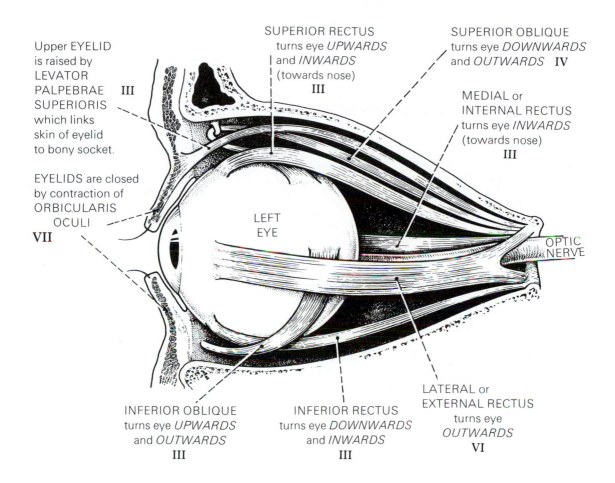

Upper EYELID
is raised by
LEVATOR
PALPEBRAE III
SUPERIORIS
which links
skin of eyelid
to bony socket.

EYELIDS are closed
by contraction of
ORBICULARIS
OCULI
VII

SUPERIOR RECTUS
turns eye *UPWARDS*
and *INWARDS*
(towards nose)
III

SUPERIOR OBLIQUE
turns eye *DOWNWARDS*
and *OUTWARDS* IV

MEDIAL or
INTERNAL RECTUS
turns eye *INWARDS*
(towards nose)
III

LEFT
EYE

OPTIC
NERVE

INFERIOR OBLIQUE
turns eye *UPWARDS*
and *OUTWARDS*
III

INFERIOR RECTUS
turns eye *DOWNWARDS*
and *INWARDS*
III

LATERAL or
EXTERNAL RECTUS
turns eye
OUTWARDS
VI

Acting together, the extrinsic muscles of the eyeballs can bring about **rotatory** movements of the eyes.

The extrinsic muscles are supplied by motor fibres from cranial nerves III, IV and VI.

CONTROL OF EYE MOVEMENTS

Both eyes normally move together so that images continue to fall on corresponding points of both retinae.

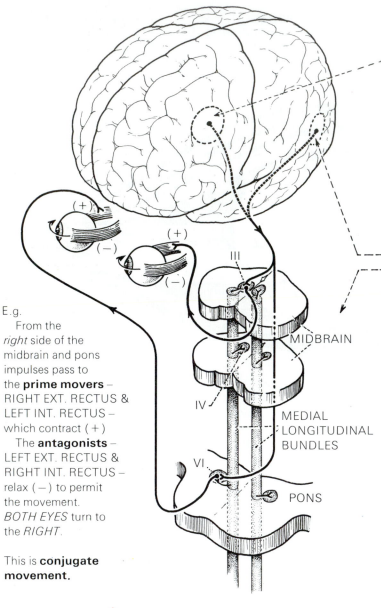

VOLUNTARY EYE MOVEMENTS are initiated in motor centres in FRONTAL LOBES.

Impulses from *one side* of the CEREBRAL CORTEX turn *both eyes* to the *other side* of visual field.

REFLEX EYE MOVEMENTS
Two groups – (1) Those in response to visual stimuli. (2) Those in response to non-visual stimuli.
In control of these are:-
CENTRES in OCCIPITAL LOBES:
CENTRES in MIDBRAIN and PONS which give rise to CRANIAL NERVES III, IV and VI.

Impulses from *one side* of the MIDBRAIN and PONS turn eyes to the *same side*.

These centres are *closely linked* with each other and with HIGHER and LOWER CENTRES in the central nervous system, so that the eyes are moved reflexly in response to many stimuli, e.g. loud noises or proprioceptive messages from vestibular organs.

E.g.
From the *right* side of the midbrain and pons impulses pass to the **prime movers** – RIGHT EXT. RECTUS & LEFT INT. RECTUS – which contract (+)
The **antagonists** – LEFT EXT. RECTUS & RIGHT INT. RECTUS – relax (−) to permit the movement.
BOTH EYES turn to the *RIGHT*.

This is **conjugate movement.**

III

MIDBRAIN

IV

MEDIAL LONGITUDINAL BUNDLES

VI

PONS

IRIS, LENS AND CILIARY BODY

The **IRIS** is a muscular diaphragm with a central opening – the PUPIL.

PUPIL
IRIS
CIRCULAR and RADIAL muscle fibres

IRIS controls amount of **light** entering the EYE

CIRCULAR smooth muscle fibres – SPHINCTER PUPILLAE – contract to make pupil smaller in bright light.

RADIAL fibres – DILATOR PUPILLAE – contract to make pupil larger with change from **light** to **dark**; **near** to **distant** vision (also with **fear** and **pain**).

ACCOMMODATION

When CILIARY MUSCLE contracts, SUSPENSORY LIGAMENT is slackened. Tension on CAPSULE of LENS is relaxed. Because lens is elastic, *ANTERIOR* surface springs forwards → LENS becomes more convex especially in its central part. This brings near objects into focus (accommodation reflex).

The **LENS** is a transparent biconvex crystalline disc.

Outer elastic CAPSULE blends with SUSPENSORY LIGAMENT which suspends LENS behind IRIS

The LENS and IRIS are attached to **CILIARY BODY** which contains fibres of SMOOTH MUSCLE

LENS

LENS brings **light rays** to a **focus** upside down on the RETINA. To do this under all conditions it must be able to alter its curvature.

CILIARY BODY

HYPOTHALAMUS

IRIS

CILIARY GANGLION

CORNEA

PUPIL   LENS

EDINGER-WESTPHAL NUCLEUS

SUSPENSORY LIGAMENT

SUPERIOR CERVICAL GANGLION

CILIARY MUSCLE

CERVICAL SYMPATHETIC CHAIN

NEAR RESPONSE   When subject looks at near objects, in addition to accommodation, visual axes converge and pupils constrict. The latter increases depth of focus.

These changes are brought about reflexly. The *ingoing* impulses travel in the optic nerves. The *outgoing* motor impulses travel in parasympathetic to ciliary body and sphincter pupillae and in sympathetic to dilator pupillae.

ACTION OF LENS

The normal lens brings light rays to a sharp focus upside down on the retina. It can do this whether we are looking at an object far away or one close at hand. The curvature increases reflexly to accommodate for near vision.

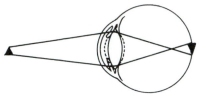

Rays of light coming from every point of a *DISTANT* object (over 20 feet away) are *PARALLEL*.

They pass through the CORNEA, AQUEOUS HUMOUR

and LENS which refract them to a sharp focus – upside down and reversed from side to side – on the retina.

The conscious mind learns to interpret the image and project it to its true position in space.

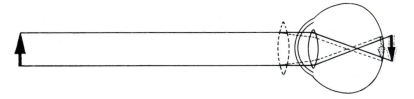

Rays of light coming from a *NEAR* object (less than 20 feet away) *DIVERGE* as they pass to the eye.

A more convex lens is required to bring these rays to a sharp focus on the retina.

If the EYEBALL is *too short*, rays from a distant object are brought into focus *BEHIND* the retina when the ciliary muscle is relaxed.

This is longsightedness or **hypermetropia**.

The longsighted eye has to accommodate even for distant vision: i.e. ciliary muscles contract to give a more convex lens and distant objects are then seen clearly. This limits amount of accommodating power left for near objects and the nearest point for sharp vision is then further away. It can be corrected by fitting spectacles with convex lenses.

If the EYEBALL is *too long*, rays from a distant object are brought into focus *IN FRONT* of the retina.

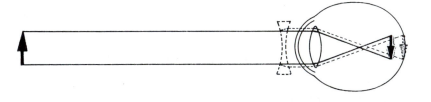

This is shortsightedness or **myopia** – only objects near the eye can be seen clearly.

It can be corrected by using a concave lens.

Part of the **retina** can be seen by means of an instrument – the **ophthalmoscope** – which shines a beam of light through the **pupil** of the eye on to the retina.

The part of the retina seen in this way is called the **fundus oculi.**

N.B. Observer uses his left eye to look into the patient's left eye and holds the ophthalmoscope in his left hand. Observer uses his right eye to look into patient's right eye and holds the ophthalmoscope in his right hand.

OPTIC DISC The nerve fibres from all parts of the retina converge on this area to leave the eyeball as the OPTIC NERVE. It has no RODS or CONES and therefore is not itself sensitive to light, hence it forms a 'BLIND SPOT' on the retina.

RETINAL BLOOD VESSELS enter or leave the eyeball here.

MACULA LUTEA – or 'YELLOW SPOT' – with

FOVEA CENTRALIS – area of acute vision – contains CONES only (the receptors stimulated in **bright** and **coloured light**). When we look at an object the eyes are directed so that the image will fall on the fovea of each eye.

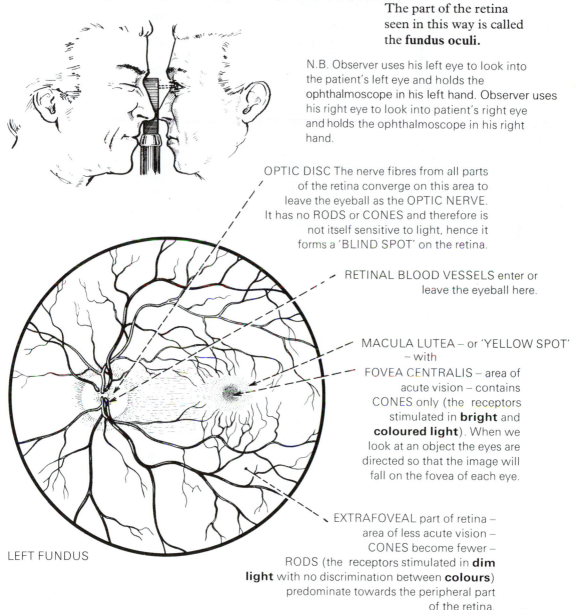

LEFT FUNDUS

EXTRAFOVEAL part of retina – area of less acute vision – CONES become fewer – RODS (the receptors stimulated in **dim light** with no discrimination between **colours**) predominate towards the peripheral part of the retina.

RETINA

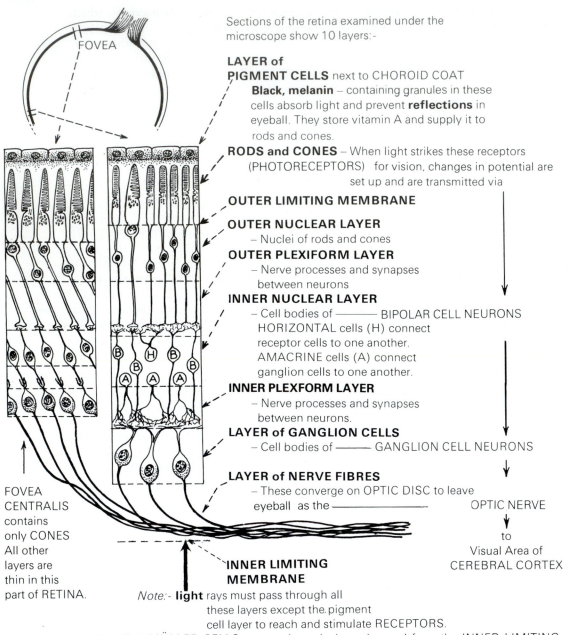

FOVEA

Sections of the retina examined under the microscope show 10 layers:-

LAYER of
PIGMENT CELLS next to CHOROID COAT
    Black, melanin – containing granules in these cells absorb light and prevent **reflections** in eyeball. They store vitamin A and supply it to rods and cones.

RODS and CONES – When light strikes these receptors (PHOTORECEPTORS) for vision, changes in potential are set up and are transmitted via

OUTER LIMITING MEMBRANE

OUTER NUCLEAR LAYER
    – Nuclei of rods and cones

OUTER PLEXIFORM LAYER
    – Nerve processes and synapses between neurons

INNER NUCLEAR LAYER
    – Cell bodies of ——— BIPOLAR CELL NEURONS
    HORIZONTAL cells (H) connect receptor cells to one another.
    AMACRINE cells (A) connect ganglion cells to one another.

INNER PLEXFORM LAYER
    – Nerve processes and synapses between neurons.

LAYER of GANGLION CELLS
    – Cell bodies of ——— GANGLION CELL NEURONS

LAYER of NERVE FIBRES
    – These converge on OPTIC DISC to leave eyeball as the ——————— OPTIC NERVE

to
Visual Area of
CEREBRAL CORTEX

FOVEA
CENTRALIS
contains
only CONES
All other
layers are
thin in this
part of RETINA.

INNER LIMITING
MEMBRANE

Note:- **light** rays must pass through all these layers except the pigment cell layer to reach and stimulate RECEPTORS.

Supporting cells called MÜLLER CELLS extend through the retina and form the INNER LIMITING MEMBRANE on the inner surface of the retina and the OUTER LIMITING MEMBRANE in the receptor layer.

White light is really due to the fusion of *coloured lights*. These coloured lights are separated by shining a beam of white light through a glass prism. This is called the **visible spectrum**.

780 nm	700 nm		600 nm			500 nm			390 nm
RED		ORANGE	YELLOW	YELL.-GR.	GREEN	BLUE-GREEN	BLUE	VIOLET	

If light is bright or intense the spectrum appears brightest to man's eye in the orange band (610 nm).

If brightness or intensity of light source is gradually reduced, colour perception is gradually lost and the spectrum appears as a luminous band with a very dark area in the red band and with its brightest part in the green band (530 nm).

Visual receptors contain **pigments** whose structure changes in the presence of light. This results in a cascade of reactions (including formation of **metarhodopsin II** and decreased intracellular cGMP) leading to the closure of Na^+ channels in the receptor membrane. This, along with Ca^{2+} release, produces **hyperpolarization** of the receptor membrane and this potential change spreads over bipolar cells and horizontal cells and induces the production of action potentials in amacrine and ganglion cells. Nerve impulses then travel to the visual cortex.

SCOTOPIC VISION is vision in *dim light*. It depends on the **rods**.

Rods are of one type ———— and give ———— **MONOCHROMATIC VISION**

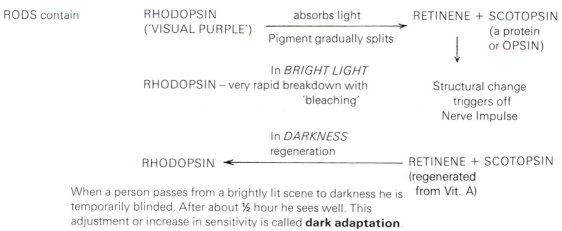

When a person passes from a brightly lit scene to darkness he is temporarily blinded. After about ½ hour he sees well. This adjustment or increase in sensitivity is called **dark adaptation**.

As light brightness or intensity increases rods lose their sensitivity and cease to respond.

MECHANISM OF VISION

Photopic vision is vision in *bright light*. It depends on the **cones.**
Cones are of three types ——— giving **TRICHROMATIC VISION.**

Each type with a   Each contains RETINENE plus a protein. Each
different   protein is like scotopsin in the rods but differs
photosensitive   slightly from it.

**VISUAL
PIGMENT**   with its own wavelength to which it is sensitive, which
it absorbs and by which its structure is changed, resulting
in receptor stimulation.

'*RED*'   responds maximally to *YELLOW-ORANGE* light (565 nm)
'*GREEN*' responds maximally to —— *GREEN* light (535 nm)
'*BLUE*'   responds maximally to —— *BLUE* light (440 nm)

All three types of CONE are stimulated in roughly equal proportions when *WHITE* light falls on retina:

The sensation of any *other colour* is determined by the relative frequency of the impulses from each of these cone systems.

The various types of colour blindness could be explained in terms of the absence or deficiency of one or more of these special receptors.

A colour sensation has three qualities:-

Hue ——   depends largely on wavelength.
Saturation – purity –
A 'saturated' colour has no white light mixed with it.
An 'unsaturated' colour has some white light mixed with it.
Intensity –   brightness – depends largely on 'strength' of the light.

As the intensity of light is reduced the cones cease to respond and the rods take over.

When a person passes from darkness to bright light he is dazzled but after a short time he sees well again.

This adjustment or decrease in sensitivity on exposure to bright light is called **light adaptation** but is, strictly speaking, the disappearance of dark adaptation.

VISUAL PATHWAYS TO THE BRAIN

The **receptors** for **vision** are linked by a chain of neurons with **receiving** and **integrating** centres in the **occipital lobes** of the **cerebral cortex**.

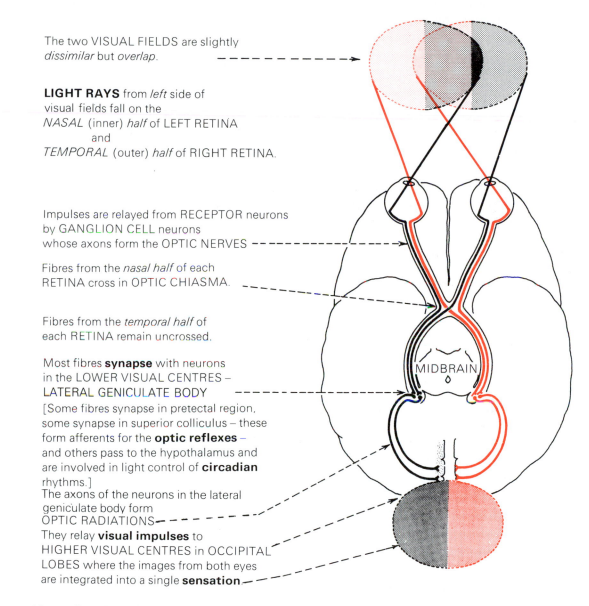

The two VISUAL FIELDS are slightly *dissimilar* but *overlap*.

LIGHT RAYS from *left* side of visual fields fall on the *NASAL* (inner) *half* of LEFT RETINA and *TEMPORAL* (outer) *half* of RIGHT RETINA.

Impulses are relayed from RECEPTOR neurons by GANGLION CELL neurons whose axons form the OPTIC NERVES

Fibres from the *nasal half* of each RETINA cross in OPTIC CHIASMA.

Fibres from the *temporal half* of each RETINA remain uncrossed.

Most fibres **synapse** with neurons in the LOWER VISUAL CENTRES – LATERAL GENICULATE BODY [Some fibres synapse in pretectal region, some synapse in superior colliculus – these form afferents for the **optic reflexes** – and others pass to the hypothalamus and are involved in light control of **circadian** rhythms.]
The axons of the neurons in the lateral geniculate body form OPTIC RADIATIONS
They relay **visual impulses** to HIGHER VISUAL CENTRES in OCCIPITAL LOBES where the images from both eyes are integrated into a single **sensation**.

MIDBRAIN

Note:- One side of the **occipital cortex** receives impressions from the **field of vision** on the opposite side.

STEREOSCOPIC VISION

When we look at some object or scene the view seen by the **right eye** is slightly different from the view seen by the **left eye**.

These two **dissimilar retinal images** are fused in the visual centres of the brain to give a 3-dimensional picture – an appreciation of *depth* as well as of *height* and *width*.

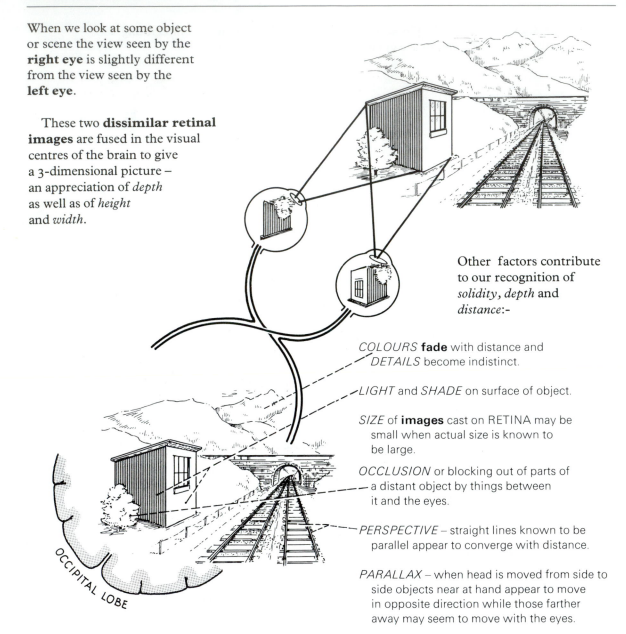

Other factors contribute to our recognition of *solidity*, *depth* and *distance*:-

COLOURS **fade** with distance and *DETAILS* become indistinct.

LIGHT and *SHADE* on surface of object.

SIZE of **images** cast on RETINA may be small when actual size is known to be large.

OCCLUSION or blocking out of parts of a distant object by things between it and the eyes.

PERSPECTIVE – straight lines known to be parallel appear to converge with distance.

PARALLAX – when head is moved from side to side objects near at hand appear to move in opposite direction while those farther away may seem to move with the eyes.

By complex mental processes these points are interpreted in terms of distance and depth.

When **light** falls on the **retina** the **pupils contract**.

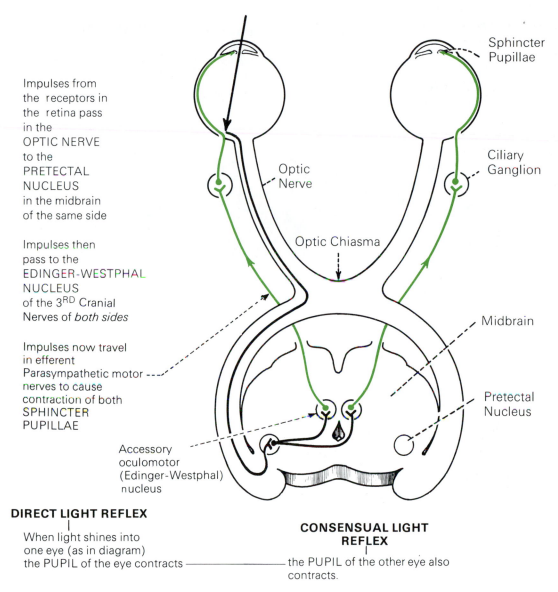

Impulses from
the receptors in
the retina pass
in the
OPTIC NERVE
to the
PRETECTAL
NUCLEUS
in the midbrain
of the same side

Impulses then
pass to the
EDINGER-WESTPHAL
NUCLEUS
of the 3RD Cranial
Nerves of *both sides*

Impulses now travel
in efferent
Parasympathetic motor
nerves to cause
contraction of both
SPHINCTER
PUPILLAE

Accessory
oculomotor
(Edinger-Westphal)
nucleus

Sphincter
Pupillae

Optic
Nerve

Ciliary
Ganglion

Optic Chiasma

Midbrain

Pretectal
Nucleus

DIRECT LIGHT REFLEX

When light shines into
one eye (as in diagram)
the PUPIL of the eye contracts —————

CONSENSUAL LIGHT REFLEX

————— the PUPIL of the other eye also
contracts.

This cuts down the amount of light entering the eyes and protects the retinae from excessive stimulation. It also increases depth of focus and improves the sharpness of the Retinal images.

The **Argyll-Robertson pupil** is one of the signs of cerebral syphilis. In this condition the pupil does *not* constrict in response to light but *does* constrict as part of the **near response** (p. 243). This indicates that the pathways for these two constrictor responses are different.

251

EAR

The ear has 3 separate parts, each with different roles in the mechanism of **hearing**:-

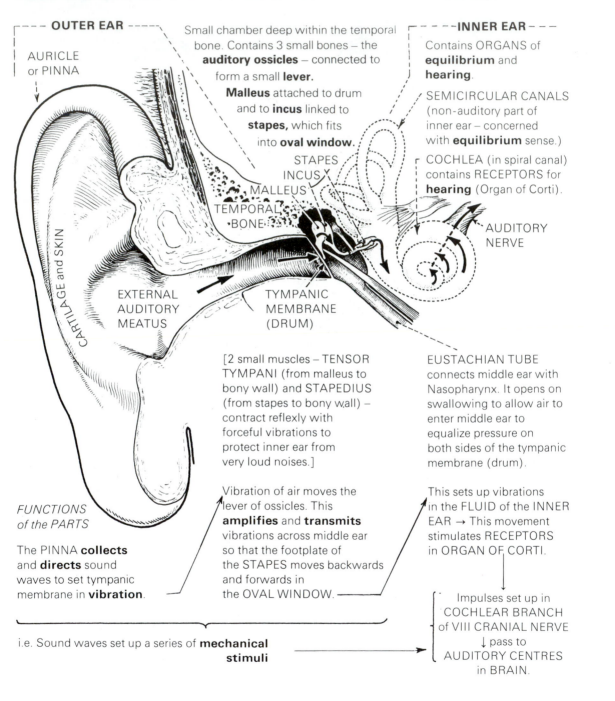

- - - OUTER EAR - - - - - -

AURICLE or PINNA

Small chamber deep within the temporal bone. Contains 3 small bones – the **auditory ossicles** – connected to form a small **lever.** **Malleus** attached to drum and to **incus** linked to **stapes,** which fits into **oval window.**

- - - -INNER EAR - - -

Contains ORGANS of **equilibrium** and **hearing.**

SEMICIRCULAR CANALS (non-auditory part of inner ear – concerned with **equilibrium** sense.)

COCHLEA (in spiral canal) contains RECEPTORS for **hearing** (Organ of Corti).

STAPES
INCUS
MALLEUS
TEMPORAL
BONE

AUDITORY NERVE

CARTILAGE and SKIN

EXTERNAL AUDITORY MEATUS

TYMPANIC MEMBRANE (DRUM)

[2 small muscles – TENSOR TYMPANI (from malleus to bony wall) and STAPEDIUS (from stapes to bony wall) – contract reflexly with forceful vibrations to protect inner ear from very loud noises.]

EUSTACHIAN TUBE connects middle ear with Nasopharynx. It opens on swallowing to allow air to enter middle ear to equalize pressure on both sides of the tympanic membrane (drum).

FUNCTIONS of the PARTS

The PINNA **collects** and **directs** sound waves to set tympanic membrane in **vibration**.

Vibration of air moves the lever of ossicles. This **amplifies** and **transmits** vibrations across middle ear so that the footplate of the STAPES moves backwards and forwards in the OVAL WINDOW.

This sets up vibrations in the FLUID of the INNER EAR → This movement stimulates RECEPTORS in ORGAN OF CORTI.

Impulses set up in COCHLEAR BRANCH of VIII CRANIAL NERVE ↓ pass to AUDITORY CENTRES in BRAIN.

i.e. Sound waves set up a series of **mechanical stimuli**

The cochlea is the essential organ of **hearing**.

It consists of:-

The BONY COCHLEA which spirals 2¾ times round central pillar of bone.

The MEMBRANOUS COCHLEA which is enclosed between the VESTIBULAR and BASILAR membranes

These spiral compartments are filled with FLUID

STRIA VASCULARIS (pigmented, granular cells with profuse blood supply) secretes **endolymph** of SCALA MEDIA

Because of ionic differences, the interior of the hair cells is 70 mV −ve to the Perilymph and 150 mV −ve to the Endolymph.

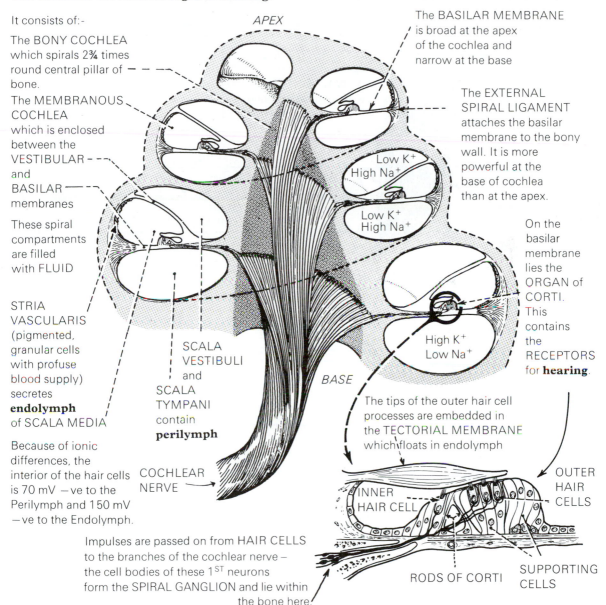

APEX

The BASILAR MEMBRANE is broad at the apex of the cochlea and narrow at the base

The EXTERNAL SPIRAL LIGAMENT attaches the basilar membrane to the bony wall. It is more powerful at the base of cochlea than at the apex.

Low K^+ High Na^+

Low K^+ High Na^+

On the basilar membrane lies the ORGAN of CORTI. This contains the RECEPTORS for **hearing**.

SCALA VESTIBULI and SCALA TYMPANI contain **perilymph**

BASE

High K^+ Low Na^+

COCHLEAR NERVE

The tips of the outer hair cell processes are embedded in the TECTORIAL MEMBRANE which floats in endolymph

INNER HAIR CELL

OUTER HAIR CELLS

Impulses are passed on from HAIR CELLS to the branches of the cochlear nerve – the cell bodies of these 1ST neurons form the SPIRAL GANGLION and lie within the bone here.

RODS OF CORTI

SUPPORTING CELLS

Inner hair cells are probably the primary sensory cells which generate action potentials. **Outer** hair cells have a parasympathetic innervation. May influence vibration pattern to improve hearing.

MECHANISM OF HEARING

This is most readily understood if the **cochlea** is imagined as straightened out:-

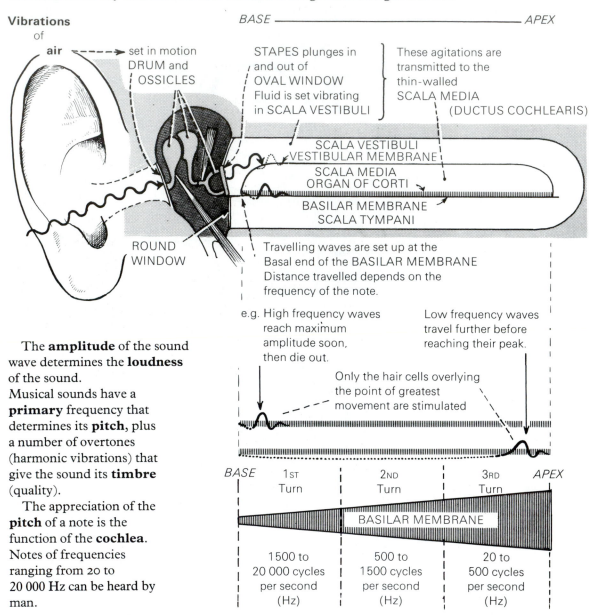

BASE ————————————————————— APEX

Vibrations of **air** ⤏ set in motion DRUM and OSSICLES

STAPES plunges in and out of OVAL WINDOW Fluid is set vibrating in SCALA VESTIBULI

These agitations are transmitted to the thin-walled SCALA MEDIA (DUCTUS COCHLEARIS)

SCALA VESTIBULI
VESTIBULAR MEMBRANE
SCALA MEDIA
ORGAN OF CORTI
BASILAR MEMBRANE
SCALA TYMPANI

ROUND WINDOW

Travelling waves are set up at the Basal end of the BASILAR MEMBRANE Distance travelled depends on the frequency of the note.

e.g. High frequency waves reach maximum amplitude soon, then die out.

Low frequency waves travel further before reaching their peak.

Only the hair cells overlying the point of greatest movement are stimulated

BASE | 1ST Turn | 2ND Turn | 3RD Turn | APEX

BASILAR MEMBRANE

1500 to 20 000 cycles per second (Hz) | 500 to 1500 cycles per second (Hz) | 20 to 500 cycles per second (Hz)

The **amplitude** of the sound wave determines the **loudness** of the sound.

Musical sounds have a **primary** frequency that determines its **pitch**, plus a number of overtones (harmonic vibrations) that give the sound its **timbre** (quality).

The appreciation of the **pitch** of a note is the function of the **cochlea**. Notes of frequencies ranging from 20 to 20 000 Hz can be heard by man.

The potential differences between the hair cells and the endolymph and perilymph are responsible for **depolarization** of the hair cells when moved in one direction and **hyperpolarization** when moved in the other.

The **receptors** for hearing are linked by a chain of **neurons** with the **receiving centres** for hearing in the **temporal lobes** of the **cerebral cortex**.

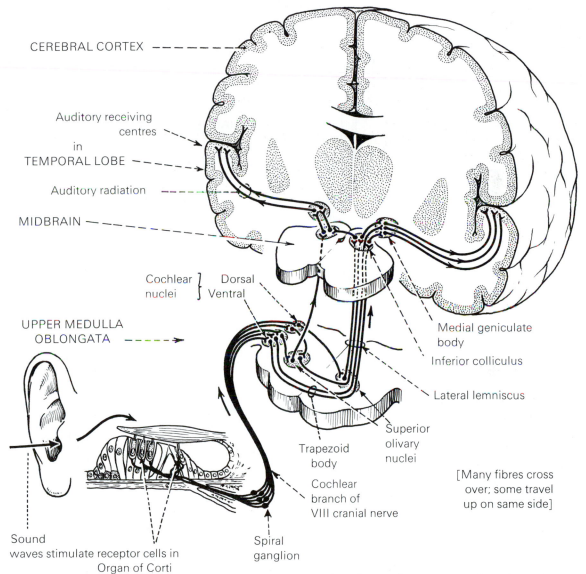

CEREBRAL CORTEX

Auditory receiving centres

in

TEMPORAL LOBE

Auditory radiation

MIDBRAIN

Cochlear nuclei } Dorsal Ventral

UPPER MEDULLA OBLONGATA

Medial geniculate body

Inferior colliculus

Lateral lemniscus

Superior olivary nuclei

Trapezoid body

Cochlear branch of VIII cranial nerve

Spiral ganglion

Sound waves stimulate receptor cells in Organ of Corti

[Many fibres cross over; some travel up on same side]

Impulses travel in vestibulocochlear nerve and are relayed as shown. Some are sent into the **reticular activating system**.

SPECIAL PROPRIOCEPTORS

The 'special' proprioceptors of the body are found in the non-auditory part of the inner ear – the labyrinth. They are stimulated by movements or change of position of the head in space enabling balance to be maintained.

Three **SEMICIRCULAR CANALS** – in each inner ear – one in each of three planes of space – contain receptors.

Membranous tubes containing endolymph embedded in bone surrounded by perilymph

These receptors, situated in the ampulla of each canal, are stimulated mechanically by the *starting or stopping of rotatory movements* of the head in space.

One end of each canal has a swelling – the AMPULLA

CRISTA AMPULLARIS

VESTIBULAR BRANCH OF VIII CRANIAL NERVE

SUPERIOR

POSTERIOR

LATERAL

COCHLEA

MACULA

SACCULE

ENDOLYMPHATIC DUCT

Both ends of each canal open into the UTRICLE

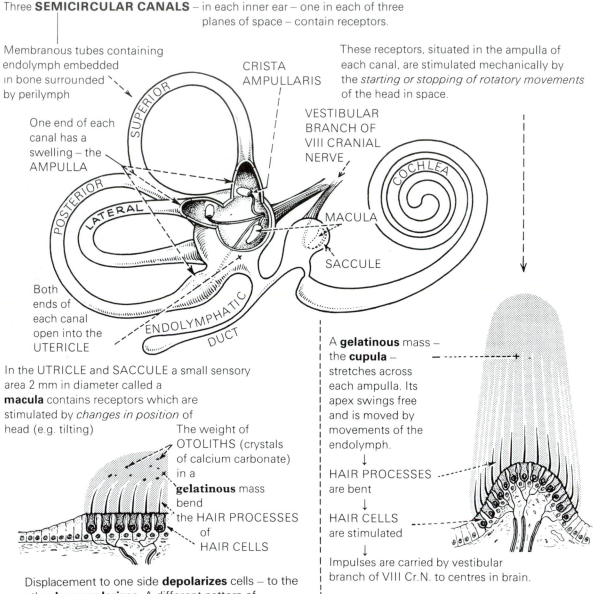

In the UTRICLE and SACCULE a small sensory area 2 mm in diameter called a **macula** contains receptors which are stimulated by *changes in position* of head (e.g. tilting)

The weight of OTOLITHS (crystals of calcium carbonate) in a **gelatinous** mass bend the HAIR PROCESSES of HAIR CELLS

A **gelatinous** mass – the **cupula** – stretches across each ampulla. Its apex swings free and is moved by movements of the endolymph.
↓
HAIR PROCESSES are bent
↓
HAIR CELLS are stimulated
↓
Impulses are carried by vestibular branch of VIII Cr.N. to centres in brain.

Displacement to one side **depolarizes** cells – to the other **hyperpolarizes**. A different pattern of excitation occurs for each position of the head. Conveyed by fibres of vestibular branch of VIII cranial nerve to brain.

ORGAN OF EQUILIBRIUM: MECHANISM OF ACTION

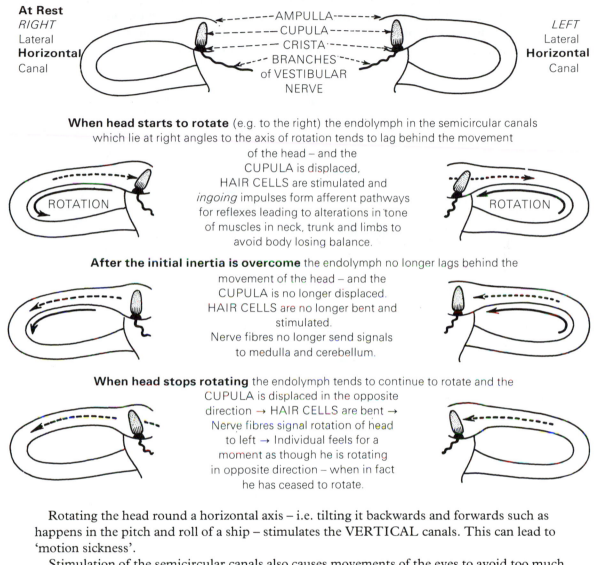

At Rest
RIGHT
Lateral
Horizontal
Canal

AMPULLA
CUPULA
CRISTA
BRANCHES
of VESTIBULAR
NERVE

LEFT
Lateral
Horizontal
Canal

When head starts to rotate (e.g. to the right) the endolymph in the semicircular canals which lie at right angles to the axis of rotation tends to lag behind the movement of the head – and the CUPULA is displaced, HAIR CELLS are stimulated and *ingoing* impulses form afferent pathways for reflexes leading to alterations in tone of muscles in neck, trunk and limbs to avoid body losing balance.

ROTATION

ROTATION

After the initial inertia is overcome the endolymph no longer lags behind the movement of the head – and the CUPULA is no longer displaced. HAIR CELLS are no longer bent and stimulated. Nerve fibres no longer send signals to medulla and cerebellum.

When head stops rotating the endolymph tends to continue to rotate and the CUPULA is displaced in the opposite direction → HAIR CELLS are bent → Nerve fibres signal rotation of head to left → Individual feels for a moment as though he is rotating in opposite direction – when in fact he has ceased to rotate.

Rotating the head round a horizontal axis – i.e. tilting it backwards and forwards such as happens in the pitch and roll of a ship – stimulates the VERTICAL canals. This can lead to 'motion sickness'.

Stimulation of the semicircular canals also causes movements of the eyes to avoid too much displacement of the image being cast on the retinae.

During rotation there is a slow movement of the eyes in the direction opposite to that of rotation, then a quick return to the normal position. This is **nystagmus**. It occurs continuously while rotating and continues for a short time after movement has ceased.

The semicircular canal mechanism predicts ahead of time that mal-equilibrium is going to occur. It allows equilibrium centres to make preventive adjustments.

VESTIBULAR PATHWAYS TO BRAIN

The special proprioceptive end-organs in the **labyrinth** are linked through the **vestibular nuclei** with **receiving** and **integrating centres** in the **cerebellum**, and with **motor centres** in the **midbrain** and **spinal cord** through which they initiate reflex muscular movements of eyes, head and neck and trunk and limb muscles to adjust balance and posture.

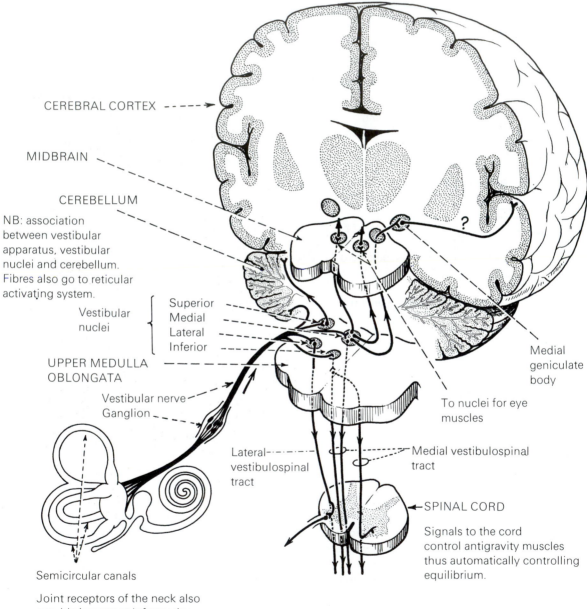

CEREBRAL CORTEX

MIDBRAIN

CEREBELLUM

NB: association between vestibular apparatus, vestibular nuclei and cerebellum. Fibres also go to reticular activating system.

Vestibular nuclei

Superior
Medial
Lateral
Inferior

UPPER MEDULLA OBLONGATA

Vestibular nerve
Ganglion

Medial geniculate body

To nuclei for eye muscles

Lateral vestibulospinal tract

Medial vestibulospinal tract

SPINAL CORD

Signals to the cord control antigravity muscles thus automatically controlling equilibrium.

Semicircular canals

Joint receptors of the neck also provide important information needed for equilibrium.

Proprioceptors are the sense organs stimulated by **movement** of the body itself. They make us aware of the movement or position of the body in space and of the various parts of the body to each other. They are important as ingoing afferent pathways in reflexes for adjusting posture and tone.

General proprioceptors are found in **skeletal muscles, tendons** and **joints**.

GOLGI ORGAN – in tendons –
stimulated by tension which occurs
when muscle is *STRETCHED* and
when it is *CONTRACTED*

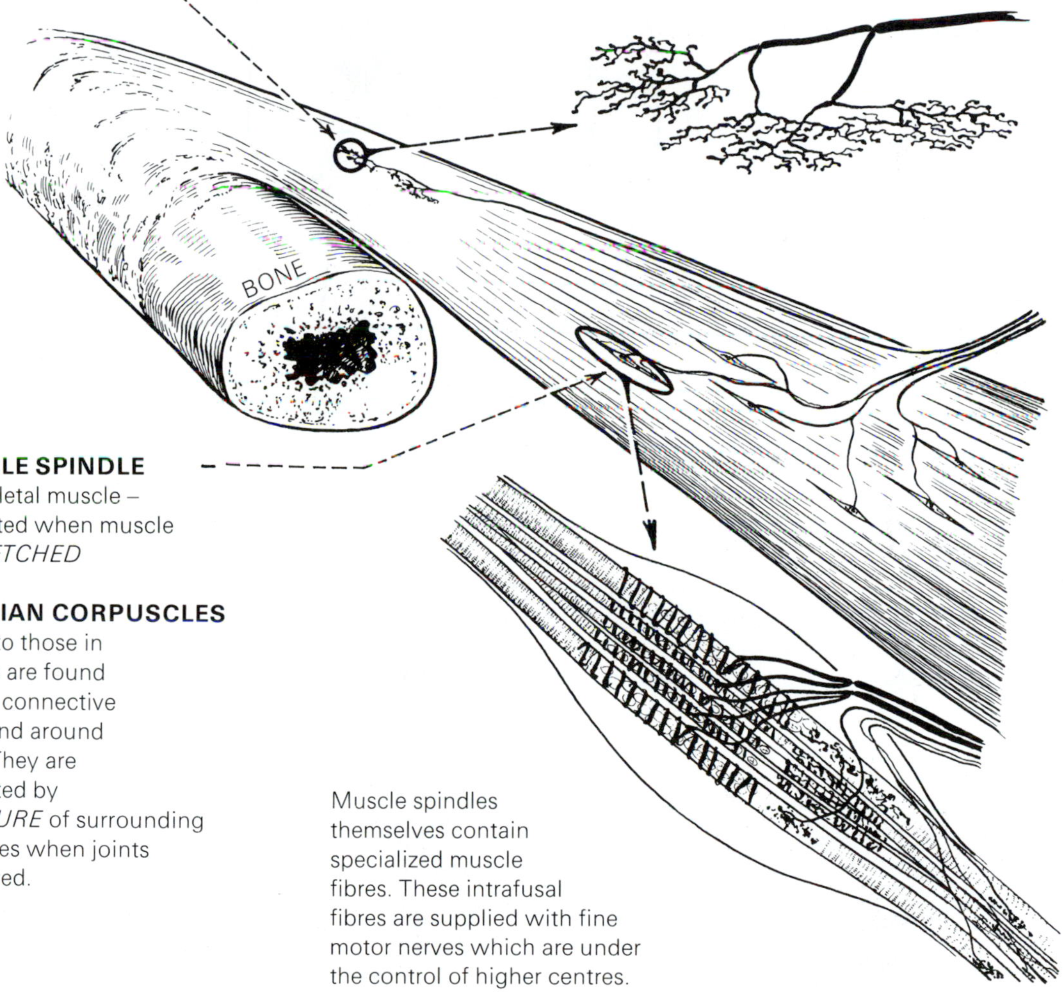

BONE

MUSCLE SPINDLE
– in skeletal muscle –
stimulated when muscle
is *STRETCHED*

PACINIAN CORPUSCLES
similar to those in
the skin are found
in deep connective
tissue and around
joints. They are
stimulated by
PRESSURE of surrounding
structures when joints
are moved.

Muscle spindles
themselves contain
specialized muscle
fibres. These intrafusal
fibres are supplied with fine
motor nerves which are under
the control of higher centres.

PROPRIOCEPTOR PATHWAYS TO BRAIN

General proprioceptor end-organs (receptors) may be linked with centres in (a) **cerebellum** or (b) **parietal lobe** of **cerebral cortex** by a chain of three neurons.

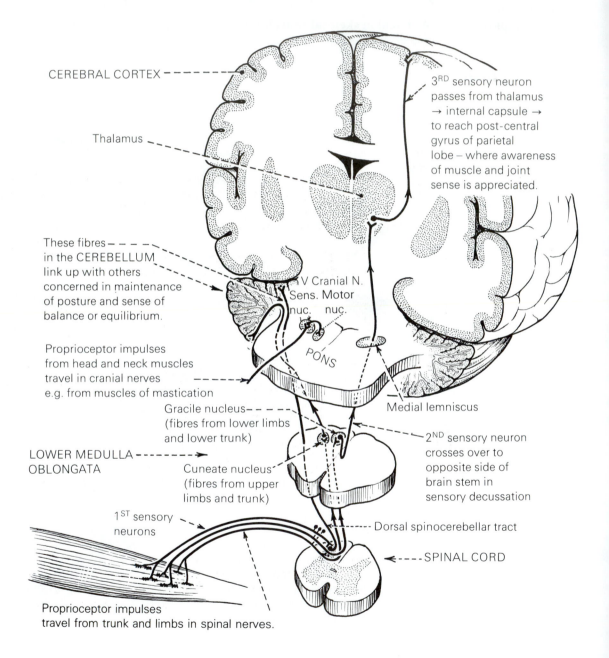

CEREBRAL CORTEX

Thalamus

3RD sensory neuron passes from thalamus → internal capsule → to reach post-central gyrus of parietal lobe – where awareness of muscle and joint sense is appreciated.

These fibres in the CEREBELLUM link up with others concerned in maintenance of posture and sense of balance or equilibrium.

Proprioceptor impulses from head and neck muscles travel in cranial nerves e.g. from muscles of mastication

V Cranial N.
Sens. Motor
nuc. nuc.

PONS

Gracile nucleus (fibres from lower limbs and lower trunk)

Medial lemniscus

LOWER MEDULLA OBLONGATA

Cuneate nucleus (fibres from upper limbs and trunk)

2ND sensory neuron crosses over to opposite side of brain stem in sensory decussation

1ST sensory neurons

Dorsal spinocerebellar tract

SPINAL CORD

Proprioceptor impulses travel from trunk and limbs in spinal nerves.

There are *five basic skin sensations* – **touch, pressure, pain, warmth,** and **cold.** There is much controversy as to how these are registered. In some areas they appear to be served by special nerve endings (sensory receptors or end-organs) in the skin. These receptors are not uniformly distributed over the whole body surface. (E.g. touch 'endings' are very numerous in hands and feet but are much less frequent in the skin of the back.)

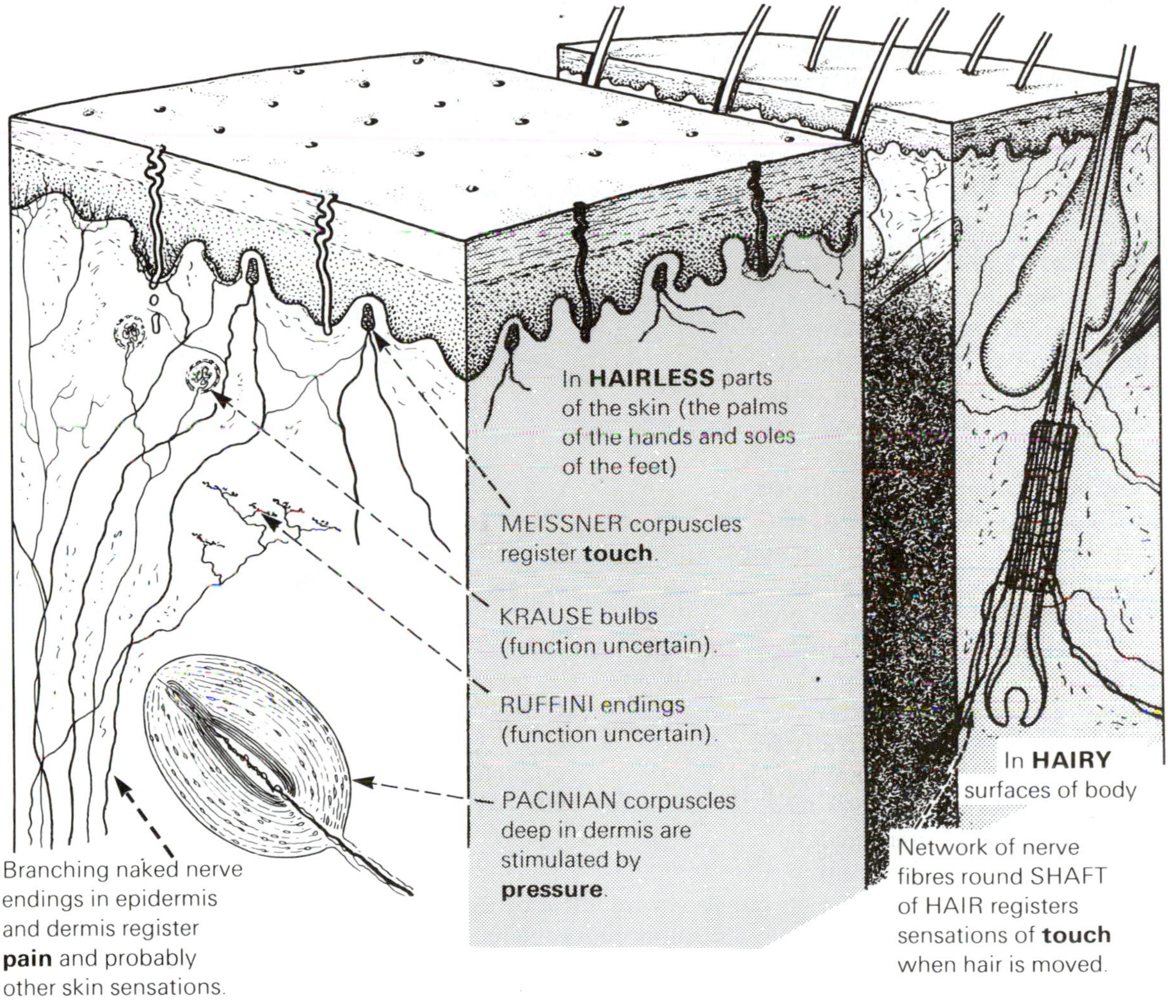

In **HAIRLESS** parts of the skin (the palms of the hands and soles of the feet)

MEISSNER corpuscles register **touch**.

KRAUSE bulbs (function uncertain).

RUFFINI endings (function uncertain).

PACINIAN corpuscles deep in dermis are stimulated by **pressure**.

Branching naked nerve endings in epidermis and dermis register **pain** and probably other skin sensations.

In **HAIRY** surfaces of body

Network of nerve fibres round SHAFT of HAIR registers sensations of **touch** when hair is moved.

Tickling, itching, softness, hardness, wetness are probably due to stimulation of two or more of these special endings and to a blending of the sensations in the brain.

Much has still to be discovered about skin receptors. Still to be explained, for example, is why in the cornea and the skin of the ear several types of sensation can be appreciated without the presence of specialized receptors.

261

PAIN

Pain is an important symptom which commonly causes a patient to consult a doctor. NB: pain is a **sensation** which is felt when **nociceptors** are stimulated by tissue damage. The terms pain and nociception are often used synonymously.

Nociceptors are free nerve endings which are stimulated by excessive heat, mechanical stimuli or chemicals, e.g. bradykinin released from γ-globulins as a result of cell damage.

Pain is either *fast* pain................or...............*slow* pain

Short, sharp, well localized, e.g. pin prick or knife cut. Conveyed by small myelinated fast Aδ or Group III fibres (45 metres/s).	Burning, aching, poorly localized, associated with tissue destruction. Conveyed by unmyelinated slower C or Group IV fibres (1 metre/s).

Nociceptive afferent fibres after synapsing in the dorsal horn ascend to the reticular formation, thalamus, sensory cortex and autonomic centres.

GATE CONTROL THEORY

In the dorsal horn of the spinal cord onward transmission of nerve impulses from nociceptive afferent fibres via **T** or **transmission cells** depends on the activity of large sensory afferent neurons from peripheral touch receptors. Impulses in these touch sensory afferents can block the pain pathway by stimulating an interneuron in the **substantia gelatinosa** which will presynaptically inhibit all input to the T-cell. If impulse traffic in the nociceptor afferents is greater than in the touch receptor afferents then afferent impulses will pass on and pain will be appreciated. This theory explains why rubbing your skin in a painful area can help to lessen the pain.

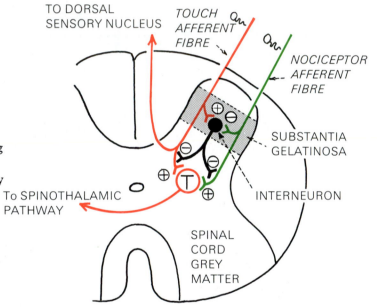

Impulses in nociceptor afferents can also be inhibited by descending fibres from the sensory cortex, the grey matter around the midbrain aqueduct and the brain stem reticular formation. These descending fibres terminate in the dorsal grey column of the spinal cord. They contain **opioid peptides** (encephalins, endorphins and dynorphins) which act as transmitters or neuromodulators.

SENSORY PATHWAYS FROM SKIN OF FACE

The receptors or nerve endings for **ordinary skin sensations** are linked by three neurons with receiving centres in the **parietal lobes**.

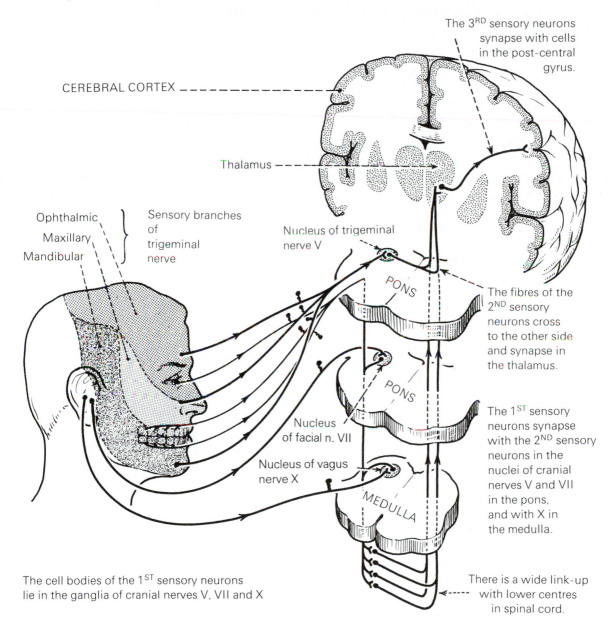

The 3RD sensory neurons synapse with cells in the post-central gyrus.

CEREBRAL CORTEX

Thalamus

Ophthalmic
Maxillary
Mandibular

Sensory branches of trigeminal nerve

Nucleus of trigeminal nerve V

PONS

The fibres of the 2ND sensory neurons cross to the other side and synapse in the thalamus.

PONS

Nucleus of facial n. VII

Nucleus of vagus nerve X

The 1ST sensory neurons synapse with the 2ND sensory neurons in the nuclei of cranial nerves V and VII in the pons, and with X in the medulla.

MEDULLA

The cell bodies of the 1ST sensory neurons lie in the ganglia of cranial nerves V, VII and X

There is a wide link-up with lower centres in spinal cord.

PAIN AND TEMPERATURE PATHWAYS FROM TRUNK AND LIMBS

The nerve endings registering pain and warmth or cold are linked by a chain of three neurons with the sensory area in the **parietal lobes** of the **cerebral cortex**. It is called the **anterolateral system**.

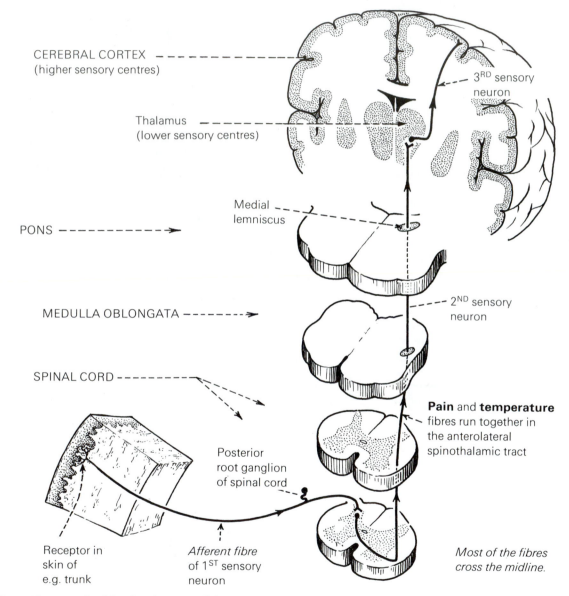

CEREBRAL CORTEX
(higher sensory centres)

3RD sensory neuron

Thalamus
(lower sensory centres)

Medial lemniscus

PONS

MEDULLA OBLONGATA

2ND sensory neuron

SPINAL CORD

Pain and **temperature** fibres run together in the anterolateral spinothalamic tract

Posterior root ganglion of spinal cord

Receptor in skin of e.g. trunk

Afferent fibre of 1ST sensory neuron

Most of the fibres cross the midline.

Pain *can* be perceived in the absence of the cerebral cortex. However the cortex is necessary to interpret the meaning of the pain and relate it to past experience.

TOUCH AND PRESSURE PATHWAYS FROM TRUNK AND LIMBS

Touch and pressure endings are linked by a chain of three neurons with the **partietal lobes.** It is called the **dorsal column (lemniscal) system**.

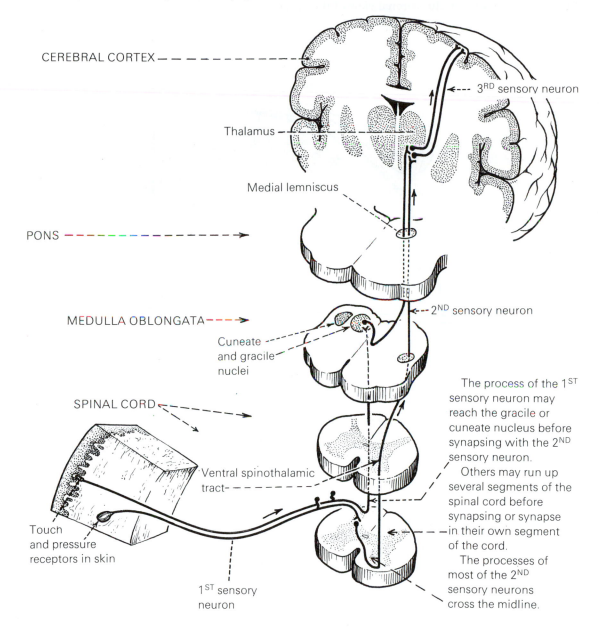

CEREBRAL CORTEX

3RD sensory neuron

Thalamus

Medial lemniscus

PONS

MEDULLA OBLONGATA

2ND sensory neuron

Cuneate and gracile nuclei

SPINAL CORD

The process of the 1ST sensory neuron may reach the gracile or cuneate nucleus before synapsing with the 2ND sensory neuron.

Others may run up several segments of the spinal cord before synapsing or synapse in their own segment of the cord.

The processes of most of the 2ND sensory neurons cross the midline.

Ventral spinothalamic tract

Touch and pressure receptors in skin

1ST sensory neuron

SENSORY CORTEX

The 3RD sensory neurons (conveying information from the *opposite side* of the body) synapse with cells in the **post-central gyrus** of the **parietal lobe** of the **cerebral cortex**. The exact points on this gyrus at which impulses coming from the different regions of the skin surface terminate are indicated on this coronal view of the gyrus.

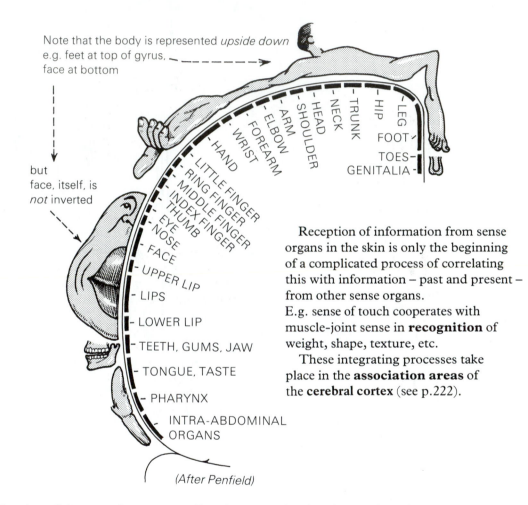

Note that the body is represented *upside down* e.g. feet at top of gyrus, face at bottom

but face, itself, is *not* inverted

HAND
LITTLE FINGER
RING FINGER
MIDDLE FINGER
INDEX FINGER
THUMB
EYE
NOSE
FACE
UPPER LIP
LIPS
LOWER LIP
TEETH, GUMS, JAW
TONGUE, TASTE
PHARYNX
INTRA-ABDOMINAL ORGANS

WRIST
FOREARM
ELBOW
ARM
SHOULDER
HEAD
NECK
TRUNK
HIP
LEG
FOOT
TOES
GENITALIA

(After Penfield)

Reception of information from sense organs in the skin is only the beginning of a complicated process of correlating this with information – past and present – from other sense organs.
E.g. sense of touch cooperates with muscle-joint sense in **recognition** of weight, shape, texture, etc.

These integrating processes take place in the **association areas** of the **cerebral cortex** (see p.222).

The *sizes* of the receptive areas are directly proportional to the number of sensory receptors in each peripheral area of the body. Note the relatively large area devoted to **face** (especially lips) and to **hand** (especially thumb and index finger) while trunk representation is very small.

The **motor nerve cells** which send out impulses to initiate **voluntary movement** of **skeletal muscles** lie in the **precentral gyrus** of each **frontal lobe** in the **cerebral cortex**.

Each cerebral hemisphere controls the muscles on the *opposite side* of the body.

The exact point in the **gyrus** where neurons controlling any one part of the body are situated is indicated in this coronal view of the gyrus.

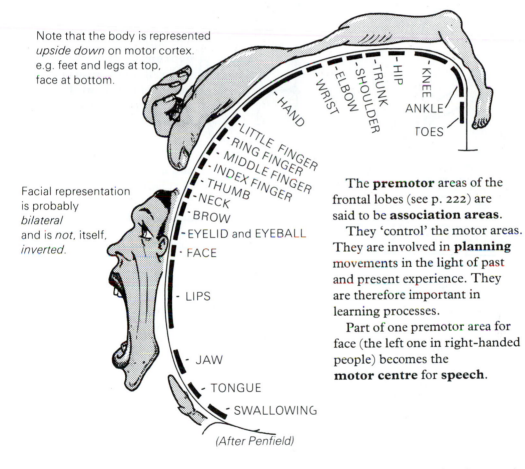

Note that the body is represented *upside down* on motor cortex. e.g. feet and legs at top, face at bottom.

Facial representation is probably *bilateral* and is *not*, itself, *inverted*.

TRUNK — SHOULDER — ELBOW — WRIST — HAND — HIP — KNEE — ANKLE — TOES — LITTLE FINGER — RING FINGER — MIDDLE FINGER — INDEX FINGER — THUMB — NECK — BROW — EYELID and EYEBALL — FACE — LIPS — JAW — TONGUE — SWALLOWING

(After Penfield)

The **premotor** areas of the frontal lobes (see p. 222) are said to be **association areas**.

They 'control' the motor areas. They are involved in **planning** movements in the light of past and present experience. They are therefore important in learning processes.

Part of one premotor area for face (the left one in right-handed people) becomes the **motor centre** for **speech**.

The amount of motor cortex devoted to a particular part of the body is related, not to its relative size, but to the **precision** with which its movements can be controlled. Note the large area of the motor cortex (and therefore the very large number of neurons) devoted to control of voluntary movements of the **hands**. This enables them to perform complicated movements and to acquire highly intricate skills: similarly with muscles of the mouth, lips, tongue and face which are used for talking.

MOTOR PATHWAYS TO HEAD AND NECK

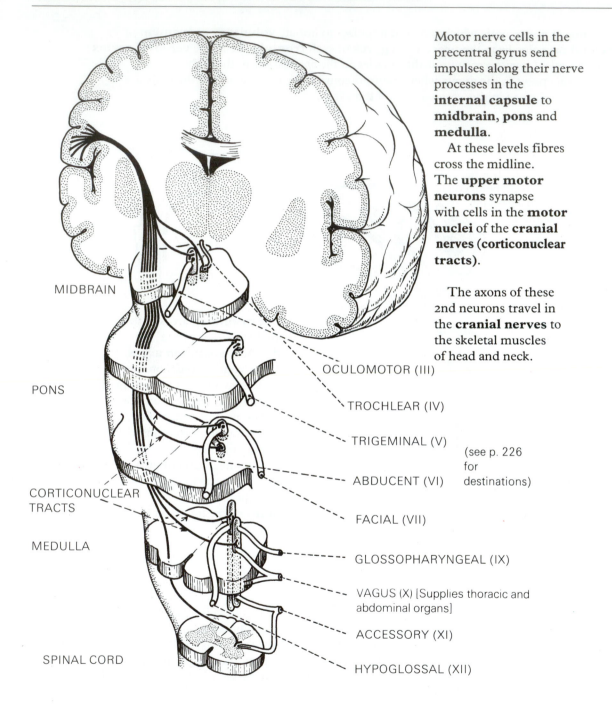

Motor nerve cells in the precentral gyrus send impulses along their nerve processes in the **internal capsule** to **midbrain**, **pons** and **medulla**.

At these levels fibres cross the midline. The **upper motor neurons** synapse with cells in the **motor nuclei** of the **cranial nerves (corticonuclear tracts)**.

The axons of these 2nd neurons travel in the **cranial nerves** to the skeletal muscles of head and neck.

MIDBRAIN

PONS

CORTICONUCLEAR TRACTS

MEDULLA

SPINAL CORD

OCULOMOTOR (III)

TROCHLEAR (IV)

TRIGEMINAL (V)

ABDUCENT (VI)

FACIAL (VII)

GLOSSOPHARYNGEAL (IX)

VAGUS (X) [Supplies thoracic and abdominal organs]

ACCESSORY (XI)

HYPOGLOSSAL (XII)

(see p. 226 for destinations)

MOTOR PATHWAYS TO EXTREMITIES

The **controlling centres** in the **motor cortex** are linked by two neurons with the **effector organs** – the **voluntary muscles**.

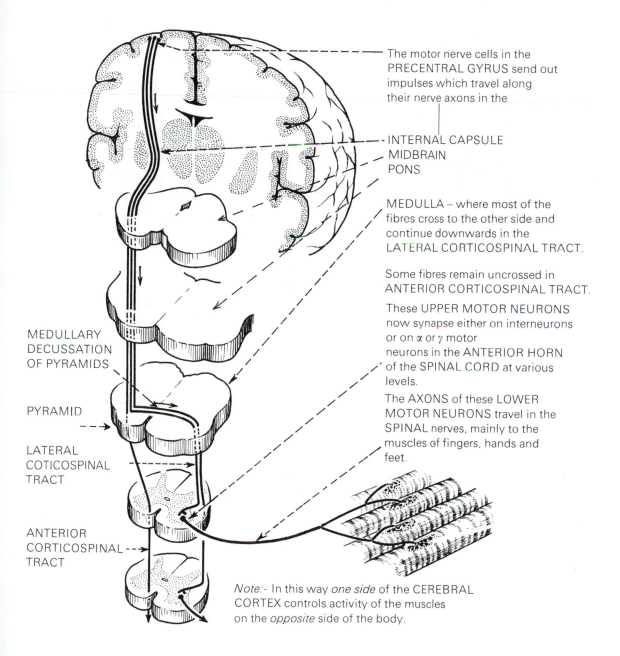

The motor nerve cells in the PRECENTRAL GYRUS send out impulses which travel along their nerve axons in the

INTERNAL CAPSULE
MIDBRAIN
PONS

MEDULLA – where most of the fibres cross to the other side and continue downwards in the LATERAL CORTICOSPINAL TRACT.

Some fibres remain uncrossed in ANTERIOR CORTICOSPINAL TRACT.

These UPPER MOTOR NEURONS now synapse either on interneurons or on α or γ motor neurons in the ANTERIOR HORN of the SPINAL CORD at various levels.

The AXONS of these LOWER MOTOR NEURONS travel in the SPINAL nerves, mainly to the muscles of fingers, hands and feet.

MEDULLARY
DECUSSATION
OF PYRAMIDS

PYRAMID

LATERAL
COTICOSPINAL
TRACT

ANTERIOR
CORTICOSPINAL
TRACT

Note:- In this way *one side* of the CEREBRAL CORTEX controls activity of the muscles on the *opposite* side of the body.

MOTOR UNIT

The **axon** of the **lower motor neuron** divides into many branches. Each branch ends at the **motor end-plate** of a single muscle fibre.

The **motor nerve** with the group of **muscle fibres** it supplies is known as a **motor unit**.

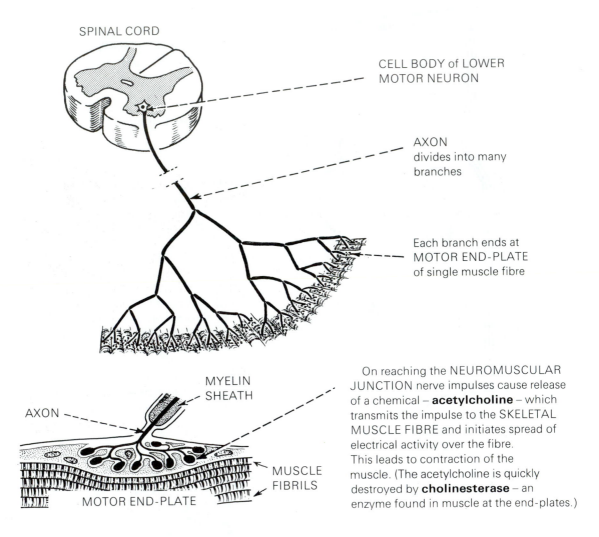

SPINAL CORD

CELL BODY of LOWER
MOTOR NEURON

AXON
divides into many
branches

Each branch ends at
MOTOR END-PLATE
of single muscle fibre

MYELIN
SHEATH

AXON

MUSCLE
FIBRILS

MOTOR END-PLATE

On reaching the NEUROMUSCULAR JUNCTION nerve impulses cause release of a chemical – **acetylcholine** – which transmits the impulse to the SKELETAL MUSCLE FIBRE and initiates spread of electrical activity over the fibre. This leads to contraction of the muscle. (The acetylcholine is quickly destroyed by **cholinesterase** – an enzyme found in muscle at the end-plates.)

In muscles requiring very fine control, e.g. extraocular muscles, *one* axon innervates only about *ten* muscle fibres. In muscles requiring less precise control *one* axon may innervate about *2000* muscle fibres.

Each **motor neuron** in the **anterior horns** of the **spinal cord** serves as the **pathway** for motor impulses initiated in:
THE CEREBRUM (*of opposite side*)
and travelling in –

1 CORTICO-SPINAL TRACT
The motor neuron also serves as the **pathway** for coordinating corrective (restraining or facilitating) impulses discharged by – NUCLEI in the BRAIN STEM (*of same or opposite side*) and travelling in –

2 RUBROSPINAL TRACT
from red nucleus (*opposite side*)
3 DORSAL VESTIBULOSPINAL TRACT
from dorsal vestibular nucleus (*same side*)
4 OLIVOSPINAL TRACT
from olivary nucleus (*same side*)
5 RETICULOSPINAL TRACT
from reticular nuclei (*same side*)
6 VENTRAL VESTIBULOSPINAL TRACT
from vestibular nuclei (*opposite side*)
7 TECTOSPINAL TRACT
from tectum (*opposite side*)

The motor neuron also receives relays of afferent impulses from other reflex centres in –

spinal cord
8 For REFLEXES of *same* segment of cord and from *same* side of cord.
9 For REFLEXES of *same* segment but *other* side of cord and body.
10 For REFLEXES from *other* segments of cord but *same* side of cord.
11 For REFLEXES from *other* segments and *other* side of cord.

'Final common pathway'

Skeletal muscle fibres

If several sources compete for the **'final common pathway'** at any one time, allied ones may reinforce each other; if incompatible, those which are initiated by **painful** stimuli (i.e. **protective reflexes**) take precedence.

MULTINEURONAL MOTOR PATHWAYS

The actual performance of a **voluntary movement** – initiated in the **cerebrum** – involves cooperation by **motor centres** in the **corpus striatum** and **brain stem**.

These form indirect
pathways to the
final common pathway

CORPUS STRIATUM
    CAUDATE NUCLEUS
    LENTIFORM NUCLEUS
    i.e. { PUTAMEN
         { GLOBUS
           PALLIDUS

These centres have important
2-way connections with
each other and with
HIGHER and LOWER MOTOR
and SENSORY Centres

CEREBELLUM
(for centres and connections
see pp. 275, 276)

MIDBRAIN and PONS
    RED NUCLEI
    SUBSTANTIA NIGRA
    RETICULAR NUCLEI
       (pontine and
       medullary)
    TECTUM (superior
       colliculus)
    VESTIBULAR NUCLEI

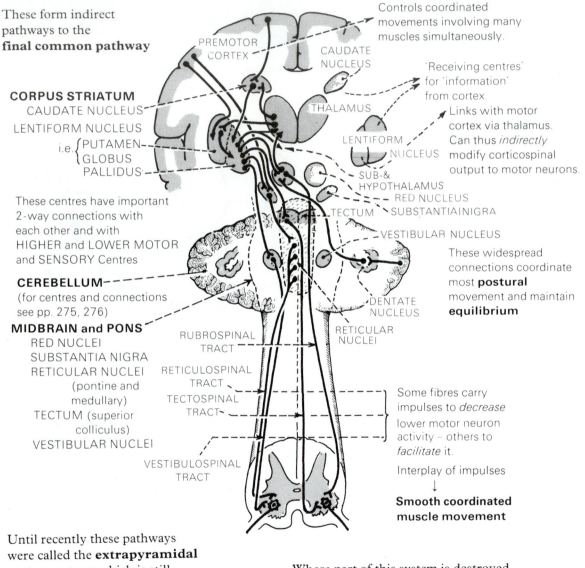

PREMOTOR CORTEX

CAUDATE NUCLEUS

Controls coordinated
movements involving many
muscles simultaneously.

THALAMUS

'Receiving centres'
for 'information'
from cortex.

LENTIFORM NUCLEUS

Links with motor
cortex via thalamus.
Can thus *indirectly*
modify corticospinal
output to motor neurons.

SUB-&
HYPOTHALAMUS

RED NUCLEUS
TECTUM   SUBSTANTIA INIGRA

VESTIBULAR NUCLEUS

These widespread
connections coordinate
most **postural**
movement and maintain
equilibrium

DENTATE
NUCLEUS

RETICULAR
NUCLEI

RUBROSPINAL
TRACT

RETICULOSPINAL
TRACT

TECTOSPINAL
TRACT

Some fibres carry
impulses to *decrease*
lower motor neuron
activity – others to
facilitate it.

Interplay of impulses
↓
**Smooth coordinated
muscle movement**

VESTIBULOSPINAL
TRACT

Until recently these pathways
were called the **extrapyramidal
system**, a term which is still
used in clinical literature
but is being abandoned by
neurophysiologists.

Where part of this system is destroyed
by disease, varying types of **rigidity**,
tremor and **uncoordinated muscle movement**
result.

PATHWAYS CONTROLLING MOTOR ACTIVITY

Many **sensory receptors, cerebral nuclei** and **integrating centres** are involved in the control of **motor activity**.

Sensory input comes from **skin, joints, muscle** and the **special senses** of **vision, hearing** and **balance**.

The **cerebellum** constantly monitors movement and regulates the range, rate and force of contraction.

The **basal ganglia** and **brain stem nuclei** are all involved in integrating the control of **motor output**.

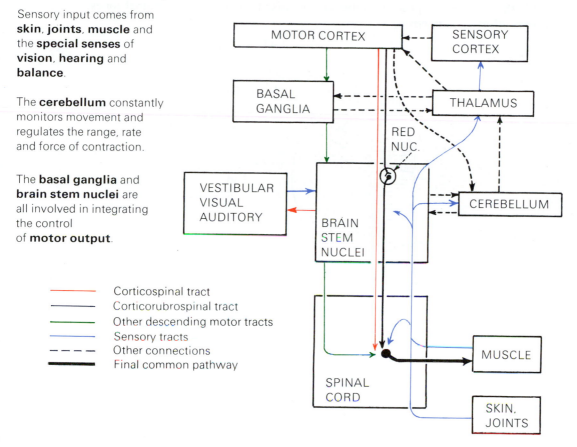

Corticospinal tract
Corticorubrospinal tract
Other descending motor tracts
Sensory tracts
Other connections
Final common pathway

The **corticospinal pathway** until recently was called the **pyramidal** tract since it runs through the **medullary pyramids**. All other descending pathways except the cerebellar were called the **extrapyramidal system**. The concept of two independent systems controlling movement is incorrect and most authors regard the terms as redundant. However the terms may persist for some time.

A more recent classification is based on the position of **termination** of the descending fibres in the grey matter of the spinal cord. The **lateral motor system** includes the **corticospinal** tracts and the **corticorubrospinal** pathway. It controls fine movements, particularly of the fingers and hands. The **medial motor system** includes the other descending tracts which originate primarily in the **brain stem**. It controls the muscles of the trunk and proximal parts of the limbs, thus it controls posture and equilibrium.

CEREBELLUM

The cerebellum has two hemispheres. Each hemisphere has three lobes, **anterior, posterior** and **flocculonodular**.

Functionally the anterior and posterior lobes are organized into three longitudinal zones, **lateral**, **intermediate** and **vermis**.

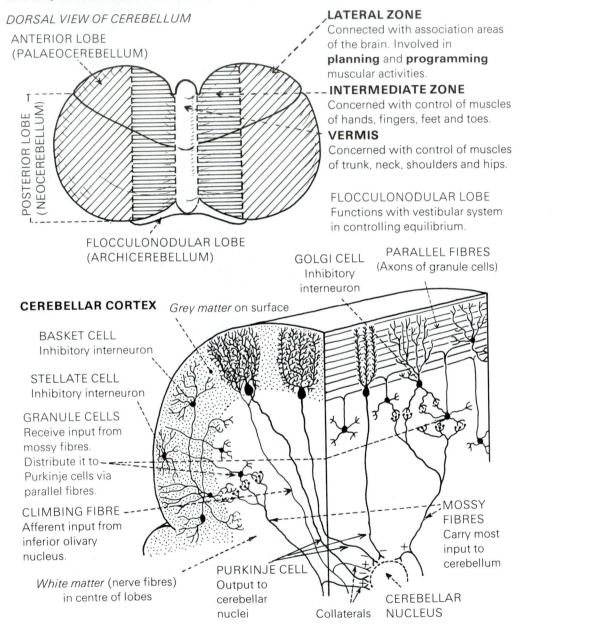

DORSAL VIEW OF CEREBELLUM

ANTERIOR LOBE
(PALAEOCEREBELLUM)

POSTERIOR LOBE
(NEOCEREBELLUM)

FLOCCULONODULAR LOBE
(ARCHICEREBELLUM)

LATERAL ZONE
Connected with association areas of the brain. Involved in **planning** and **programming** muscular activities.

INTERMEDIATE ZONE
Concerned with control of muscles of hands, fingers, feet and toes.

VERMIS
Concerned with control of muscles of trunk, neck, shoulders and hips.

FLOCCULONODULAR LOBE
Functions with vestibular system in controlling equilibrium.

CEREBELLAR CORTEX *Grey matter on surface*

GOLGI CELL
Inhibitory interneuron

PARALLEL FIBRES
(Axons of granule cells)

BASKET CELL
Inhibitory interneuron

STELLATE CELL
Inhibitory interneuron

GRANULE CELLS
Receive input from mossy fibres. Distribute it to Purkinje cells via parallel fibres.

CLIMBING FIBRE
Afferent input from inferior olivary nucleus.

White matter (nerve fibres) in centre of lobes

PURKINJE CELL
Output to cerebellar nuclei

Collaterals

MOSSY FIBRES
Carry most input to cerebellum

CEREBELLAR NUCLEUS

All afferent fibres send impulses to the cerebellar nuclei via collateral fibres.

Each cerebellar hemisphere is linked with the rest of the nervous system through three bundles of nerve fibres – the superior, middle and inferior peduncles.

All sensory ingoing fibres to cerebellum send collateral branches to deep cerebellar nuclei.

INGOING PATHWAYS

The cerebellum receives information...

...from SPINAL CORD
(Ventral spinocerebellar tract) – information about the arrival and strength of motor signals in spinal cord.

...from EYES and EARS
(Tectocerebellar tracts) –

...from CEREBRAL CORTEX
(Corticopontocerebellar tract) – information about muscle movements 'planned' by cortex.

...from OLIVARY NUCLEUS
Receives proprioceptive and cutaneous information from spinal cord. Also connections from motor cortex, basal ganglia and reticular formation. Sole source of **climbing fibres**.

SUPERIOR
PEDUNCLE

MIDLINE

NUCLEI
PONTIS

MIDDLE
PEDUNCLE

INFERIOR
PEDUNCLE

MIDLINE

OLIVARY
NUCLEUS

ACCESSORY
CUNEATE
NUCLEUS

...from GENERAL PROPRIOCEPTORS (Dorsal spinocerebellar tract) in sacral, lumbar and thoracic regions – information about tension and contraction of muscle, movement of joints; forces acting on body surface.

...from SPECIAL PROPRIOCEPTORS (Vestibulocerebellar tract)
Semicircular canals – information about **movement** of head in space.

Utricle
Saccule } – information about **position** of head in space.

...from CUNEATE NUCLEUS
(Cuneocerebellar tract) – same information as dorsal spinocerebellar tract but from neck and upper limbs.

The cerebellum continuously receives information about the exact position of all parts of the body in space and what movements are 'planned'. Since the information is received below the level of consciousness it gives no sensation.

CEREBELLUM

OUTGOING PATHWAYS
Originate in VERMIS,
LATERAL and INTERMEDIATE
ZONES and FLOCCULONODULAR LOBE.

In addition to the grey matter on the surface of the cerebellar cortex, there are masses of grey matter – called **cerebellar nuclei** – deep within the cerebellum.

The **intermediate zone** sends signals to GLOBOSE and EMBOLIFORM nuclei. Thence to RED NUCLEUS of opposite side to influence activity of rubrospinal motor pathway.

The **vermis** sends impulses direct to vestibular nucleus. Also via FASTIGIAL NUCLEUS to pontine and medullary reticular formation. Helps to control posture.

The **flocculonodular lobe** sends signals direct to VESTIBULAR NUCLEUS. Thence to vestibulospinal tract — — — — to coordinate movements and position of head with postural tone of limb muscles.

The **lateral zone** sends impulses to DENTATE nucleus, thence to CEREBRAL CORTEX via THALAMUS. Coordinates corticospinal motor activities.
Also a small projection to RED NUCLEUS of opposite side to influence activity of the rubrospinal motor pathway.

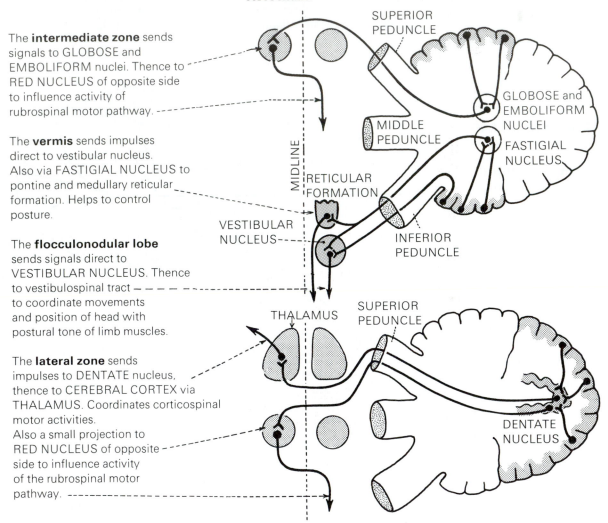

The cerebral cortex probably initiates purposeful movements. During such movements proprioceptors are continually supplying information to the cerebellum about the changing positions of muscles and joints. The cerebellum is then responsible for collating this information and sending out impulses to regulate the rate, range, force and direction of movements.

The muscle spindle is the key structure in the complex self-regulating mechanism for the control of movement of skeletal muscle.

In **voluntary movement** a muscle can be made to contract by impulses reaching it by one of *two routes:-*

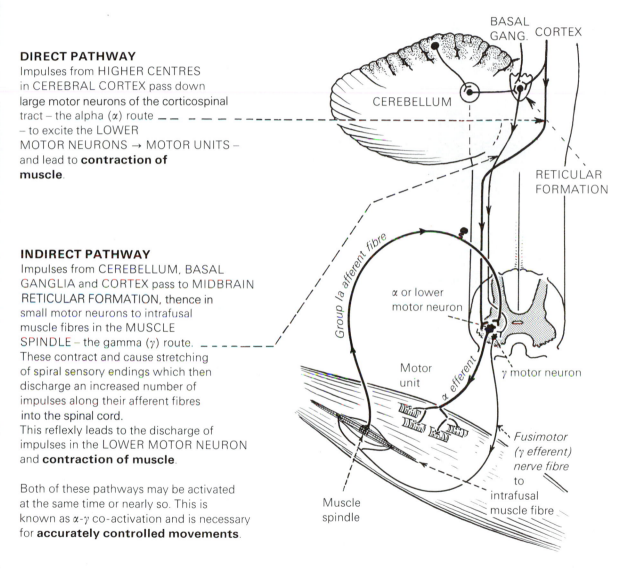

DIRECT PATHWAY
Impulses from HIGHER CENTRES in CEREBRAL CORTEX pass down large motor neurons of the corticospinal tract – the alpha (α) route – to excite the LOWER MOTOR NEURONS → MOTOR UNITS – and lead to **contraction of muscle**.

INDIRECT PATHWAY
Impulses from CEREBELLUM, BASAL GANGLIA and CORTEX pass to MIDBRAIN RETICULAR FORMATION, thence in small motor neurons to intrafusal muscle fibres in the MUSCLE SPINDLE – the gamma (γ) route.
These contract and cause stretching of spiral sensory endings which then discharge an increased number of impulses along their afferent fibres into the spinal cord.
This reflexly leads to the discharge of impulses in the LOWER MOTOR NEURON and **contraction of muscle**.

Both of these pathways may be activated at the same time or nearly so. This is known as α-γ co-activation and is necessary for **accurately controlled movements**.

It is probable that the **cerebral cortex** decides what is to be done, and the **cerebellum** constantly monitors and controls the movement.

277

SKELETAL SYSTEM

BONES and MUSCLES — concerned with **movement** of the body.

SKELETON — RIGID FRAMEWORK gives **shape** and **support** to body.
 is JOINTED to permit **movement**.

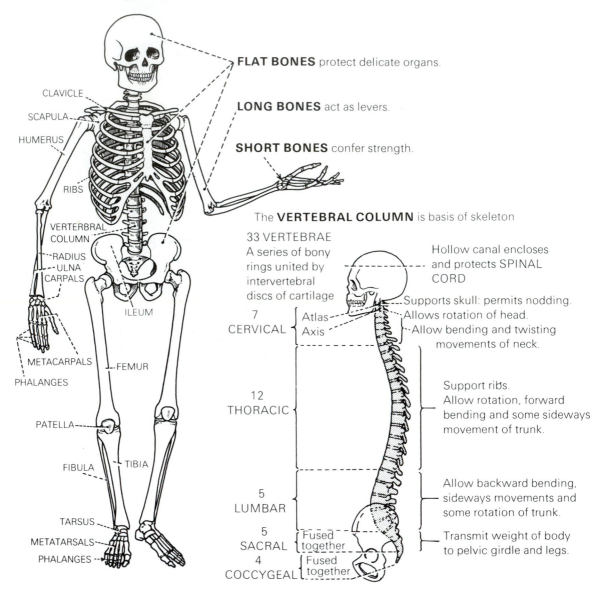

FLAT BONES protect delicate organs.

LONG BONES act as levers.

SHORT BONES confer strength.

CLAVICLE
SCAPULA
HUMERUS
RIBS
VERTERBRAL
COLUMN
RADIUS
ULNA
CARPALS
ILEUM
METACARPALS
PHALANGES
FEMUR
PATELLA
FIBULA
TIBIA
TARSUS
METATARSALS
PHALANGES

The **VERTEBRAL COLUMN** is basis of skeleton

33 VERTEBRAE
A series of bony
rings united by
intervertebral
discs of cartilage

Hollow canal encloses
and protects SPINAL
CORD

7
CERVICAL

Atlas
Axis

Supports skull: permits nodding.
Allows rotation of head.
Allow bending and twisting
 movements of neck.

12
THORACIC

Support ribs.
Allow rotation, forward
bending and some sideways
movement of trunk.

5
LUMBAR

Allow backward bending,
sideways movements and
some rotation of trunk.

5
SACRAL

Fused
together

Transmit weight of body
to pelvic girdle and legs.

4
COCCYGEAL

Fused
together

All bones give attachment to muscles.

Bones are moved at joints by the *contraction* and *relaxation* of **muscles** attached to them.

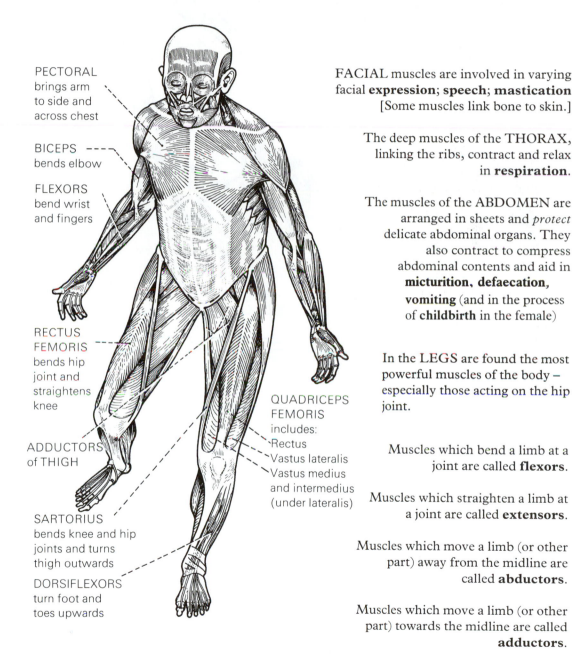

PECTORAL
brings arm
to side and
across chest

BICEPS
bends elbow

FLEXORS
bend wrist
and fingers

RECTUS
FEMORIS
bends hip
joint and
straightens
knee

ADDUCTORS
of THIGH

SARTORIUS
bends knee and hip
joints and turns
thigh outwards

DORSIFLEXORS
turn foot and
toes upwards

QUADRICEPS
FEMORIS
includes:
Rectus
Vastus lateralis
Vastus medius
and intermedius
(under lateralis)

FACIAL muscles are involved in varying facial **expression; speech; mastication** [Some muscles link bone to skin.]

The deep muscles of the THORAX, linking the ribs, contract and relax in **respiration**.

The muscles of the ABDOMEN are arranged in sheets and *protect* delicate abdominal organs. They also contract to compress abdominal contents and aid in **micturition, defaecation, vomiting** (and in the process of **childbirth** in the female)

In the LEGS are found the most powerful muscles of the body – especially those acting on the hip joint.

Muscles which bend a limb at a joint are called **flexors**.

Muscles which straighten a limb at a joint are called **extensors**.

Muscles which move a limb (or other part) away from the midline are called **abductors**.

Muscles which move a limb (or other part) towards the midline are called **adductors**.

279

SKELETAL MUSCLES

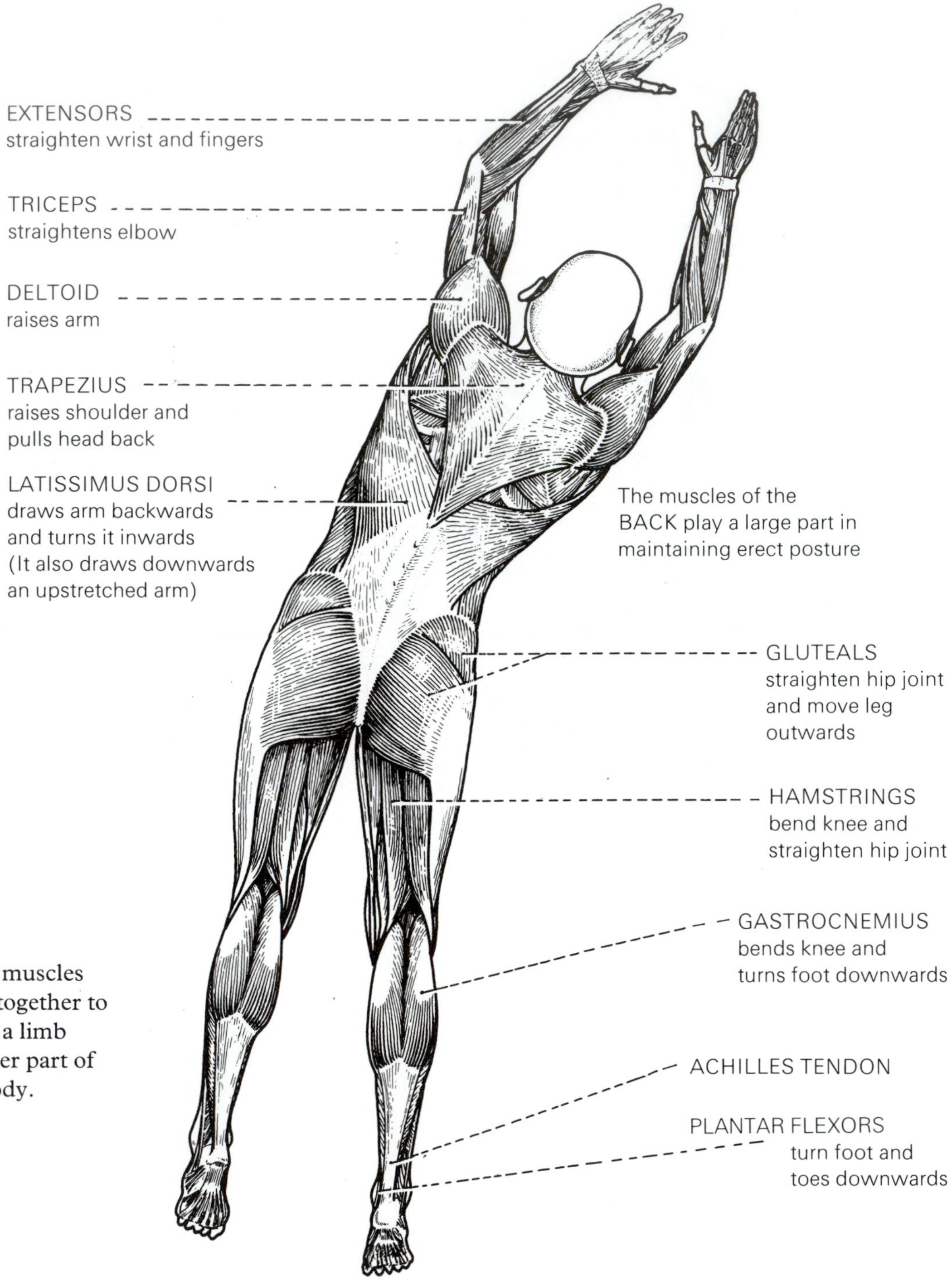

EXTENSORS
straighten wrist and fingers

TRICEPS
straightens elbow

DELTOID
raises arm

TRAPEZIUS
raises shoulder and
pulls head back

LATISSIMUS DORSI
draws arm backwards
and turns it inwards
(It also draws downwards
an upstretched arm)

The muscles of the
BACK play a large part in
maintaining erect posture

GLUTEALS
straighten hip joint
and move leg
outwards

HAMSTRINGS
bend knee and
straighten hip joint

GASTROCNEMIUS
bends knee and
turns foot downwards

Some muscles
work together to
rotate a limb
or other part of
the body.

ACHILLES TENDON

PLANTAR FLEXORS
turn foot and
toes downwards

The long bones particularly form a light framework of **levers**.
The skeletal muscles attached to them contract to operate these levers.

When a muscle contracts it shortens.

This brings its two ends closer together.

Since the two ends are attached to different bones by **tendons** one or other of the bones must move.

Two bones meet or articulate at a **joint**.

Joint surfaces are covered with a layer of smooth **cartilage**.

To avoid friction when the two surfaces move on one another a **synovial membrane** secretes a **lubricating fluid**.

BICEPS

TRICEPS

FLEXION

EXTENSION

The muscle which *contracts* to move the joint is called the prime mover or **agonist** (the biceps in flexion of elbow).

To allow the movement to take place, however, other muscles near the joint must cooperate:-
The oppositely acting muscles gradually *relax* – these are called the **antagonists** and exercise a 'braking' control on the movement. (E.g. the extensors – chiefly triceps – in flexion of the elbow.)
Other muscles steady the bone giving 'origin' to the prime mover so that only the 'insertion' will move – these muscles are called **fixators**.
Still other muscles help to steady, for most efficient movement, the joint being moved – called **synergists**.

When the elbow is straightened the reverse occurs:-
Triceps, the prime mover, *contracts*; biceps, the antagonist, *relaxes*.

RECIPROCAL INNERVATION

The coordinated group action of muscles is made possible by the many synaptic connections between interneurons of the *ingoing* or **proprioceptive neurons** of one muscle group and the *outgoing* or **motor neurons** of the functionally opposite group of muscles.

    This is shown diagrammatically for the reciprocal contraction and relaxation of the extensors and flexors during the **stretch reflex**. See page 230.

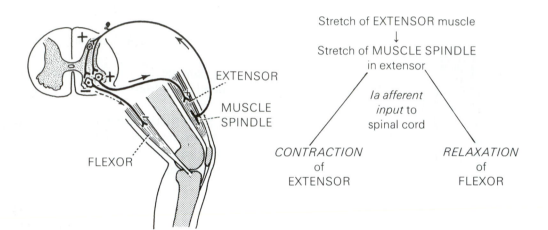

Stretch of EXTENSOR muscle
↓
Stretch of MUSCLE SPINDLE
in extensor

Ia afferent input to spinal cord

EXTENSOR

MUSCLE SPINDLE

FLEXOR

CONTRACTION of EXTENSOR

RELAXATION of FLEXOR

Reciprocal innervation is due to an **inhibitory** interneuron (within the spinal cord) interposed between the sensory nerve fibre and the α-motor neuron of the **flexor** muscle.

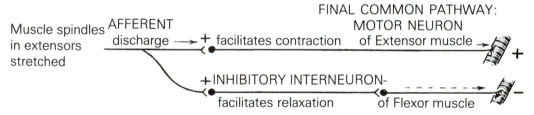

Muscle spindles in extensors stretched

AFFERENT discharge → + facilitates contraction

FINAL COMMON PATHWAY: MOTOR NEURON of Extensor muscle +

+INHIBITORY INTERNEURON- facilitates relaxation

of Flexor muscle −

Reciprocal **inhibition** of the flexor muscle is mediated by a **disynaptic** (two synapses) pathway. Contraction of the **extensor** muscle is mediated by a **monosynaptic** pathway.

SKELETAL MUSCLE AND THE MECHANISM OF CONTRACTION

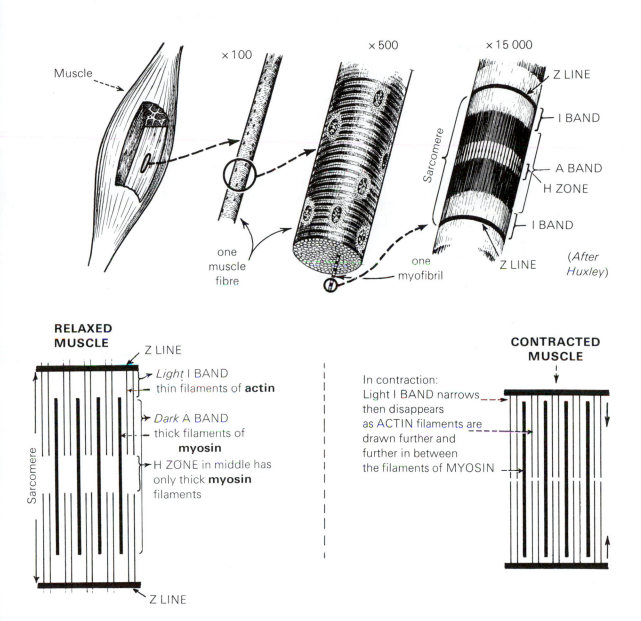

Muscle

×100

×500

×15 000

Z LINE

I BAND

A BAND

H ZONE

I BAND

Z LINE

(*After Huxley*)

Sarcomere

one muscle fibre

one myofibril

RELAXED MUSCLE

Z LINE

Light I BAND
thin filaments of **actin**

Dark A BAND
thick filaments of **myosin**

H ZONE in middle has only thick **myosin** filaments

Sarcomere

Z LINE

CONTRACTED MUSCLE

In contraction:
Light I BAND narrows then disappears
as ACTIN filaments are drawn further and further in between the filaments of MYOSIN

Energy for contraction is derived from glucose and fat in the mitochondria (page 8). The energy is transported in ATP from the mitochondria to the contractile filaments. ATP splits readily into ADP and phosphate, releasing its trapped energy where needed (page 42).

283

SKELETAL MUSCLE – MOLECULAR BASIS OF CONTRACTION

Actin and **myosin** filaments *slide* past each other during *contraction* of skeletal muscle.

SARCOMERE

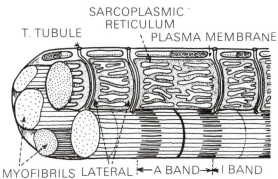

ACTIN — MYOSIN MOLECULE — CROSS-BRIDGES — Z-LINE

CHAINS OF ACTIN MOLECULES

TROPONIN — Ca^{2+} BINDING SITE — TROPOMYOSIN — Ca^{2+}

COVERED ← BINDING SITES → EXPOSED

RELAXED **CONTRACTED**

T. TUBULE — SARCOPLASMIC RETICULUM — PLASMA MEMBRANE

MYOFIBRILS   LATERAL SACS   ← A BAND → ← I BAND

The globular heads of **myosin** molecules (**cross-bridges**) bind to special binding sites on **actin** filaments, then move in an arc, like the oars of a boat, and pull the actin towards the centre of the sarcomere.

ATP molecules are necessary, to link the cross-bridges to actin, energize their movement and break the links at the end of each cycle, to allow further cycling.

In *relaxed* muscle, cross-bridge binding is inhibited by regulatory proteins, **troponin** and **tropomyosin**. Tropomyosin *covers* the actin binding sites. Troponin *holds* the tropomyosin in this blocking position. To initiate cross-bridge cycling Ca^{2+} attaches to troponin and changes its *shape*. This change moves tropomyosin away from and thus exposes the binding sites, allowing cross-bridge cycling and contraction to proceed. Removal of Ca^{2+} *reverses* the process and the muscle *relaxes*.

To initiate *contraction* Ca^{2+} is released from the **lateral sacs** of the **sarcoplasmic reticulum**, segments of which are wrapped round the myofibrils covering each A and I band. Between each segment a **transverse** or **T-tubule** runs right round the myofibril and connects with the plasma membrane. Its lumen is continuous with the extracellular fluid. Action potentials travelling along the muscle membrane pass down into the **T-tubules** and cause release of Ca^{2+} from the sarcoplasmic reticulum to initiate contraction. At the end of contraction Ca^{2+} is pumped back into the sarcoplasmic reticulum, removing it from the troponin and thus *relaxation* occurs.

CHAPTER 11

AUTONOMIC NERVOUS SYSTEM AND CHEMICAL TRANSMISSION AT NERVE ENDINGS

AUTONOMIC NERVOUS SYSTEM

The autonomic nervous system innervates three types of tissue – **cardiac muscle**, **smooth muscle** and **gland cells**. Its functions are normally reflexly controlled and are carried out below the level of consciousness.

It has two divisions – **parasympathetic** and **sympathetic**.

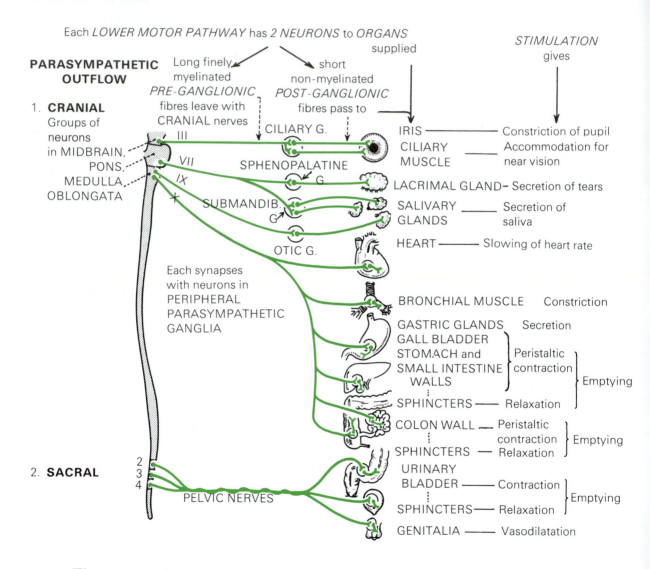

Each *LOWER MOTOR PATHWAY* has *2 NEURONS* to *ORGANS* supplied

STIMULATION gives

PARASYMPATHETIC OUTFLOW

Long finely myelinated *PRE-GANGLIONIC* fibres leave with CRANIAL nerves

short non-myelinated *POST-GANGLIONIC* fibres pass to

1. **CRANIAL** Groups of neurons in MIDBRAIN, PONS, MEDULLA OBLONGATA

III — CILIARY G.

IRIS ———— Constriction of pupil

CILIARY MUSCLE ____ Accommodation for near vision

VII — SPHENOPALATINE G.

LACRIMAL GLAND – Secretion of tears

IX — SUBMANDIB. G.

SALIVARY ____ Secretion of GLANDS saliva

OTIC G.

HEART ———— Slowing of heart rate

Each synapses with neurons in PERIPHERAL PARASYMPATHETIC GANGLIA

BRONCHIAL MUSCLE   Constriction

GASTRIC GLANDS   Secretion
GALL BLADDER
STOMACH and } Peristaltic
SMALL INTESTINE contraction } Emptying
WALLS
SPHINCTERS ——— Relaxation

COLON WALL — Peristaltic contraction } Emptying
SPHINCTERS — Relaxation

2. **SACRAL** 2 3 4

PELVIC NERVES

URINARY BLADDER ———— Contraction } Emptying
SPHINCTERS ——— Relaxation

GENITALIA ——— Vasodilatation

The parasympathetic system is concerned mainly with the production and conservation of energy, e.g. it promotes reabsorption from the gut, slows the heart, etc.

The parasympathetic and sympathetic systems usually act in balanced reciprocal fashion. The activity of an organ at any one time is the result of the two opposing influences. However this is not *always* true, e.g. most blood vessels have only a sympathetic innervation.

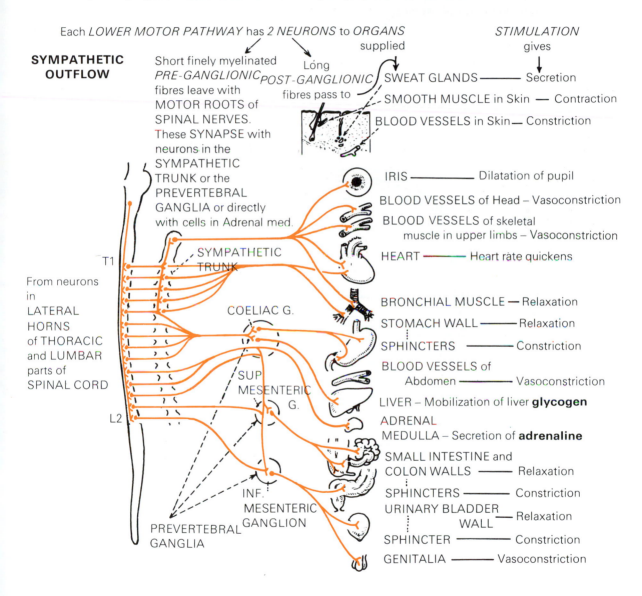

Each *LOWER MOTOR PATHWAY* has *2 NEURONS* to *ORGANS* supplied   *STIMULATION* gives

SYMPATHETIC OUTFLOW

Short finely myelinated *PRE-GANGLIONIC* fibres leave with MOTOR ROOTS of SPINAL NERVES. These SYNAPSE with neurons in the SYMPATHETIC TRUNK or the PREVERTEBRAL GANGLIA or directly with cells in Adrenal med.

Long *POST-GANGLIONIC* fibres pass to

SWEAT GLANDS —— Secretion

SMOOTH MUSCLE in Skin — Contraction

BLOOD VESSELS in Skin — Constriction

IRIS —— Dilatation of pupil

BLOOD VESSELS of Head – Vasoconstriction

BLOOD VESSELS of skeletal muscle in upper limbs – Vasoconstriction

HEART —— Heart rate quickens

BRONCHIAL MUSCLE — Relaxation

STOMACH WALL —— Relaxation

SPHINCTERS —— Constriction

BLOOD VESSELS of Abdomen —— Vasoconstriction

LIVER – Mobilization of liver **glycogen**

ADRENAL MEDULLA – Secretion of **adrenaline**

SMALL INTESTINE and COLON WALLS —— Relaxation

SPHINCTERS —— Constriction

URINARY BLADDER WALL — Relaxation

SPHINCTER —— Constriction

GENITALIA —— Vasoconstriction

T1

From neurons in LATERAL HORNS of THORACIC and LUMBAR parts of SPINAL CORD

L2

SYMPATHETIC TRUNK

COELIAC G.

SUP MESENTERIC G.

INF. MESENTERIC GANGLION

PREVERTEBRAL GANGLIA

The sympathetic system is regarded as preparing the animal for 'fight' or 'flight'.

AUTONOMIC REFLEX

Autonomic centres in the **brain** and **spinal cord** receive *sensory inflows* from the **viscera**. (Less is known about their exact pathways than about *motor outflows*.) Some of the *sensory neurons* convey information about events in the viscera to **higher autonomic centres** which send impulses to modify the activity of

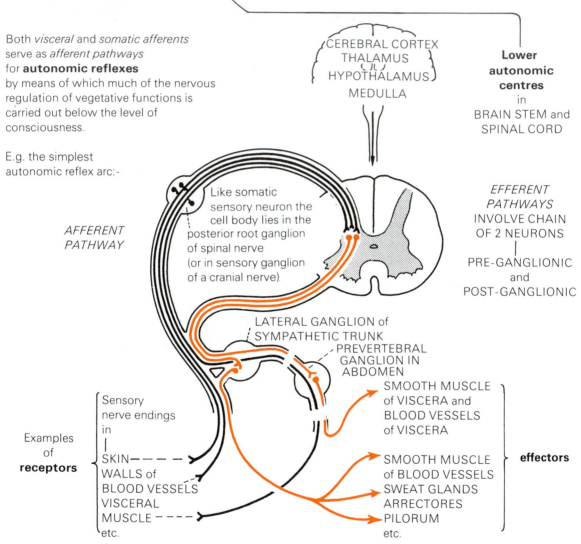

Both *visceral* and *somatic afferents* serve as *afferent pathways* for **autonomic reflexes** by means of which much of the nervous regulation of vegetative functions is carried out below the level of consciousness.

E.g. the simplest autonomic reflex arc:-

CEREBRAL CORTEX
THALAMUS
HYPOTHALAMUS
MEDULLA

Lower autonomic centres in BRAIN STEM and SPINAL CORD

AFFERENT PATHWAY

Like somatic sensory neuron the cell body lies in the posterior root ganglion of spinal nerve (or in sensory ganglion of a cranial nerve).

EFFERENT PATHWAYS INVOLVE CHAIN OF 2 NEURONS
|
PRE-GANGLIONIC and POST-GANGLIONIC

LATERAL GANGLION of SYMPATHETIC TRUNK
PREVERTEBRAL GANGLION IN ABDOMEN

Examples of **receptors**

Sensory nerve endings in
|
SKIN – – –
WALLS of BLOOD VESSELS
VISCERAL MUSCLE – – –
etc.

SMOOTH MUSCLE of VISCERA and BLOOD VESSELS of VISCERA

SMOOTH MUSCLE of BLOOD VESSELS
SWEAT GLANDS
ARRECTORES PILORUM
etc.

effectors

The autonomic reflex arc differs from the somatic reflex arc mainly in that it has *two efferent neurons*. Transmission of impulse from *afferent* to *efferent* probably involves one or more interneurons.

Reflex control of blood pressure is a more complex and an important example of an autonomic reflex (see pp. 108, 109).

CHEMICAL TRANSMISSION AT NERVE ENDINGS

When an **action potential** reaches the endings of a nerve, a **neurotransmitter** is liberated. It diffuses across the gap between the nerve endings and the next neuron or effector cell and attaches itself to **receptors** on the membrane of the cell. This attachment alters the permeability of the post-synaptic membrane to Na^+ ions and results in onward spread of the action potential over the post-synaptic cell. Some transmitters can **inhibit** the post-synaptic membrane by altering its permeability to Cl^- or K^+.

Some nerves when stimulated liberate acetylcholine. These are called *cholinergic* nerves. Others liberate noradrenaline. These are called *adrenergic* nerves.

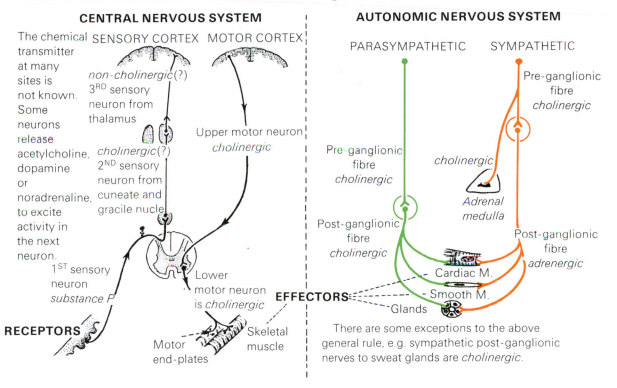

There are many other chemical substances secreted by neurons which may act as synaptic transmitters to regulate certain aspects of cerebral activity.

Many neurons within the CNS release inhibitory, *not* excitatory, transmitters. One such substance is GABA (gamma aminobutyric acid). Another is the amino acid, glycine.

Cholinergic receptors are of two types, **nicotinic** and **muscarinic**. Adrenergic receptors are classified as α_1, α_2, β_1, or β_2.

INDEX

Abdomen
  muscles of, 279
  reflex vasoconstriction, 108
  reflex vasodilatation, 109
Abducent nerve (cranial VI), 224, 226, 241, 242, 268
Absorption
  calcium and phosphorus role of parathormone, 170–172
  large intestine, 64, 65, 85
  small intestine, 38–40, 64, 65, 82–84
  stomach, 70
  striated border epithelium, 11
  vitamin B$_{12}$, 70, 118
Accessory
  cuneate nucleus, 275
  nerve (cranial XI), 224, 226, 268
  oculomolor (Edinger-Westphal) nucleus, 243, 251
  sex organs, 192, 196
    at menopause, 216
    during menstrual cycle, 201
    female, 196, 197, 200
    hormonal control of, 198
    male, 192–195
Acclimatization
  to altitude, 119
Accommodation
  for near vision, 239, 243, 244
A cells of pancreas, 189
Acetylcholine
  adrenal medulla, 176
  autonomic nervous system, 289
  chemical transmission, 289
  gut wall, 87
  neuromuscular junction, 270
Acetylcholinesterase, 270
Acetylcoenzyme A, 40–42
Achilles tendon, 230, 280
Acid base balance, 157, 158
  definition, 157
Acidity
  effect on O$_2$ combination with haemoglobin, 139, 141
Acidosis, 174, 189
Acinus
  pancreatic, 74, 75
Acquired immune deficiency syndrome (AIDS), 125
Acromegaly, 181
ACTH *see* Adrenocorticotrophic hormone
Actin, 283, 284
Action
  potential, 58
  propagation of, 59
  trigger for transmitter release, 289
Active transport, 52
  primary, 53
  secondary, 53
Adaptation
  in eye, 247, 248
  in receptors, 235
Addison's disease, 174
Adenine, 28
Adenohypophysis, 178

Adenosine diphosphate (ADP), 42, 117, 283
Adenosine triphosphate (ATP), 8, 40–42, 53, 87, 283, 284
Adenylate cyclase, 61, 186
Adipose tissue, 14
  deposition in growth, 47
  in breast, 213
Adolescence
  energy requirements in, 48, 49
  growth, 47
  puberty in female, 200
  puberty in male, 195
  reproductive organs, 202, 213
Adrenal cortex, 173
  disease of, in man, 174, 175, 182
  effect of anterior pituitary on, 179
  in carbohydrate metabolism, 39, 189
  in water balance, 152, 160, 173, 188
  overactivity, 175
  underactivity, 174
  vitamin C in, 36
Adrenal gland (of fetus), 208
Adrenaline, 176, 177
  body temperature control, 45
  role in intermediate carbohydrate metabolism, 41
  vasoconstrictor effects, 108, 176
  vasodilatation effects, 109
Adrenal medulla, 173, 176, 287
  actions of adrenaline, 177
  effect of metabolism, 39
  innervation, 289
  role in body temperature control, 45
Adrenergic nerves
  chemical transmission, 289
  in gut, 87, 88, 287
Adrenergic receptors, 289
Adrenocorticotrophic hormone (ACTH, corticotrophin), 173, 175, 179, 182
  overproduction of, 182
Adrenogenital syndrome, 175
Afferent
  arteriole (kidney), 148–150
  nerves, 229–231, 288
  pathway, 233
Afferents
  in cardiac reflexes, 100
Afterbirth, 211
Agglutination, 120–122
Agglutinin α, β, 120, 121
Agglutinogen A, A$_1$, and B, 120, 121
Agglutinogen D, 122
Agonist (prime mover) of joint, 281
Air
  atmospheric, 137, 138
  capacity of lungs, 136
  composition of respired air, 137, 138
  conducting passages, 131, 137
Alanine, 27
Albumin
  digestion of, 37
  human blood, 116

serum, 37
Aldosterone, 152, 160, 173–175, 188
Aldosteronism, 175
Alimentary tract, 64
  basic structural pattern, 79
  heat loss, food and faces, 43–45
  innervation, 87
  nervous control, 88, 286, 287
  progress of food along, 65
Alkalosis, 175
Alpha-gamma (α–γ) coactivation, 277
Alpha (α) motor neurons, 269, 277, 282
Alveolar
  duct, 132
  membrane, 132
  pressure, 134
  sac, 132
Alveolar air, 132, 137, 138
  partial pressure of gases, 137–139, 142
Alveoli
  of breasts, 213–215
  of lungs, 132, 134, 136–138
Amacrine cells (retina), 246, 247
Amino acids, 27
  absorption and fate, 37, 38, 82–84
  dietary sources of, 34
  energy from, 41
  essential, 32
  in blood, 116
  reabsorption from glomerular filtrate, 151
Ammonia, 41, 116
  deamination in liver, 38
  excretion in faeces, 38
  production by kidneys, 38, 151, 158, 163
Ammonium salts, 31
Amniocentesis, 208
Amnioic
  cavity, 21, 208
  fluid, 208, 210
  membranes, 208
Amoeba, 2, 3
Ampulla
  of semicircular canals, 256, 257
  of Vater, 74
Amygdala, 225
Amylase, 37, 39, 66, 74
Anabolic steroids, 173
Anaemia, 119
Anal canal, 64, 65, 85
  defaecation, 86
  innervation, 88
  sphincters, 65, 85, 86, 88
Anaphase, 9, 20
Androgen binding protein (ABP), 195
Androgens, 196, 200
  androstenedione, 173
  dehydroepiandrosterone, 173
  of adrenal cortex, 173–175, 182
  overproduction, 175
  underproduction, 174

Angiotensin I, 160, 188
Angiotensin II, 160, 173, 188
Angiotensinogen, 160, 188
Anion, 54
Ankle jerk, 230
Antagonist (joint), 281
Anterior commissure, 223
Anterior horn (spinal cord), 227, 232, 269, 271
Anterior lobe (cerebellum), 274
Antibodies, 13, 116, 123–126
  blood group antibodies, 120–122
Antidiuretic hormone (ADH, Vasopressin), 152, 159, 160,
    184, 186–188
Antigens, 123, 126
  blood group antigens, 120–122
  self antigens, 125
Antithrombin III, 117
Antigravity muscles, 258
Aorta, 93–95, 106
  baroreceptors, 101, 108, 109, 145
  blood flow in, 112
  blood pressure in, 104
  chemoreceptors, 108, 109, 144, 145
  cross-section and blood flow, 112
  valve, 96, 106
Aortic bodies
  chemoreceptors
    in regulation of arteriolar tone, 108, 109
    in regulation of respiration, 144, 145
Aortic arch
  baroreceptors
    in cardiac reflexes, 101
    in regulation of arteriolar tone, 108, 109
    in regulation of respiration, 145
Apneustic centre (respiration), 143
Appendix, 65, 84, 85
Aqueous humour, 239, 244
Arachnoid, 127
Archicerebellum *see* Flocculondular lobe of cerebellum
Arcuate arteries of kidney, 148
Areola of breast, 213, 214
Argyll-Robertson pupil, 251
Arrectores pilorum (smooth muscle of skin), 45, 177
  in autonomic reflex, 288
Arterial elasticity, 106
Arterioles
  blood pressure in, 104
  histology, 103
  nervous regulation of, 107
  of kidney, 148–150, 188
  of pulmonary circulation, 113
  reflex and chemical regulation of tone, 108, 109
Artery, 90
  blood flow in, 112
  blood pressure in, 104, 105
  histology, 103
  mesenteric, 82
  muscular, 103, 112
  pulmonary, 94, 113, 132
  renal, 148, 149

INDEX

retina, central artery of, 239
splenic, 124
superior hypophyseal, 179
supra-optic, 184
umbilical, 207, 209
Artificial respiration, 135
Ascending limb (of Henle's loop), 149
Ascorbic acid (vitamin C), 36
Assimilation of food, 64, 65
Association areas of cerebral cortex, 222, 265–267
    functions of, 266, 267
    pathways, 67, 71, 75, 77
Aster, 9
Atherosclerosis
    after menopause, 216
    and growth hormone overproduction, 181
    in diabetes mellitus, 189
Atria
    of heart, 91–95, 97, 98, 109, 113, 160
Atrial natriuretic peptide, 160, 188
    volume receptors, 160
Atrio
    –ventricular groove, 92, 93
    –ventricular (AV) node, 97–99
Auditory
    cortical representation, 255
    ossicles, 252, 254
    pathways, 255
    receptors, 19, 253, 255
    stimuli, 235, 254
Auerbach's plexus, 79, 87
    large intestine, 85
    oesophagus, 68, 69
    small intestine, 77, 80, 82
    stomach, 71
Auricle (ear), 252
Autocrines, 60
Autoimmune disease, 126
Autonomic nervous system, 286, 287
    centres connected to nociceptor afferents, 262
    chemical transmission in, 289
    in control of gastro-intestinal movements, 72, 77, 86–88
        adrenal medulla, 176
        gastro-intestinal secretions, 71, 75, 80
        micturition, 162
        respiration, 145
    in eye reflexes, 243, 251
    in regulation of arteriolar tone, 107–109
        heart action, 99–101
    outflows, 286, 287
    reflex, 288
    salivary centres, 67
Axon, 17–19, 228, 270
    hillock, 228
    reflex, 109

Baby
    birth, 210
    development, 21, 204–210

energy expenditure, 49
    growth, 46
Bacteria
    defence against, 13, 123, 125, 126
    nitrogen cycle, 31
    of intestine, 36, 38, 76, 85
    synthesis of vitamins K and B, 35, 36
Bainbridge reflex, 100
Balance, sense of (see Equilibrium)
Baroreceptors
    and vasopressin, 188
    in cardiac reflexes, 100, 101
    in respiratory reflexes, 145
    in vasomotor reflexes, 108, 109
Basal ganglia, 223, 225, 232, 273, 275, 277
Basal metabolism, 48, 49
    anterior pituitary activity and, 183
    thyroid activity and, 167–169
Base (definition), 157
Base, reabsorption of, 157
Basilar membrane of cochlea, 253, 254
Basophil cells
    of blood, 13, 119
B cells of pancreas, 189
Beri-beri, 36
Betz cells (giant pyramidal), 182, 222
    functions of, 267–269, 271, 277
Bicarbonate
    buffer system, 157
    carriage in blood, 114, 116, 140–142
    conservation by kidney, 157
    pancreatic, 75
    protection of stomach mucosa, 70
Biceps, 279, 281
Bile, 76, 78, 115
    acids, 76, 77, 83
    canaliculi, 76
    concentration of, 153
    duct, 65, 74, 76
    expulsion of, 77
    pigments, 76, 85, 116, 118
    salts, 76, 77
    secretion, 77
    storage, 77
Bilirubin, 76, 116
    glucuronide, 76
Biliverdin, 76
Binding sites, 53
    high affinity, 53
    low affinity, 53
Binocular vision, 249, 250
Biotin, 36
Bipolar cells, 19
    auditory, 253, 255
    retinal, 246
    vestibular, 258
Birth
    body proportions at, 46
    change in circulation at, 209
    energy requirements, 49

normal, 210
Bladder (urinary), 148, 161, 162, 194
  innervation, 162, 286, 287
Blastocyst, 21, 204, 205
Blastulation, 204
Blind spot, 245
Blindness
  colour, 248
  night, 33
Blood, 13, 90, 116
  acid-base balance, 157, 158
  carbon dioxide, 42, 137–142
  cells, 13, 116, 118, 119, 123, 124
  cholesterol, 168
  circulation, 90, 91
    fetal, 209
    kidney, 148, 149
    placental, 207
    portal, 84
    pulmonary, 113, 132
    splenic, 124
  coagulation, 13, 35, 117, 177
  corpuscles, 13
  exchange of water and electrolytes, 110, 114
  flow in liver, 76
    kidney, 150
    pulmonary, 113
    systemic, 112
  gases, 137–142
    in placenta, 207
  groups, 120–122
  haemopoiesis, 118, 119
  haemostasis, 117
  heat dispersal, 43–45
  oxygen, 42, 137–142
  plasma, 114–116
    clearance, 154–156
    proteins, 116, 117, 123
  platelets, 13, 117, 119
  pressure
    and adrenaline, 177
    and angiotensin, 188
    and noradrenaline, 176
    and posterior pituitary, 184
    arterial, 104–109
    capillaries, 110
    in underactivity of adrenal cortex, 174
    in underactivity of anterior pituitary, 183
    kidney glomeruli, 150
    measurement, 105
    pulmonary, 113
  regulation of water balance, 115, 160
  reservoir, spleen, 124
  sinusoids, liver, 76
  specific gravity, 116
  sugar, 39, 40, 84, 116
    adrenaline and, 177
    anterior pituitary and, 181–183
    role of glucagon, 189
    role of insulin, 189

role of adrenal cortex, 174, 175, 182
  transfusion, 120, 122
  transport
    foodstuffs, 84
    haemopoietic factor, 118
    heat, 43–45
    hormones
      endocrine glands, 166–189
      gastro-intestinal, 71, 72, 75, 77, 80, 82
      reproductive system, male, 192–195
        female, 196–217
      respiratory gases, 130, 132, 137–142, 144, 207
    waste products (for excretion)
      in faeces, 38
      in kidney, 148–155
  velocity, quantity (of flow), 112
  viscosity, 104, 116
  volume, 102, 114, 116, 174, 188
Blood-brain barrier, 127
Blood vessels, 90, 91, 103
  blood pressure in, 104, 105, 113
  capillaries, 103, 110, 113
  control of, 107–109
  cross-section area, 112
  development, 21, 206
  effects of adrenaline, 177
  elastic arteries, 103, 106
  flow in, 112, 113
  in heat dispersal, 43–45
  innervation of, 107, 286–288
  median eminence, 179
  role of vitamin C, 36
  structure of, 11, 14, 103
  veins, 103, 111
B-lymphocytes, 123–126
B-memory cells, 126
Body-building compounds, 32, 34, 50
Body cavities, 11
  development of, 21
  pericardial, 92
  peritoneum, 79
  pleural, 132
Body fluids
  composition, 24, 114
    control of, 110, 114
    defence of tonicity, 159
    defence of volume, 160
    osmolarity, 159, 160
    role of ADH, 159, 160, 184, 186, 187
    role of adrenal cortex, 173
    role of kidney, 151–153, 157, 158
  intake, 115
  output, 115
  volume and sodium, 160
  water balance, 115
Body of stomach, 65, 70, 73
Body temperature, 43–45
  effect of thyroid, 168, 169
  in panhypopituitarism, 183
Böhr effect, 139

INDEX

Bolus of food, 66, 69
Bones, 15, 34, 35, 278, 279, 281
  action of growth hormone, 179–183
  action of calcitonin, 167, 170
  action of parathormone, 170–172
  as levers, 281
  chemical constituents of, 24, 34
  marrow, 13
  haemopoiesis, 118, 119
  skeleton, 278
  vitamin D deficiency, effect of, 35
Bowman's
  capsule, 149, 150, 153
  glands (nose), 236
Brachial artery, 105
Bradykinin, 262
Brain, 220, 222
  cells of, 18
  centres and pathways (*see* Cerebral cortex)
  cerebrospinal fluid, 127
  control of respiratory movements, 143–145
  coronal section, 225
  cranial nerves, 226
  development of, 221
  horizontal section, 223
  links with cerebellum, 273, 275, 276
  links with multineuronal motor pathways, 272
  link-up of neurons, 232–234
  lobes, 222
  membranes of, 127
  motor cortex, 267
  role in control of muscle movement, 277
  sensory cortex, 266
  ventricles of, 127, 223–225
  vertical section, 224
  Wernicke's area, 222
Brain stem, 224, 225, 232, 272, 273, 288
Breast (*see* Mammary gland)
Breathing
  mechanism of, 134
Broad ligament (uterine), 197
Bronchi, 130–132
Bronchioles, 131, 132
  action of adrenaline on, 177
  muscle, innervation of, 286, 287
  respiratory, 132
  stretch receptors in, 145
  terminal, 132
Brunner's glands, 78, 80
  sympathetic innervation of, 80
Brush border, 83
Buffer
  nerves (heart), 101
  salts in urine, 157, 158
  systems, 157, 158
Bulbo-urethral glands, 194
Bundle of His, 97

Caecum, 65, 81, 84, 85
  control of movements, 86, 87, 286, 287

Calcitonin, 167, 170
Calcium, 24, 50
  absorption, 84, 170–172
  and calcitonin, 167
  and muscle contraction, 284
  content of food, 34
  excretion, 163
  in blood, 116, 170
  in blood clotting, 117
  ion (Ca^{2+}), 61, 170
  metabolism and parathormone, 35, 170–172
  muscle excitability, 171
  second messenger system, 61
  urinary, 172
Calmodulin, 61
Calorie
  intake, 43
  requirements, 48, 49
  values of foods, 32
Calyx (kidney), 149
Cannon's point, 86
Capacitation (sperm), 194
Capillaries, 90, 91
  and tissue fluid exchange, 110, 114
  blood (hydrostatic) pressure in, 104, 110, 113
  cross-section area/blood flow
    in pulmonary, 113, 132
    in systemic, 112
  crystalloid osmotic forces across, 110
  kidney, 149–156
  protein osmotic forces across, 110, 113
  structure, 103
  vitamin C and, 36
Capsule of kidney, 149
  pressure in, 150
Carbamino compound, 141, 142
Carbohydrate, 24, 25
  absorption, 82–84
  body fat and, 40
  caloric value of, 32
  carbon cycle, 29, 30
  dietary sources, 33, 50
  digestion, 37, 66, 74, 78
  energy release from, 41, 42
  metabolism, 39
    adrenal cortex and, 173
    adrenaline and, 177
    islets of Langerhans and, 189
    pituitary and, 179–183
  plant sources, 33
  release of energy from, 41, 42
  requirements, 48, 50
  specific dynamic action of, 48
Carbon, 24, 32
Carbon cycle, 29, 30
Carbon dioxide
  in blood, 94, 116
    carriage and transfer, 140–142
    exchange of respiratory gases, 137–140
    fetal-maternal, 207

in regulation of arteriolar tone, 108, 109
    blood pressure, 108
    respiration, 144, 145
in carbon cycle, 29, 30
in respired air, 137 ·
Carbonic anhydrase
  in blood (RBC), 141, 142
  kidney, 157, 158
Carboxypeptidase, 38, 74
Cardiac
  centres, 99–101
  cycle, 95, 96
  fibrous-tissue ring, 92, 97
  murmurs, 96
  muscle, 92, 93, 97
    chemical transmission, 289
    histology, 16
  nerves, 99
  output, 102, 104, 188
    to kidneys, 148, 150, 188
  pump (in venous return), 111
  reflexes, 100, 101
  sphincter (stomach), 65, 68–70
    control of, 88
Cardio-acceleratory centre, 99
Cardio-inhibitory centre, 99
Cardiovascular system, 22, 90
Carotene, 35
Carotid body
  and blood pressure, 108, 109
  in regulation of respiration, 144, 145
  in vasomotor reflexes, 108, 109
Carotid sinus, 101, 145
  and blood pressure, 108, 109
  and respiration, 145
  and vasomotor reflexes, 108, 109
  nerves, 226
Carrier proteins, 52
Cartilage, 15
  at joints, 281
  in ear, 252
  in trachea, 131
Catalyst, 37
Cation, 54
Cauda equina, 227
Caudate nucleus, 223, 225
  links with other centres, 272
Cell, 2, 5
  anterior horn, 18
  Betz, 18, 222
  brush border, 11
  chemical constituents, 24–28
  ciliated, 11
  columnar, 11
  control of, role of thyroid, 167–169
  cornified, 12
  cubical, 11
  differentiation, 10
  division, 9, 20
    meiosis, 20

    mitosis, 9
  enzyme systems, 42
  fat, 10
  fluid in, 114, 159
  goblet, 11
  inclusions, 5
  Leydig, 193
  membrane, 2, 6
  muscle, 10
  mucous, 11
  nerve, 10
  neuroglial, 18
  organelles, 5, 6
  oxidations, 42
  pancreas A and B, 189
  permeability, 6, 52, 53, 114
  Purkinje, 18, 274
  satellite, 17, 19
  Schwann, 17, 18
  secretory, 10, 11
  serous, 11
  Sertoli, 193
  squamous, 11
  sustentacular, 236
  (see also Tissues)
Cell mediated immunity, 126
Cellular absorption, small intestine, 83
Cellulose, 25
Central artery of retina, 239, 245
Central canal of spinal cord, 127, 221, 227
Central integrating centres, 220
Central nervous system, 22, 220, 229, 232, 289
Centre, swallowing, 69
Centriole, 5, 8, 9, 20
Centromere, 9
Centrosome, 5, 8, 9, 20
Cerebellum, 224, 225, 232, 274–276
  basket cells, 274
  climbing fibres, 274, 275
  cortex, 274
  development of, 221
  Golgi cells, 274
  granule cells, 274
  histology, 18, 274
  in control of muscle movements, 273, 277
  links with multineuronal motor pathways, 272, 273
  lobes, 274
  mossy fibres, 274
  nucleus, 274
  parallel fibres, 274
  peduncles, 275, 276
  proprioceptor pathways, 260
  Purkinje cells, 274
  vestibular pathways and centres, 258
  white matter, 274
  zones, 274
Cerebral cortex
  association areas, 222, 266
  centres and pathways
    auditory, 255

autonomic, 288
eye movements, 241
motor to head and neck, 268
motor to trunk and limbs, 269
olfactory, 236
pain and temperature from trunk and limbs, 262, 264
proprioceptor, 260
sensory, 263
taste, 238
touch and pressure, 265
vestibular, 258
vision, 249
convolutions, 222
cranial nerves, 226
development of, 221
'edifice', 232
grey matter, 222
histology, 18, 222
links with cerebellum, 273, 275, 276
links with multineuronal motor pathways, 272
motor areas, 222, 267
neurons, 222
columns, 222
modules, 222
role in control of muscle movement, 277
sensory areas, 222, 266
structure, 222
coronal section, 225
horizontal section, 223
vertical section, 224
Cerebral hemispheres, 222, 224
Cerebral nuclei, 273
Cerebral peduncles, 225
Cerebrospinal fluids, 114, 127, 223
in development, 221
in respiration, 144
Cerebrum, 222, 271, 272
development of, 221
Cervical enlargement of spinal cord, 227
Cervical mucus (uterus), 201
Cervical nerves, 227
Cervix (see also under Uterus), 197, 201, 210
Channel proteins, 56
Channels, 52, 60
K$^+$, 58
ligand gated, 52, 61
Na$^+$, 58
voltage gated, 52, 58, 59, 61
Chemical messengers, 166
Chemical senses, 235, 236
smell, 236
taste, 237, 238
Chemical transmission, 60, 228, 289
in gut, 87
Chemoreceptors
and arteriolar tone, 108, 109
in regulation of respiration, 144, 145
Chenodeoxycholic acid, 76
Chest (see Thorax)
Chewing, 66

Chiasma, optic, 226, 249
Chief or peptic cells, stomach, 70
Childbirth, 210
effect of oxytocin, 185
Children
anterior pituitary in, 179–181
effect of thyroid deficiency in, 168
energy requirements in, 48, 49
growth, 46, 47
effect of vitamins on, 35, 36
mammary glands in, 213
overactivity of adrenal cortex in, 175
reproductive system in, 192, 196
thymus in, 125
uterus in, 202
vitamin D deficiency in, 35
Chloride, 24
and membrane potential, 56
in blood, 116
in body fluids, 114
in kidney, 151, 152
Chloride shift, 141, 142
Chlorine, 24
Chlorophyll, 29
Cholecystokinin/pancreozymin (CCK-PZ), 72, 75, 77, 166
Cholera toxin, 80
Choleretic action, 76
Cholesterol, 173
absorption, 83
in bile, 76
in blood, 116
in myxoedema, 168
Cholic acid, 76
Choline, 36
Cholinergic nerves, 289
Cholinesterase (acetylcholinesterase), 270
Chordae tendineae, 92, 93
Chorionic
gonadotrophins (see Placenta)
membranes, 208
villi, 205–207
Choroid
coat of eye, 239, 246
plexuses, 127, 223
Chromaffin granules
adrenal medulla, 176
Chromatids, 9, 20
Chromatin, 5, 6, 9
in meiosis, 20
Chromosomes, 5, 6, 9, 21
DNA in, 5, 6, 28
in meiosis, 20
in mitosis, 9
nucleotides in, 28
Chvostek's sign, 171
Chylomicrons, 83
Chyme, 71, 72, 75, 78, 81
Chymotrypsin, 37, 38, 74
Cilia, 4, 11, 12
in bronchi, 132

in trachea, 131
in uterine (Fallopian) tube, 197
Ciliary
arteries, 239
body, 239, 243
ganglion, 243, 251, 286
muscle, 239, 243, 244
nerves, 226, 286
Cimetidine, 71
Circadian rhythms, 249
Circulation
course of, 91
fetal
changes at birth, 209
placental, 207
portal, 84
pulmonary, 91, 113
splenic, 124
Citric acid cycle, 8, 40–42
Claustrum, 223, 225
Clearance tests
creatinine, 154
diodone, 156
inulin, 154
PAH, 156
urea, 155
Cleavage of zygote, 21, 204
Clot formation, 117
Clot retraction, 117
Clotting (see Blood coagulation)
Clotting factors, 35, 117
Coccygeal nerves, 227
Cochlea, 252–254, 256
centres and nerve pathways, 226, 255
mechanism of hearing, 254
Cochlear branch of vestibulocochlear nerve (VIII), 226, 252, 255
Coeliac ganglion, 88, 287
Coenzyme A, 36, 42
Coenzyme systems, 42
vitamin B complex in, 36
Cold
as skin sensation, 261
effect on arteriolar tone, 108
effect on respiration, 145
effects on skin, 45
pathways from skin, 263, 264
Collagen fibres, 14, 117
Collateral ganglion in abdomen, 288
Collecting tubules (ducts), 149, 151, 152, 186, 187
Colliculi, 224, 225, 249, 255, 272
development of, 221
Colloid
osmotic pressure, 110
(see under Osmotic pressure)
Colon (ascending, transverse, descending and pelvic), 65, 85, 86
absorption in, 84
functions, 85
innervation, 88, 286, 287

movements in, 86
secretion into, 80
Colour
blindness, 248
sensation, 248
vision, 245, 247, 248
Coma, 225
diabetic, 189
Common bile duct, 76
Communication between cells, 60
Complement system, 126
Concentration of urine, 151, 152
Conditioned reflexes, 234
expulsion of bile from gall bladder, 77
salivary, 67
secretion of gastric juice, 71
secretion of pancreatic juice, 75
Conducting arteries, 103, 104, 106, 113
Conduction
in heat loss, 43–45
in nerve, 17, 59, 220, 228
of heart beat, 97, 98
Cones (retina), 245, 246
mechanism of vision, 248
Conjugate deviation (eyes), 242
Conjunctiva, 240
Connective tissues, 11, 13–15, 182
development of, 21
of oesophagus, 68
role of vitamin C, 36
Consciousness and reticular activating system, 225
Consensual light reflex, 251
Contractile filaments, 283, 284
Contractile vacuole, 2, 3, 5
Contractility, 3, 10, 16, 22, 283, 284
Contraction, muscle, 281–284
control of, 270, 277
Convection
in heat loss from skin, 43–45
Converting enzyme, 160
Convoluted tubules, 149–154, 157–160
action of ADH, 184, 186, 187
action of adrenal corticoids, 173–175
action of parathormone, 170–172
Copper
in blood, 116
in haemopoiesis, 118
Cornea, 239, 243, 244
Coronary blood vessels
vasodilatation, 109
Corpus albicans, 198, 201
Corpus callosum, 224, 225
Corpus luteum, 198–201, 203, 205, 206, 208, 212–215, 217
control of progesterone
secretion by anterior pituitary, 179, 200, 217
Corpus striatum, 223–225, 272
links with other centres, 272
Cortex, cerebral (see under Cerebral cortex and Brain)
Cortex of kidney, 149
Corti, organ of, 252–255

Corticoids, 173–175
  and anterior pituitary, 179, 182
Corticospinal tract (formerly pyramidal tract), 224, 232, 273, 277
  chemical transmission in, 289
Corticosteroid binding globulin, 173
Corticosterone, 173, 175
Corticotroph cells, 179
  in haemopoiesis, 118
  overproduction of, 182
Corticotrophin (see Adrenocorticotrophic hormone)
  releasing hormone, 179
Cortisol (hydrocortisone), 173, 175, 179
Cortisone, 175
Costal cartilages, 133
Cotransport, 53
Coughing, respiration in, 145
Counter-current system (kidney), 152
Counter transport, 53
Cranial nerves, 226, 268
  sensory ganglia of, 288
Creatine, 38
Creatinine
  clearance, 154
  in blood, 116
  urinary, 151, 163
Cretin, 168
Crista (of semicircular canals), 256, 257
Cross bridges (muscle), 284
Crypts of Lieberkühn, 78, 80
Crystalline lens, 239, 243, 244
Cubical epithelium, 11
Cumulus oophorus, 198
Cuneate nucleus
  chemical transmission in, 289
  links with cerebellum, 275
  proprioceptor pathways, 260
  touch and pressure pathways from trunk and limbs, 265
Cupula of semicircular canals, 256, 257
Cushing's syndrome, 175, 182
Cutaneous sensation
  centres, 222, 266
  nature of stimuli, 235
  pathways, 226, 263–265
  receptors, 19, 161
    links with cerebellum, 275
Cyanocobalamine (vitamin B_{12}), 36
Cyclic 3, 5-adenosine monophosphate (cAMP), 61, 186
Cyclic 3, 5-guanosine monophosphate (cGMP), 61
Cystic duct, 76
Cystine, 27
Cytochrome system, 42
Cytoplasm, 2, 3, 4, 5, 9, 10
Cytosine, 28
Cytosol, 5, 83

Dark adaptation of eye, 247
D cells of pancreas, 189
Dead space air, 137
Deamination, 38

Decidua, 206
  after childbirth, 211
Defaecation, 65, 85, 86, 88
Defence
  digestive tract in, 70, 78, 79, 85
  immunity, 13, 126
  lymphatic tissue, 123–125
  white blood cells, 13, 116, 118, 119
Dehydration, tissues, causes, 160
Dendrites, 17–19, 228
Dense fibrous tissue, 14
Dentate nucleus (cerebellum), 276
Deoxycorticosterone, 173
Deoxyribonucleic acid (DNA), 5, 6, 28
Deoxyribose, 28
Depolarization, 57
Depot fat, 39, 40
Depth perception, 250
Descending limb (of Henle's loop), 149
Development of the individual, 21
  growth, 46, 47
  intrauterine, 204–206
Dextrinase (iso-maltase), 37, 39, 78
Diabetes
  insipidus, 187
  mellitus, 189
Diaphragm, 68, 69, 130, 133, 134, 143
  innervation, 143
  in venous return, 111
  in vomiting, 73
Diarrhoea, 159, 188
Diastasis, 95
Diastole, 95
  blood pressure in, 104, 105
  elastic arteries in, 103, 106
  heart sounds in, 96
Diastolic blood pressure, 105
Diencephalon, 221
Diet
  balanced, 32, 50
  body-building requirements, 32, 34
  effect on haemoglobin regeneration, 118
  energy requirements, 32, 33, 48–50
  influence on growth, 46, 47
  minerals, 50
  vitamins, 32, 35, 36, 50
Diffusion
  facilitated, 52
  simple, 52
Digestion, 37–40, 64
  in intestine, 74, 78
  in mouth, 66
  in stomach, 70
Digestive system, 22
  blood circulation and, 91
Diglycerides, 40, 61
Dihydrotestosterone, 192
Dihydroxycholecalciferol (1,25-DHCC), 166, 170–172
Diiodotyrosine, 167
Dilator pupillae, 243

action of adrenaline, 177
Diodone
  renal clearance, 154, 156
Dipeptides, 27, 83
Diphosphoglycerate (DPG)
  and O_2-haemoglobin combination, 139, 141
Diploid number, 20
Disaccharide, 25, 78
Dissociation curves, 139, 140
Disynaptic pathway, 282
Diuresis, 160, 187
Donor (blood), 120–122
Dopamine, 179, 213, 289
Dorsal column system, 265
Dorsal horn (spinal cord), 262
Dorsal vestibulo-spinal tract, 271, 272
Down regulation (receptors), 60
Drum membrane, ear, 252, 254
Duct
  breast, 213–215
  ejaculatory, 194
  lymphatic, 123
  thoracic, 123
Ductless glands, 166
Ductus arteriosus, 209
Ductus venosus, 209
Duodenum, 65, 74, 78
  absorption in, 82
    vitamin B_{12}, 118
  bile in, 76, 77
  digestion in, 78
  emptying, 72
  hormones of, 72, 75, 77, 80, 82
  movements of, 81
  secretion, 78
    control of, 80
Dup (second heart sound), 96
Dura mater, 127
Dwarfing
  in pituitary deficiency, 180
  in thyroid deficiency, 168
Dynorphins, 262

Ear, 252, 253
  drum, 252, 254
  mechanism of hearing, 254
  nature of stimuli, 235
  pathways to brain, 255
  receptors, 19, 253, 255
Echocardiography, cardiac output, 102
Ectoderm, 21
  development of nervous system, 221
Ectoplasm, 2
Edinger-Westphal (accessory oculomotor) nucleus, 243, 251
Effective filtration pressure, kidney, 150
Effector neurons, 233, 234
Effector organs
  autonomic nervous system, 286, 288
  central nervous system, 269
  chemical transmission to, 54, 289
  in reflex action, 229–233

Efferent
  arterioles, (kidney), 149, 150
  nerves, 229, 231
  pathway, 232
Ejaculatory duct, 194
Elastic arteries, 183
  function of, 106
  in pulmonary circulation, 113
Elastic fibres, 14
  in arteries, 103
  in cartilage, 15
  in respiratory system, 14, 131, 132, 134
Elasticity, arterial, 106
Electric charges and ions, 54
Electrocardiogram, 98
Electrocardiograph, 98
Electrolytes, 114
  control of distribution, 110
    in blood, 116
    in urine, 163
  control of excretion
    role of kidney, 148, 150, 151, 157–160
    role of parathyroid, 170–172
    role of posterior pituitary, 186, 187
    role of adrenal cortex, 173–175
  distribution in body, 114
Electromagnetic waves, 235
Electron transport chain, 42
Electrotonic potentials, 57
Embryo
  development, 21, 204, 221
  implantation, 21, 205, 206
  placenta, 207
Embryonic germ disc, 21
Emetic drugs, 73
Emotion
  and ADH, 186
  and adrenaline, 176, 177
  and lactation, 214
  and limbic system, 225
  and respiration, 145
  (see also Stress)
Endocardium, 92, 97
  histology, 92
Endocrine glands, 60, 166–189
Endocrine system, 22, 166–189, 220
Endocytosis, 53
Endoderm, 21
Endolymph
  of ear, 253, 254
  of semicircular canals, 256, 257
Endolymphatic duct, 256
Endometrium, 197, 198, 201–207, 211, 212, 217
Endoplasm, 2
Endoplasmic reticulum, 5, 7, 8
  rough, 5, 8
  smooth, 5, 8
Endorphins, 262
Endothelium, 11
  heart valves, 93

of blood vessels, 103
permeability, 114
Energy
balance, 43
cellular, 39, 41, 42
foods, 33, 50
for growth, 46
for muscle contraction, 283
muscular, 39
release, 32, 41, 42
requirements, 32, 48–50
role of thyroid, 167–169
sources, 29
uses, 42–47, 283
work, 48, 49
Enkephalins, 87, 262
Enteric nervous system, 87
Enterogastric reflex, 71
Enterogastrone, 166
Enteropeptidase (enterokinase), 74
Environment
influence on growth, 46
internal, stability of
acid-base balance, 157, 158
role of endocrines, 166, 170–175, 184, 186, 187
role of kidneys, 148, 159
water balance, 115, 160
stimuli from, 235
Enzymes, 3, 6, 11, 37–42, 64, 116
α amylase, 66
carbonic anhydrase
in kidney, 157, 158
in red blood cells, 141, 142
glutaminase, 158
intrinsic factor, 118
lysozyme, 240
mouth, 37, 39, 66
pancreas, 37, 40, 74, 75
role of vitamin B complex, 36
small intestine, 37–39, 78
stomach, 37–39, 70, 71, 118
Eosinophils
adrenal cortex, 174
blood, 13, 119
parathyroid, 170
Epididymis, 192, 193
Epiglottis, 69
Epithelia, 11, 12
development of, 21
stratified squamous, 12
transitional, 12
(for sites see under Organs)
Equilibrium (balance)
and cerebellum, 274
mechanism of action, 235, 257
medial motor system, 273
organs of, 252, 256, 257
vestibular connections, 272
vestibular pathways to brain, 226, 258, 260
Equilibrium potential, 55, 58

Erythrocytes (red blood corpuscles), 13
agglutinogens, 120–122
carriage and transfer of O_2 and CO_2, 141, 142
destruction of, 124
electrolytic composition, 114
formation, 118, 119
functions, 116
Erythropoietin, 118, 166
Erythropoiesis, 118
Ethnoid bone, 236
cribriform plate of, 236
Eustachian tube, 236, 252
Evaporation in heat loss, 43, 44
Excretion, 3, 10, 22, 148
digestive system, 64
in bile, 76
large intestine, 85
salivary glands, 66
kidneys, 148–163
clearance of,
creatinine, 154, 156
diodone, 154, 156
inulin, 154
PAH, 156
urea, 154, 155
respiratory system, 130–145
Exercise
blood elements in, 119
body temperature in, 43–45
carbon dioxide in, 144
cardiac output in, 102
energy requirements for, 48, 49
pulmonary ventilation in, 113, 136
urine formation in, 163
Exocytosis, 53
Exophthalmos, 169
Expiration, 130, 132
in artificial respiration, 135
mechanism of, 134
nervous control of, 143
Expiratory
capacity, 136
centres, 143
muscles, 134
reserve volume, 136
Extension of joints, 281
reciprocal innervation in, 282
Extensor reflexes
ankle jerk, 230
knee jerk, 230
External auditory meatus, 252
External genitalia
female, 196, 200, 216
innervation, 286, 287
male, 192–195
reflex vasodilatation, 107, 109
External (lateral) rectus muscle of eye, 241, 242
in conjugate deviation, 242
innervation, 226
External respiration, 130, 138

External spiral ligament of cochlea, 253
Exteroceptors, 235
Extracellular fluid, 113
  aldosterone and, 173, 188
  distribution, control of, 110
Extracellular ions, 114
Extraembryonic coelom, 21
Extrapyramidal system, *see* Multineuronal motor pathways
Extrinsic factor (vitamin B$_{12}$), 118
Extrinsic muscles of eye, 241
  innervation, 226
  reflex movements of, 242
Eyelids, 240, 241
Eyes, 239
  accommodation, 243, 244
  autonomic fibres to, 243, 286, 287
  conjugate movement, 242
  control of movements, 242
  cortical centres, 242, 249
  effects of adrenaline on, 177
  extrinsic muscles of, 241
  fundus oculi, 245
  innervation, 226
  light reflex, 251
  mechanism of vision, 246–248
  movements, 241
  nature of stimulus, 235
  nerve pathways from, 249
  protective mechanisms, 240
  stereoscopic vision, 250
  visual purple (rhodopsin), 35, 247

Facial muscles, 279
  nerve supply, 226, 268
Facial nerve (cranial VII), 224, 226, 241, 242, 263, 286
  motor pathways to head and neck, 268
  parasympathetic fibres, 240, 286
    to lacrimal gland, 240
  pathways for taste, 238
  sensory pathways from face, 263
Factors, clotting, 35, 117
Faeces, 85
  bile pigments in, 76
  calcium in, 170
  expulsion, 86
  fluid loss in (or in water balance), 115
  heat loss in, 43
Fallopian tubes (uterine tubes), 194, 196, 197, 201–204
  at maturity, 203
  at menopause, 216
  at puberty, 200
  fertilization in, 204
  motility of, 194
Fat, 24, 26, 29, 30
  absorption of, 76, 82–84
  caloric value of, 32
  cells, 14
  dietary, 33, 50
    specific dynamic action, 48
  digestion of, 37, 70, 74, 76, 78

energy release from, 41, 42
  from carbohydrate, 39
  globules, 5
  heat insulation, 44, 45
  in blood, 116
  in growth, 47
  metabolism, 40
    endocrines affecting, 168, 169, 173, 175, 179, 181, 182, 189
  storage of, 10, 40
  transport of, 84
  (*see also* Lipids)
Fat-soluble vitamins, 35
  absorption, 84
Fatty acids, 26, 37, 40, 41, 78
  absorption and transport, 82, 84
  essential, 32
  polyunsaturated, 26
  role in emulsion of fats, 76
  saturated, 26
  unsaturated, 26
F cells, of pancreas, 189
Fear, 225
  and respiration, 145
Feeding, 225
Female reproductive system and sex hormones (*see under* Reproductive system)
Fertilization, 21, 203, 204, 217
Fetal circulation, 209
Fetus
  birth weight, 46
  circulation in, 209
  development of, 21, 204–208
  haemolytic disease, 122
  haemopoiesis in, 124
  placental circulation, 207
  Rh incompatibility, 122
Fibres
  myelinated, 59
  unmyelinated, 59
Fibrin, 116, 117
Fibrinogen, 116, 117
Fibrinolysin (plasmin), 177
Fibrous tissue, 14
  in blood vessels, 103
Fick principle, 102
'Fight or flight' mechanism, 177, 287
Filtration (*see under* Kidney)
Final common pathway, 271–273
  in reciprocal innervations, 282
Firing level, 57
First heart sound ('lubb'), 96
First messenger, 61
Fissures (brain), 222
Fixators, 281
Flavoproteins, 42
Flexion, of joints, 281
  reciprocal innervation, 282
Flocculonodular lobe of cerebellum, 274, 276
Fluids of body (*see under* Body fluids)

INDEX

Folic acid, 36
  in haemopoiesis, 118
Follicle-stimulating hormone (FSH) *see also* Gonadotrophins, 179, 195, 198, 200, 202, 203, 205, 206, 212–217
Follicular cells (ovary), 198
Follicular fluid (ovary), 198
Food
  absorption of, 82, 84
  balanced diet, 50
  body-building, 32, 34, 50
  bolus, 66, 81
    and respiration, 145
  digestion of, 37, 66, 70, 74, 78
  energy balance, 43–46
  energy-giving, 32, 33
  energy requirements, 32, 48–50
  metabolism, 38–40
  oxidation of, 42
  progress along alimentary canal or digestive tract, 65
  specific dynamic action, 48
  vitamins, 35, 36
Food vacuole, 2–4
Foramen of Munro, 127
Foramen ovale, 209
Forced breathing, 134, 143
Forebrain, 224, 225
  development of, 221
Fornix (brain), 224
Fourth heart sound, 96
Fourth ventricle, 127
Fovea centralis, 245, 246
Free fatty acids, 189
Fröhlich's dwarf, 180
Frontal lobes, 221–224, 267
  centres for eye movements, 242
Frontal sinus, 236
Fructose, 25, 33, 39, 83
Functional residual capacity (lungs), 136, 137
Fundus
  of eye, 245
  of stomach, 65, 70
    movements of, 72
Fusimotor fibres, 277

GABA (gamma aminobutyric acid), 289
Galactogogue action of oxytocin, 185
Galactose, 25, 39, 83
Gall bladder, 64, 65, 76, 78
  expulsion of bile, control of, 77
  parasympathetic innervation, 286
Gametes, maturation of, 20
  (*see also under* Ovum and Spermatozoa)
Gamma aminobutyric acid, 289
Gamma globulin
  in plasma, 116
Gamma motor neurons (efferent fibres), 269, 277
Ganglion cells (retina), 246, 247, 249
Gap junctions, 228
Gases, in blood (*see under* Blood)
Gastric glands, 70, 71

  parasympathetic innervation, 286
Gastric hormones, 71
Gastric inhibitory peptide (GIP), 71, 72
Gastric juice, 70, 115
  control of secretion, 71
Gastric mucosa, 118
  in haemopoiesis, 118
Gastric phase of gastric secretion, 71
Gastric secretion, phases, 71
Gastrin, 70, 71, 75, 166
Gastrin releasing peptide (GRP), 87
Gastrocnemius muscle, 280
Gastro-colic reflex, 86
Gastro-ileal reflex, 81
Gastro-intestinal hormones, 71, 72, 75, 77, 80, 82, 166
Gastro-intestinal secretion in water balance, 115
Gastro-intestinal tract, 64, 65
  absorption – calcium and phosphorus – role of parathormone, 170–172
  actions of adrenaline on, 177
  in Addison's disease, 174
Gate control theory (pain), 262
G cells (stomach), 70
Genes, 5, 6, 7, 9, 28
Germinal epithelium, 198
Giantism, 181
Glands
  Brunner, 78, 80
  bulbo-urethral, 194
  cells, innervation, 286
  chemical transmission at, 289
  endocrine
    adrenal cortex, 173
      medulla, 176
    islets of Langerhans, 189
    ovary, 198, 199
    parathyroid, 170
    pituitary, 178, 179, 184
    testis, 193–195
    thymus, 125
    thyroid, 167–169
  endometrial, 197, 201–203, 205
  gastric, 70
  intestinal, 78
  lacrimal, 240
  liver, 76
  mammary, 213, 215
  mucous
    oesophageal, 68
    tracheal, 131
  pancreatic, 74
  prostate, 194
  salivary, 11, 66
  sweat, 43–45
  tarsal, 240
Globulin, 116, 123
Globus pallidus, 223
  links with other centres, 232, 272
Glomerular
  capillaries, 148–150, 153–156

filtrate, 150, 151, 154, 155
filtration rate, 154
Glomerulus, kidney, 148–152, 154–156
  hydrostatic pressure, 150
  osmotic pressure, 150
  volume of filtrate, 187
Glossopharyngeal nerve (cranial IX), 69, 224, 226, 286
  carotid sinus nerve, 101, 108, 109, 144, 145
  motor pathways to head and neck, 268
  parasympathetic fibres, 286
  pathway for taste, 238
Glottis, 131
  in sneezing, 145
Glucagon, 40, 74, 189
Glucoamylase, 37, 39, 78
Glucocorticoids, 39, 173–175, 182
  corticosterone, 173
  cortisol, 173
  in fat and protein metabolism, 174, 175
Gluconeogenesis, 189
Glucose, 25, 37
  absorption and transport, 83, 84
  conversion to fat, 39
  formation from protein, 172–175
  kidney reabsorption, 151
  metabolism, 39, 41
  oxidation of, 42
  role of glucagon and insulin, 189
Glucuronic acid, 76
Glutamic acid (transmitter CNS), 289
Glutaminase in kidney, 158
Glutamine in kidney, 158
Glycerides, 26, 37
  absorption in small intestine, 82, 84
  digestion in intestine, 40, 78
  role in emulsifying fats, 7, 6
Glycerol, 26, 37, 40, 78, 82, 84
Glycine, 27, 76, 289
Glycocholic acid, 76
Glycogen, 37, 39, 41
  action of adrenaline on, 177
  and adrenal cortical hormones, 173–175
  and energy release, 41, 42
  and thyroid, 168, 169
  granules, 5
  in liver, 33, 38–40
  in muscles, 33, 39
  in vaginal epithelium, 197
  Krebs citric acid cycle, 42
  role of glucagon and insulin, 189
Glycogenesis, 189
Glycogenolysis, 189
Glycolipids, 28
Glycolysis, 41
Glycoproteins, 28
Goblet cells, 11, 12, 80, 85
  in trachea, 131
Goitre, 169
Golgi apparatus, 5, 7, 8
Golgi organs (in tendons), 259

Gonadotroph cells, 179
Gonadotrophins (luteinizing LH and follicle stimulating FSH
    hormones), 179, 180, 205–207, 215–217
  at menopause, 212, 215, 216
  at puberty, 195, 200
  in regulation of female
    reproductive cycle, 200–217
  lack of, 167
  placental, 207, 208
Gonadotrophin releasing hormone (see also luteinizing
    hormone releasing hormone LHRH), 179, 195, 200,
    202, 203, 205, 206, 212, 213, 215
Gonads, 192, 196
  at puberty, 195, 200
  role of anterior pituitary in control of, 179
Goose-flesh, 45
G protein, 61
Graafian follicle, 198, 200–203, 205, 212, 213, 215, 217
  role of anterior pituitary, 217
Gracile nucleus
  chemical transmission in, 289
  links with cerebellum, 275
  proprioceptor pathways, 260
  touch and pressure pathways from trunk and limbs, 265
Granular leucocytes, 13, 119
Granules, secretion, 7, 8
Granulosa cells (ovary), 198, 200
Graves' disease, 169
Gravity and venous pressure, 104, 111
Greater curvature, stomach, 70
Greater splanchnic nerve to adrenal medulla, 176
Grey matter
  brain, 222–224
  spinal cord, 227, 262
Groups, blood, 120–122
Growth, 3, 22, 30, 46, 47
  adolescence, 47
  adulthood, 47
  at puberty, 195, 200
  childhood, 46
  dietary requirements for, 34, 48, 49
  hormone, 38–40, 46, 179–182
  hormone releasing hormone, 179
  infancy, 46
  inhibiting hormone (somatostatin), 87, 179, 189
  phases, 46, 47
  role of endocrines in control of, 166
    adrenal cortex, 173–175
    anterior pituitary, 46, 47, 179–182
    gonads, 195, 200
    thyroid, 167
  role of vitamins, 35, 36
  spurt, 46, 47
Guanine, 28
Gustatory cells, 237
Gut
  calcium absorption in, 170–172
  movements, 88
  structure, 79
Gyrus, 222, 266, 267

INDEX

Haematocrit, 156
Haemocytoblast, 119
Haemoglobin
  appearance in red blood corpuscle, 119
  breakdown products, 76
  carbamino compound, 141, 142
  carriage of respiratory gases, 139–142
  iron, 24, 34
Haemolysis, 120–122
Haemolytic disease of newborn, 122
Haemophilia, 117
Haemopoiesis, 118, 119
  in lymph nodes, 123
  in spleen, 124
  nutritional factors for, 118
Haemopoietic factor, 118
Haemostasis, 117
Hair
  adrenaline and, 177
  body temperature regulation, 45
  in panhypopituitarism, 183
  in thyroid deficiency, 168
  touch sensation, 261
  (see also under Hirsutism and Secondary sex characteristics)
Hair cells
  of cochlea, 253
  of labyrinth, 256, 257
Haploid number, 20
Hassall's corpuscles, 125
Haustrations see Segmentation
Haversian canal (in bone), 15
Health, 22
  role of vitamins, 35, 36
Hearing
  auditory receptors, 19, 253, 255
  centres for, 222, 255
  mechanism of, 254
  nature of stimulus, 235
  nerve pathways, 226, 255
  organ of, 252, 253
Heart, 92–102
  blood flow through, 94
  blood pressure, 104
  cardiac muscle, 16, 92
  cardiac output, 102
  cardiac reflexes, 100, 101
  effects of adrenaline on, 177
  effects of thyroid on, 168, 169
  ejection phase, 95
  electrocardiogram, 98
  fetal, 209
  histology, 11, 16, 92, 97
  innervation, 99, 226, 286, 287
  in panhypopituitarism, 183
  murmurs, 96
  origin and conduction of heart beat, 97
  rapid filling, 95
  rate, 99
  sounds, 96
  valves, 14, 93, 95, 96

venous return to, 111
Heat, 30–32
  balance, 43–45
  effect of thyroid, 168, 169
  effect on blood vessels, 109
  effect on respiration, 145
  loss, 43–45
  nerve endings, 44, 45
  production, 43–45
  skin sensation, 261
Height, 46, 47
  (see also under Growth)
Henderson-Hasselbalch equations, 157
Henle's loop (kidney), 149, 151, 152
Heparin, 13, 117
Hepatic
  artery, 76
  circulation, 91
  duct, 76
  flexure, colon, 85
Hepatocytes, 76
Herbivorous animals, 30, 31
Heredity, 5, 6, 7, 9, 28
  influence on growth, 46
  interchange of hereditary material in meiosis, 20
  nucleic acids, 28
Hering-Breuer reflex, 145
Herring bodies, 184
Hertz, 235
Hexose, 25
Hiccoughing, 145
High altitude
  acclimatization to (red blood corpuscles in), 119
  effect on respiration, 144
Hindbrain, 224, 225
  aqueduct, 262
  development of, 221
Hippocampus (brain), 225
Histamine
  and gastric secretion, 71
Hirsutism
  in adrenogenital syndrome, 175
  in Cushing's syndrome, 182
Histone, 6
Horizontal cells (retina), 246
Hormones, 38, 60, 116, 166
  (see under Endocrines, Gastro-intestinal and Reproduction)
Hot flushes, 216
H_2 receptor blocking agents, 71
Humidity
  and control of body temperature, 44
Human immunodeficiency virus (HIV), 125
Humoral mediated immunity, 126
Humoral transmission of nerve impulses, 60, 228, 289
Hyaline cartilage, 15
Hyaloplasm, 5
Hydrochloric acid
  in gastric juice, 70, 71
Hydrocortisone, see Cortisol, 173
Hydrogen, 24

carriers, 42
ions, kidney, 157, 158
  on respiration, 144
  on vasomotor centre, 108, 109
Hydrogen peroxide, 7
Hydrostatic pressure
  of blood, glomeruli (kidneys), 150
    systemic capillaries, 110
  of cerebrospinal fluid, 127
  tissue interstitial fluids, 110
5, hydroxytryptamine (serotonin), 87, 117
Hyperinsulinism, 189
Hyperglycaemia, 189
Hypermetropia, 244
Hyperpolarization, 57, 58
Hyperthyroid, 169
Hyperventilation, 137
Hypoglossal nerve (cranial XII), 69, 224, 226, 252
Hypoglycaemia, 189
Hyponatraemia, 174
Hypophysis, 178
  (see also under Pituitary)
Hypothalamus, 166, 184, 193, 224
  autonomic centres in, 288
  corticotrophin releasing hormone, 179
  development of, 221
  dopamine (prolactin inhibiting hormone), 179
  effect on anterior pituitary, 179
  growth hormone-inhibiting hormone (somatostatin), 179
    releasing hormone, 179
  in control of adrenal cortex, 173
  in control of thyroid, 167
  in eye reflexes, 243
  influence on heart action, 99
  in function of ovary, 200
  in release of oxytocin, 184, 185, 211, 212
  in suckling reflexes, 211, 212, 214
  links with multineuronal motor pathways, 272
  osmoreceptors in, 159, 184, 186–188
  prolactin inhibiting hormone, 179
  prolactin releasing hormone, 179
  releasing hormones, 167, 173, 179, 213–215, 217
  role in body temperature regulation, 44, 45
  role in control of adrenal medulla, 176
  role in control of heart and blood vessels, 99, 107
  see Luteinizing hormone rleasing hormone
  somatostatin (see Growth hormone inhibiting hormone)
  thyrotrophin-releasing
    see Hormone (TRH), 167, 168, 179
Hypothyroid, 168

Ileo-caecal valve, 65, 78, 81, 85
  innervation and control, 86, 88
Ileum, 65, 78
Immune system, 126
Immune response, 13
Immunity, 13, 125
  antibody formation, 123, 125
    cell mediated, 126
    gamma globulin in, 123, 126

humoral mediated, 126
Immunoglobulins, 126
Immunological reactions, 123
Implantation, 21, 204–206, 217
Incisura angularis, 72
Incus (hearing ossicle), 252
Infancy (see under Baby and Growth)
Inferior colliculus, 255
Inferior mesenteric ganglion, 88, 287
Inferior oblique muscle of eye, 241
  innervation, 226, 241
Inferior rectus muscle of eye, 241
  innervation, 226, 241
Infundibular stem, 178
Infundibulum (see Pituitary stalk)
Ingestion, 64
Inheritance (see under Heredity)
Inhibin, 193, 195
Inhibition
  of gastric motility and secretion, 71, 72
Inhibitory
  interneuron, 282
  transmitters, 289
Initial segment (nerve), 228
Inner ear, cochlea, 252–254, 256
  special proprioceptors, 256
Innervation of gut wall, 87
Inositol triphosphate (IP$_3$), 61
Insensible perspiration, 43
Inspiration, 130, 132
  air in, 138
  control of, 143–145
  mechanism of, 134
Inspiratory
  capacity, 136
  centres, 143
  reserve volume, 136
Insula, 225
Insulation of skin, 43–45
Insulin, 39, 40, 74, 189
Integration in nervous system, 220
Interatrial septum, 92
Intercalated discs, 16
Intercellular cement, 11, 13
  role of vitamin C, 36
Intercostal
  muscles, 69, 130, 133, 134, 143, 145
  nerves, 69, 143
Interferons, 126
Interleukins, 126
Interlobular artery (kidney), 148
Intermediate metabolism, 38–42
Intermediate zone (cerebellum), 274, 276
Internal capsule, 223, 225, 260, 263–265, 268, 269, 272
Internal (medial) rectus muscle of eye, 241, 242
  innervation of, 226, 241, 242
Internal respiration, 130, 138
Interneurons, 231, 232, 234, 269, 282, 288
  inhibitory, 282
Interoceptors, 235

INDEX

Interphase, 9
Interstitial cells of Leydig (testis), 193, 195
Interstitial fluid, 110, 114
  pressures, 110
Interventricular septum, 92
Intervertebral discs, 15, 278
Intestinal
  absorption, 82–84
    role of parathormone, 170
  bacteria, 35, 36
  digestion, 78
    functions of bile, 76
    in epithelial cells, 83
  enzymes, 37, 78
  epithelium, 11, 79, 83
  functions, 79
  hormones, 71, 72, 75, 77, 80, 82, 166
  innervation, 81, 87, 88, 286, 287
  juice, 78, 80, 115
  motility, 81, 88
  muscle, 78, 79
  phase of gastric secretion, 71
  secretion, 80
  slow waves of depolarization, 81
  tract, general structure of, 65, 79
  villi, 82
  wall, 79
Intracellular
  ions, 114
  water, 114, 115
Intrafusal muscle fibres, 259, 277
Intrapleural pressure, 133, 134
Intrathoracic pressure, 111, 134
Intrinsic factor
  formation in stomach, 70
  function, 118
Intrinsic nerve plexus, 80, 81
Inulin clearance, 154
Involuntary muscle, 16
Involution of uterus, 212
Iodide trapping, 167
Iodine
  dietary, 34, 50
  in blood, 116
  thyroid hormone, 167, 169
Ions and charges, 54
Ion transport, 52
Iris, 239, 243
  effect of adrenaline on, 177
  in light reflexes, 251
  nerve supply, 226, 286, 287
Iron, 24, 34, 50
  in blood, 116
  in haemopoiesis, 118
  storage in liver, 76
Irritability in nervous system, 220 (see also Phenomena of life)
Irritants, effects on respiration, 145
Islets of Langerhans, 74, 166, 189
  role in carbohydrate metabolism, 39

Isometric
  contraction of heart, 95
  relaxation of heart, 95

Jejunum, 65, 78
  (see also Small Intestine)
Joints
  muscular movements at, 281, 282
  proprioceptors, 259
Juxtaglomerular apparatus, 159, 160, 188

Ketogenic amino-acids, 40
Ketone bodies, 40
  in diabetes mellitus, 189
Kidney
  ammonia formation in, 158
  anatomy and histology, 11, 12, 148, 149
  and acid base balance, 157, 158
  and adrenal cortex hormones, 173–175
  and insulin, 189
  and parathyroid, 170–172
  and posterior pituitary, 186, 187
  and water balance, 115, 152, 159, 160
  clearance tests, 154–156
  concentration of filtrate, 151, 152
  counter-current
    theory/mechanism, 152
  deamination of amino-acids, 38
  effective filtration pressure, 150
  enzymes in, 157, 158
  excretion of ketone bodies, 40, 189
  filtration of blood, 149–151
  functions, 148
  proximal tubule epithelium, 153
Kininogen (high molecular weight), 117
Knee jerk, 230
Krause bulbs, in skin, 261
Krebs citric acid cycle, 8, 40–42
Kupffer cells (liver), 76

Labour, 210
Labyrinth, 235, 256, 257
  pathways to brain, 258
Lacrimal ducts, 240
Lacrimal glands, 240
  parasympathetic innervation, 286
Lacrimal sac, 240
Lactase, intestinal, 37, 39, 78
Lactation, 199, 214
  calorific requirements for, 49
  role of anterior pituitary in, 179
Lacteals, 82, 83
Lactic acid, in muscle, 39, 41, 109
Lactose, 25, 33, 83
Lamellae (of bone), 15
Large intestine, 64, 65, 85
  absorption in, 84
  bacteria in, 36, 38, 76, 85
  deamination in, 38
  innervation, 88, 286, 287

movements, 86
Larynx, 69, 130, 131
  nerves to, 226
Lateral cerebral ventricle, 127, 223
Lateral cortico-spinal tract, 269
Lateral ganglion of sympathetic chain, 288
Lateral geniculate body, 249
Lateral horns of spinal cord, 227
  sympathetic outflow, 287
Lateral intercellular spaces of kidney, 153
Lateral lemniscus, 255
Lateral motor system, 273
Lateral sulcus of cerebrum, 206
Lateral ventricle (brain), 223–225
Lateral vestibulo-spinal tract, 258
Lateral zone of cerebellum, 274, 276
Laughter and respiration, 145
Learning, 225
Lecithin, 76
Lemniscal or dorsal column system, 265
Lemnisci, 224
Lenses for vision correction, 244
Lens of eye, 239, 243, 244
Lentiform nucleus, 223, 272
Lesser curvature of stomach, 70
Leucocytes, 13, 119, 126
Leucocytosis, 119
Leucopenia, 119
Leucopoiesis, 118, 119
Levator palpebrae superioris (eyes), 241
Leydig, cells of, 193, 195
Ligaments, histology of, 14
Ligand, 61
Light adaptation, 248
Light reflex, 251
Limbic system (brain), 225
Limbs
  muscles, 279–281
    innervation, 227
α Limit dextrins, 74
Limiting membranes (retina), 246
Lipase, 37, 40
  gastric, 70
  intestinal, 78
  pancreatic, 40, 74
Lipids, 24, 26, 29
  (see also Fat)
Lipogenesis, 189
Lipolysis, 189
Lipoproteins, 28
Lips, 131
Liver, 64, 65, 76
  action of insulin, 189
  and adrenaline, 177
  and anterior pituitary, 179
  and glucocorticoids, 173–175
  bile formation and excretion, 76
  development of, 21
  glycogen in, food value of, 33–36
  in fetal circulation, 209

innervation, 287
intermediate metabolism, 38–41
lobule, 76
sinusoids, 76
storage of vitamin B_{12}, 118
transport of food from gut to, 84
Local anaesthetic on Na^+ channels, 58
Locomotor system, 22, 235, 279–284
Longitudinal fissure of cerebrum, 222, 224
Loop of Henle, 149, 151, 152
Loose fibrous tissue, 14
Lorain dwarf, 180
Loudness, 254
Lower motor neurons, 232, 269–272, 277
  chemical transmission in, 289
  in reciprocal innervation, 282
Lower visual centres, 249
Lubb (first heart sound), 96
Lumbar enlargement of spinal cord, 227
Lumbar nerves, 227
Lungs, 130, 132, 133
  and acid-base balance, 157
  capacity of, 136
  composition of respired air, 137
  exchange of respiratory gases, 139–142
  heat loss, 43
  Hering-Breuer reflex, 145
  movement of respiratory gases, 138
  nerves to, 226
  stretch receptors in, 145
  volumes of, 136
  water loss from, 115
Luteal phase (endometrium), 205
Luteinizing hormone (LH) (see also Gonadotrophin), 179,
      195, 198, 200, 202, 203–217
  releasing hormone (LHRH), 179, 195, 198, 200, 202, 203,
      205, 206, 212, 213, 215, 217
Luteolysis, 200
Lymph, 114, 123
  absorption of glycerides, 37, 40
  in haemopoiesis, 119, 125, 126
  nodes, 119, 123
  reticular tissue of, 13
Lymphatic vessels, 123
  small intestine, 82, 84
  transport of absorbed foodstuffs, 40, 84
Lymphoblast, 119
Lymphocytes, 13, 123
  B-lymphocytes, 123–126
  formation of, 119, 123–126
  of gastro-intestinal tract, 78, 79, 85
  of thymus, 125
  role in immunity, 13, 125, 126
  T-lymphocytes, 123–126
Lymphoid tissue
  growth of, 46
  in digestive tract, 78, 79, 82, 84, 85
    in large intestine, 85
    Peyer's patch in ileum, 78
  in spleen, 124

INDEX

in thymus, 125
Lysosome, 5, 7, 35
Lysozyme, 240

Macrophage cells, 123, 124
  (*see also* Reticulo-endothelial cells)
Macula
  lutea (retina), 245
  of utricle, 256
Macula densa of kidney, 188
Magnesium, 24, 114
Male
  hormones, adrenal, 173, 175
    testis, 193–196
  organs, 193–196
Malleus (hearing ossicle), 252, 254
Maltose, 66, 74
Maltriose, 66, 74
Mammary glands, 196, 213–215
  action of anterior pituitary on, 179
  action of oxytocin, 184, 185
  after childbirth, 214
  after menopause, 215, 216
  after placental progesterone, 199
  at lactation, 214
  at puberty, 200, 213
  by maturity, 213
  hormonal control of changes in menstrual cycle, 201
  in childhood, 213
  in pregnancy, 214
  suckling reflexes, 211, 212, 214
Mammotroph cells, 179
Manganese, 24, 118
Mass peristalsis, 85, 86
Mastication, 66
  nerve of, 226
  tongue and, 237
Maturation
  of ovum, 20, 204
  of sperm, 20
Maxillary sinus, 236
Meal, rate of passage, 65
Mean arterial blood pressure, 104
Measurement of arterial blood pressure, 105
Medial geniculate body, 255, 258
Medial lemniscus, 260, 264, 265
Medial longitudinal bundles, 242, 258
Medial motor system, 273
Median eminence, 178
  capillary plexus, 179
  long portal veins, 179
Medulla oblongata, 67, 224–227
  centres
    autonomic, 286, 288
    cardiac, 99–101
    expulsion of bile, 77
    gastric secretion, 71
    pancreatic secretion, 75
    respiratory, 143–145
    salivation, 67

    swallowing, 69
    vasomotor, 100–109
    vomiting, 73
  development of, 221
  pathways
    hearing, 255
    links with cerebellum, 275
    links with other centres, 232
    motor, 268, 269
    pain and temperature, 264
    parasympathetic outflows from, 286
    pressure and touch, 265
    proprioceptor, 260
    sensory from head and neck, 263
    taste, 238
    vestibular, 258
  role in control of adrenal medullae, 176
Medulla of kidney, 149
Megakaryoblast, 119
Megakaryocyte, 119
Meiosis, 20
  in oogenesis, 198
  in spermatogenesis, 193
Meissner corpuscles, in skin, 19, 261
Meissner's nerve plexus, in gut, 79, 82, 87
Melanophores, 174
Membrane
  cells, 2, 4, 52
  channels, 6
  nuclear, 2, 5, 6, 9
  permeability to
    Ca^{2+}, 58
    Cl^-, 56
    K^+, 56
    Na^+, 56, 58
  plasma, 5, 6, 35
  potential, 52, 56, 57, 59, 228
  pumps, 6, 53
  transport, 52, 53
Menopause, 216
  changes with, 216
  mammary glands, 215
  uterus after, 212
Menstrual cycle, 197, 201, 203, 217
  at puberty, 200
  cessation at menopause, 212, 216
  endometrial changes in, 203
  length of, 197
  ovarian hormones in, 201
  ovary in, 198
  phases of, 197
  restoration after childbirth, 212
Mercury manometers, 105
Mesencephalon, 221
Mesenchyme, 13
Mesenteric vein, 84
Mesentery, of gut, 79, 82, 85
Mesoderm, 21
Metabolic rate
  basal, 48, 49

resting, 48
Metabolism, 3, 10, 22, 42, 43, 45
 basal, 48, 49
 carbohydrate, 39
 excretion of waste products of, 148, 151, 155
 fat, 40
 heat production in, 43–45
 protein, 38
 resting, 48
 role of endocrines in control of, 166
  adrenal, 173–175
  adrenaline, 177
  anterior pituitary, 179–183
  insulin, 173
  parathyroid, 170–172
  thyroid, 167–169
 specific dynamic action of protein, 48
 vitamins and, 35, 36
 water of, 115
Metaphase, 9, 20
Metarhodopsin II, 247
Metencephalon, 221
Micelle formation, 76, 83
Microfilaments, 8
Microtubules, 8
Microvilli, 78
 enzymes in, 78
Micturition, 162
Midbrain, 224, 225
 centres
  for eye reflexes, 242, 249, 251
  parasympathetic, 286
  vestibular, 258
 development of, 221
 pathways
  for hearing, 255
  links with multineuronal motor pathways, 272
  motor, 268, 269
  vestibular, 258
Middle ear, 252, 254
Milk
 dietary, 33, 34
 expression of, 185 (see also under Lactation)
Milk 'let down' reflex, 185, 214
Mineralocorticoids, 173–175
 aldosterone, 173
 and pituitary, 179, 182
 deoxycorticosterone, 173
Minerals, 24, 70
 absorption in large intestine, 85, 86
 absorption in small intestine, 82, 84
 dietary, 34, 50
 excretion (see under Kidney)
Mitochondria, 5, 8, 35, 42, 283
 cristae of, 8, 42
 matrix of, 8, 42
Mitosis, 9, 21
 in haemopoiesis, 119
 in spermatogenesis, 193
Mitral cells (olfaction), 236

Mitral value, 93, 95, 96
Mixed venous blood, 102
Monoblast, 119
Monochromatic vision, 247
Monocyte, 13, 119
Monoglycerides, 26, 40, 76, 82, 83
Monoiodotyrosine, 167
Monosaccharide, 25, 37, 78
Monosynaptic pathway, 282
Morula, 204
Mossy fibres of cerebellum, 274
Motion sickness, 73, 257
Motivation, 225
Motor cortex, 222, 232, 267
 chemical transmission in, 289
 in control of skeletal muscle movement, 273, 277
 pathways from, 268, 269, 272, 273, 275
Motor end-plate, 17, 230, 270
 chemical transmission at, 298
α Motor neuron, 282
Motor neurons, 17, 18, 220
 in cerebral cortex, 267–269, 277
 spinal cord, 227, 229–233, 270–273, 277
Motor nuclei of cranial nerves, 268
Motor pathways
 in control of voluntary movements, 273, 277
 to head and neck, 268
 to trunk and limbs, 269
Motor pool, 232
Motor systems, 273
Motor unit, 270, 271, 277
Mouth, 64–66, 131
 lining, 12
Mouth-to-mouth resuscitation (artificial respiration), 135
Mucosa
 in digestive tract, 79, 80, 85
  large intestine, 80, 85
  oesophagus, 68
  small intestine, 78, 80
  stomach, 70
  muscularis mucosa, 79
Mucous acini, 66
Mucous glands, 11
 in trachea, 131
 oesophagus, 68
 salivary glands, 66
Mucus (mucin), 11, 12
 in gall bladder, 76
  intestine, 78, 80, 85
  mouth, 66
  oesophagus, 68
  stomach, 70
  trachea, 131
Müller cells (retina), 246
Multineuronal motor pathways (formerly extrapyramidal system), 232, 271, 272
Multipolar neurons, 17, 18
Muscarinic receptors, 289
Muscles
 antigravity, 258

INDEX

action of insulin, 189
chemical transmission in, 289
contraction of, 281, 283, 284
  molecular basis, 284
development, 21
effect of adrenaline on, 177
eye, 239–241
heat production in, 43–45
histology, 14, 16, 283, 284
in diet, 33, 34
in intestine, 78, 79, 85
joint sense, 235, 259
  pathways, 260, 273, 275
of mastication, nerve supply to, 226
of oesophagus, 68
proprioceptors, 235
reciprocal innervation of, 282
release of energy in, 42, 283, 284
respiratory, 133, 134, 143
  innervation of, 143
skeletal, 279, 280, 283, 284
  contraction, 283, 284
  names, 279, 280
  movement of, 281, 282
spasm and calcium, 171
spindle, 259
  in reciprocal innervation, 282
  role in control of muscle movement, 277
  stretch reflexes, 230
stomach, 70
tone, 44, 45
  role of extraneuronal motor pathways, 272
  role of proprioceptors, 259
utilization of glucose by, 39
Muscle 'pump' (venous return), 111
Muscular arteries, 103
  blood pressure in, 104
  nervous regulation of, 107
Muscularis mucosa, 79, 82, 87
  in oesophagus, 68
  'villus pump', 82
Muscular veins, 103
  blood pressure in, 104
  'tone' in, 111
Myelencephalon, 221
Myelin, 17, 18
  and conduction of nerve impulse, 59, 228, 270
Myeloblast, 118, 119
Myelocytes, 119
Myocardium, 92
Myoepithelial cells, mammary glands, 184, 185
Myofibril, 16, 283, 284
Myogenic movement, 81, 86, 88
Myometrium, 197, 202, 203, 206, 208, 210–212
Myopia, 244
Myosin, 283, 284
Myxoedema, 168

Na^+-K^+ pump, 53, 56, 58
Nasal passages, 130, 236

Naso-lacrimal duct, 236, 240
Naso-pharynx, 69, 252
Natural killer cells, 126
Near response (eyes), 243
Negative intra-thoracic pressure, 133, 134
Neocerebellum
  (*see* Posterior lobe of cerebellum)
Nephrons, 149
  clearance of creatinine, 154
    diodone, 256
    inulin, 154
    PAH, 156
    urea, 155
Nerve cell body, 17, 19
Nerve ending, 17, 19
  temperature, 44, 45
Nerve plexus
  Auerbach's, 87
  Meissner's, 87
  oesophagus, 68
Nerves
  adrenergic, 289
  association, 17
  cholinergic, 289
  cranial, 226
  histology, 17–19
  impulse propagation, 59, 228
  motor, 17, 220
  nature of impulse, 228
  sensory, 17, 200
  spinal, 227
Nervous regulation of heart, 99
Nervous tissues, 17–19, 220
  development of, 21, 46, 221
Neurilemma, 17, 18
Neurogenic movement, 81, 86, 88
Neuroglia, 17, 18
Neurohormones, 60
  antidiuretic hormone (vasopressin), 159, 160, 184, 186–188
  oxytocin, 184
Neurohypophysis (*see* Pituitary, posterior)
Neuromuscular excitability
  role of blood calcium and parathyroid, 170, 171
Neuromuscular junction, 270
  chemical transmission at, 289
Neuromuscular transmission, 270, 289
Neurons, 17–19
  la afferent, 230, 282
  arrangement of, 234
  bipolar, 19
  α motor, 230, 282
  multipolar, 18
  synapse, 228
  unipolar, 19
Neurophysin, 184
Neurosecretion
  hypothalamus, 159, 160, 184, 186–188
Neurotransmitter, 60, 289
Neutral (unsplit) fat, 26, 40

(triglycerides), 26
absorption of, 40, 82–84
digestion of, 74, 76, 78, 83
role of bile, 76
Neutrophil, 13, 119
Newborn
circulation, 209
energy requirements of, 49
haemolytic disease of, 122
haemopoiesis in, 118
thymus in, 125
urine in, 163
Niacin (nicotinic acid), 36
Nicotinic acid (niacin), 36
Nicotinic receptors, 289
Night blindness, 35
Nitrate, 29, 31
Nitrogen, 24, 27, 28, 32
cycle, 31
in respired gases, 137, 138
retention, role of growth hormone, 179–181
Nitrogenous compounds, 31
Nociceptors, 262
Nodal tissue (heart), 97, 98
Node of Ranvier, 17, 18
in conduction of nerve impulse, 59
Noise, effect on blood vessels, 108
Non-granular leucocytes, 13
formation of, 119, 123–125
Non-myelinated fibres, 59
Noradrenaline, 176
in chemical transmission, 289
in gut, 87
Normoblast, 118, 119
Nose, 131, 226, 235, 236
Nuclear layers (retina), 246
Nucleases, 74
Nucleic acids, 28, 29
digestion of, 74
Nucleolus, 5, 7, 9
Nucleoplasm, 2, 5
Nucleoproteins, 28
Nucleotides, 28
Nucleus
caudate, 223, 272
cell, 2, 4–6, 9, 10
cerebellar, 274, 276
cochlear, 255
cranial nerve nuclei, 260, 263, 268
cuneate, 224, 260, 265, 275
dentate, 276
division of, 9, 20
Edinger-Westphal (accessory oculomotor), 243, 251
emboliform, 276
fastigial, 276
globose, 276
gracile, 224, 260, 265
histology, 5–9
lentiform, 223, 272
nucleic acids, 6, 7, 28

olivary, 225, 271, 275
paraventricular, 184, 185
pontine, 275
red, 271, 272, 276
solitarius, 238
subthalamic, 223
supra-optic, 159, 160, 184, 186
vestibular, 258, 271, 272
Nutrition, 3, 4, 10, 22, 32, 33–36
factors in haemopoiesis, 118
influence in growth, 46, 47
Nystagmus, 257

Obesity
in Cushing's syndrome, 175, 182
in Fröhlich's syndrome, 180
Occipital lobes, 222–224
centres for eye reflexes, 242
centres for vision, 239, 249, 250
Oculomotor nerve (cranial III), 224, 226, 241, 242, 251, 268, 286
parasympathetic outflow to ciliary muscle, 243, 286
Oddi, sphincter of, 74, 65
Oedema
in overactivity of adrenal cortex, 175
Oesophagus, 64, 65, 68, 70, 170
glands of, 68
innervation, 68, 88
lining, 12, 68
muscle of, 68
swallowing, 69
vomiting, 73
Oestrogen, 196–198, 200–203, 205–208, 211–217
action on mammary glands, 212–214
adrenal cortex, 173–175
after childbirth, 211
and anterior pituitary, 179
at menopause, 212, 216
at puberty, 200
effect on anterior pituitary, 179
in male, 173, 195
estradiol, 201
oestrone, 201, 216
oestriol, 201, 208
receptors, 199
Oils, 33
Olfactory bulb, 226, 236
development of, 221
cells, 236
centres, 236
epithelium, 236
nerve (cranial I), 226, 236
receptors, 236
Olivary nucleus, 255, 271, 275
Olivo-spinal tract, 271
Omnivorous animals, 30, 31
Oocyte, 198
Oogenesis, 198
hormonal control of, 200
Oogonium, 198

INDEX

Ophthalmoscope, 245
Opioid peptides, 262
Optic
  chiasma, 249–251
  disc, 245, 246
  nerve (cranial II), 226, 239, 241, 245, 246, 249, 251
  radiations, 249
  reflexes, 243, 249
Orbicularis oculi (eye), 240, 241
Organ of Corti, 252, 253, 255
  mechanism of hearing, 254
    pathways of brain, 255
Organelles, 5–8
Organs, 22
Osmoreceptors, 159, 184–186, 188
Osmotic pressure
  adjustment in stomach, 70
  crystalloids and, 110
  filtration force,
    kidney glomeruli, 150
    systemic capillaries, 110
    pulmonary capillaries, 113
  in regulation of water balance, 152, 159
    role of ADH, 159, 186, 188
    role of corticoids, 173–175, 188
  plasma proteins and, 110, 127, 150
Osteitis fibrosa cystica, 172
Osteoblasts, 170
Osteomalacia, 35
Osteoporosis
  after menopause, 216
  in adrenal overactivity, 182
Otic ganglion, 286
Otoliths, 256
Oval window, 252, 254
Ovarian cycle, 198, 201, 217
  after lactation, 215
  at menopause, 216
  at puberty, 200
  in pregnancy, 199
Ovarian hormones, 201, 217
  at involution, 212
  at menopause, 216
  at puberty, 200
  in pregnancy, 205, 206
Ovary, 166, 196–199
  action of anterior pituitary on, 179, 217
  at menopause, 216
  at puberty, 200, 202
  hormones of, 201
  influence of adrenals, 173
  in menstrual cycle, 203, 217
Ovulation, 197, 198, 200, 201, 203, 204, 217
Ovum, 196–198, 200, 203
  fertilization and maturation of, 20, 204
Oxaloacetic acid, 42
Oxidation (tissue), 39, 40, 42
  heat production, 43–45
  role of thyroid in, 167–169
Oxidative phosphorylation, 41, 42

Oxygen, 24–30
  carriage and transfer, 13, 94, 116, 139, 141, 142
  consumption, 102
  dissociation from haemoglobin, 139
  effect on uptake and release of CO_2 by blood, 140
  exchange of respiratory gases, 137, 138
  fetal-maternal, 207
  in regulation of arteriolar tone, 108, 109
  in regulation of respiration, 144, 145
  in release of energy, 39, 40, 42
  in respired air, 137, 138
  role in adrenaline secretion, 176
  transfer in placenta, 207
  uptake by blood in lungs, 102
Oxyhaemoglobin, 139–142
Oxyntic cells, 70
Oxytocin, 184, 185, 199, 210–212, 214

Pacemaker (S.A. node), 97–99
Pacinian corpuscles, 19
  in joints, 259
  in skin, 261
PAH clearance, 156
Pain, 223, 224, 235, 262
  effect on blood vessels, 108
  effect on respiration, 145
  fast, 262
  fibres (fast and slow), 262
  gate control theory, 262
  pathways from skin, 264
  skin sensation, 260
  slow, 262
Palaeocerebellum
  see Anterior lobe of cerebellum
Palate, 65
  hard, 69
  soft palate in swallowing, 69, 226
Pancreas, 64, 65, 74, 75, 78, 84, 166
  development of, 21
  digestive secretions, 74, 75
  hormonal secretions, 74
  islets of Langerhans, 74, 189
Pancreatic duct, 65, 74
Pancreatic juice, 75, 78, 115
Pancreatic polypeptide, 189
Pancreatic secretion
  control of, 75
  phases of, 75
Panhypopituitarism, 183
Pantothenic acid, 36
Papillae of tongue, 237
Papillary muscles, 92, 93
Para-aminohippuric acid (PAH) clearance, 156
Paracellular pathway
  kidney, 153
Paracrines, 60
Parafollicular cells (C cells), thyroid, 167
Parallax, 250
Paramecium, 4
Parasympathetic outflow, 286, 287

chemical transmission, 298
cranial, 286
effects of, 286
in control of expulsion of bile, 77
gastric secretion, 71
in movements of small intestine, 81
salivary secretion, 67
secretion of pancreatic juice, 75
secretion of small intestine, 80
swallowing, 69
vomiting, 73
organs supplied, 286
sacral, 286
to blood vessels, 107
gall bladder, 77
gut, 86–88
heart, 99–101
iris, 243, 251
lacrimal glands, 240
urinary bladder, 162
Parathormone, 170–172
Parathyroid, 166, 170–172
overactivity, 172
underactivity, 171
Paraventricular nucleus, 184, 185
Paraventriculo-hypophyseal tract, 184, 185
Paravertebral ganglia, 88, 99–101, 107, 287
Parietal
pericardium, 92
pleura, 133
Parietal lobes, 222, 224, 225
proprioceptor centres, 260
sensory centres, 263–266
taste centres, 238
Parotid salivary gland, 66
innervation, 226
Pars anterior (glandularis, distalis), 178
development of, 178
Pars intermedia, 178
Pars posterior (nervosa), 184, 187
development of, 178
Pars tuberalis, 178
Partial pressure of refined gases, 138–140
Parturition, 210, 211
Patellar tendon, 230
Peduncles (cerebellar), 276
Pellagra, 36
Pelvic floor muscles, 86
Pelvic nerves, 86, 88, 107, 162, 286
Pelvis (kidney), 149
Penfield's
motor cortex, 267
sensory cortex, 266
Penis, 192, 194, 195
Pepsin, 37, 38, 70
Pepsinogen, 38, 70, 71
Peptic cells, 70
Peptidase, 83
Peptide bond, 27
Peptides, 27, 38, 70, 78

Periarterial lymphoid sheaths (PALS) of spleen, 124
Periarterial macrophage sheaths (PAMS) of spleen, 124
Pericardial fluid, 114
Pericardial sac, 92
Pericardium, 92
Perilymph
cochlear, 253, 254
round semicircular canals, 256
Perineurium, 194
Peripheral nervous system, 220
Peripheral resistance, 104, 108, 109
pulmonary circulation, 113
Peristalsis
in large intestine, 85, 86
oesophagus, 68, 69
small intestine, 81
stomach, 72
uterine tubes, 197, 204, 205
Peritoneal fluid, 114
Peritoneum, 78, 79, 85
Peritubular capillaries (kidney), 151, 153
Pernicious anaemia, 36, 70
Peroxisomes, 5, 7
Perspective (vision), 250
Perspiration, 43–45
Peyer's patch, 78
pH (hydrogen-ion concentration)
of blood, 116
of urine, 157, 158, 163
Phagocytes, 13
in lymph nodes, 123
in spleen, 124
Pharynx, 64, 65, 68, 69
innervation, 226
in swallowing, 69
regulation of water intake, 159
Phenomena of life, 3, 10, 22
Phosphate metabolism, 24, 29, 34
and acid-base balance, 158
energy rich bond, 42, 283, 284
excretion, 151, 158
role of parathormone, 170–172
vitamin D and, 35, 170–172
Phosphatidylinositol 4, 5-diphosphate (PIP$_2$), 61
Phospholipase C, 61
Phospholipid, 6, 26, 52, 83, 117
Phosphorus, 22, 34
Photopic vision, 248
Photosynthesis, 29, 30
Phrenic nerve, 69, 73, 143
Physiological haemostasis, 117
Pia mater, 127
Pigment,
areola of the breast, 214
choroid coat of eye, 239
in skin, Addison's disease, 174
retina, 246
Pitch
appreciation of, 254
Pituicytes, 184

INDEX

Pituitary, 166–187
  anterior, 178–183
    adrenocorticotrophic hormone (ACTH corticotrophin), 173
    and the ovarian and endometrial cycles, 217
    and thyroid, 167–169
    at menopause, 216
    control of testes, 193, 195
    effect on mammary glands, 214, 215
    gonadotrophic hormones, 47, 195, 200, 205, 206, 208, 212
    growth hormone, 46, 179–181
    role at puberty, 195, 200, 213
    role in fat metabolism, 40
    role in intermediate carbohydrate metabolism, 39
    role in lactation, 214
    role in protein metabolism, 38
  development of, 178, 221
  posterior, 178, 184–187
    antidiuretic hormone (vasopressin), 152, 159, 160, 186–188
    in pregnancy, 210
    in suckling reflexes, 184, 185, 211, 212, 214
    in water balance, 152, 159, 160, 186–188
    oxytocin after childbirth, 211, 212, 214
    role in lactation, 214
    vasopressor effect, 184
  stalk (infundibulum), 178
Placenta, 166, 206–208, 217
  after childbirth, 21
  effect on lactation, 214
  fetal-maternal circulation, 207
  hormones, 207, 208, 211
    human chorionic gonadotrophin, 199, 208
    human chorionic somatotrophin, 208, 214
    human placental lactogen, 208
  in fetal circulation, 209
  oestrogen, 119
  placentation, 206
  progesterone, 199
  relaxin, 199
  transfer of antibodies in haemolytic disease of the newborn, 122
Plants, 30, 31
Plasma, 114, 116
  agglutinins in, 120, 121
  albumin, 116
  clearance, 154–156
    (see also under Plasma proteins)
  fibrinogen, 116
  globulin, 116
Plasma cells, 123, 124, 126
Plasma proteins, 38, 116
  as weak acids, 140
  osmotic pressure, 110, 127, 133, 150
  role in carriage and transfer of O_2 and CO_2, 140–142
Plasma volume, 114
Plasmin (fibrinolysin), 117
Plasminogen, 117
Plasminogen activators, 117

Platelet phospholipids, 117
Platelet plug, 117
Platelets, 13, 116, 124
  formation of, 119
  role in blood clotting, 117
Plates of liver cells, 76
Pleura, 132–134
  intrapleural pressure, 133, 134
  parietal, 133
  visceral, 133
Pleural fluid, 114
Pleural sac, 132
Plexiform layers (retina), 246
Pneumotaxic centres, 143
Polar bodies, 20, 204
Polycythaemia, 119, 175
Polydipsia, 189
Polymers, 25, 27
Polymorphonuclear granular leucocytes, 13
  formation of, 118, 119
Polypeptides, 27, 38, 83
Polyphagia, 189
Polysaccharide, 25, 37
Polyuria, 172, 174, 189
Pons, 224, 225, 272
  centres
    apneustic, 143
    for eye reflexes, 242
    pneumotaxic, 143
  development of, 221
  pathways through
    motor, 268, 269, 272, 273
    proprioceptor, 260
    sensory, 263–265
    taste, 238
  links with cerebellum, 273, 275
  parasympathetic outflow from, 226, 286
Pores
  nuclear, 6
  membrane, 52
Portal canal, 76
Portal vein, 76, 77, 84, 91
  absorption amino-acids, 38
    fat, 40
    glucose, 39
Post-central gyrus, 260, 263, 266
Posterior horns of spinal cord, 227, 232
Posterior lobe (cerebellum), 274
Posterior pituitary gland, 159
Posterior root ganglion, 19, 227, 229, 264, 288
Post-ganglionic fibres, 87, 286–289
Postsynaptic cells, 60
  excitation, 289
  inhibition, 289
Postural reflexes, 258, 260
  role of multineuronal motor pathways, 272
Posture
  muscles maintaining, 280
  role of cerebellum, 276, 277
    medial motor system, 273

multineuronal motor pathways, 272, 273
proprioceptors, 259, 260
vestibular mechanism, 258
Potassium, 24, 50, 55, 56, 57, 114
depletion, 175
in blood, 116, 140–142
intoxication, 174
membrane permeability, 55, 58
role in
action potential, 58
equilibrium potential, 55
resting potential, 56
tubular excretion in urine, 173–175
Potential
difference, 54, 55
electrotonic, 57
membrane, 52, 56, 58, 59
spread of change, 57
threshold, 57, 59
Precentral gyrus, 267–269
Precision of motor control, 267
Preganglionic fibres, 87, 286–289
Pregnancy
anterior pituitary in, 179, 217
energy requirements in, 49
fertilization in uterine tube, 204
mammary glands in, 214
menstrual cycle ending in, 217
ovary in, 199, 201
Rh factor in, 122
uterus in, 205–212
Prekallikrein, 117
Premotor cortex, 222, 267, 272, 276
Pressure receptors in
joints, 259
skin sensation, 19, 235, 261, 265
(see under Blood pressure, Hydrostatic pressure and Osmotic pressure)
Pretectal nucleus of midbrain, 249, 251
Prevertebral ganglia, 87, 287, 288
Primitive foregut, 21
Primitive neural tube, 221
Primordial follicle, 198
Prime movers (agonists) of joints, 281
P-R interval (ECG), 98
Procarboxypeptidases, 74
Proerythroblast, 118, 119
Progesterone, 196–201, 203, 205–208, 211–217
action on mammary glands, 213, 214
adrenal cortex, 173
and anterior pituitary, 179, 217
Prolactin, 179, 213–215
Prolactin inhibiting hormone, 179, 213, 215
Prolactin releasing hormone, 179, 214
Prolymphocyte, 119
Promyelocyte, 119
Propagation of
action potential, 59
nerve impulse, 59
Prophase, 9, 20

Proprioceptors, 235, 256, 259
effects on respiration, 145
links with cerebellum, 275, 276
pathways to brain, 258, 260
role in reciprocal innervation, 282
Prosencephalon, 221
Prostaglandins, 117, 194, 198, 200, 206, 210
Prostate gland, 192, 194, 195
Protein, 7, 24, 27, 29, 30, 33, 38, 40, 42
absorption, 82, 83, 84
and body temperature regulation, 44, 45
and growth, 46, 47
and haemopoiesis, 118
caloric values, 32
carbon cycle, 30
carrier, 52, 53
channel, 56
dietary sources, 34, 50
specific dynamic action, 45, 48
digestion, 37, 70, 74, 78
energy release from, 41, 42
formation of, 7
G protein, 61
in membrane, 52
kinase, 61, 186
masses, 5
membrane, 6
metabolism, 38
and anterior pituitary, 179–183
and glucocorticoids, 173–175
nitrogen cycle, 31
regulator, 61, 284
see also Plasma proteins
Prothrombin, 35, 117
Prothrombin activator, 117
Proton donor, 157
Protoplasm, 2, 5, 6, 27
carbohydrates in, 25, 33, 39
constituents of, 24–28, 33, 34
essential materials for building and repair, 32
fats in, 26, 33, 40
proteins in, 27, 34, 38
Protozoa, 123
Pseudopodia, 2, 3
Psychic phase
bile secretion, 77
gastric secretion, 71
pancreatic secretion, 75
Ptyalin, 39, 66
action in stomach, 70
Puberty, 192, 195, 196, 200
growth in, 47
haemopoiesis after, 118
mammary glands at, 213
thymus regression, 125
uterus at, 202
Pudendal nerves, 86, 88, 109, 162, 286
Puerperium, 211
Pulmonary

arterioles, 113
artery, 93, 94, 95, 132
  blood pressure, 113
  capillaries, 113, 142
  circulation, 91, 113, 132
  exchange in respiratory gases, 142
  valve, 96
  veins, 93, 113, 132
  ventilation, 136
Pulse (radial), 106
  pressure, 104
  wave, 106
Pumps
  Ca^{2+}, 53
  H^+, 53
  Na^+-K^+, 53, 56
Pupil of eye, 239, 243, 245, 251, 286, 287
Purines, 28, 38
Purkinje
  cell, 18, 274
  tissue, 97
Putamen, 223, 272
P-wave (ECG), 98
Pyloric
  antrum, 70, 71, 73
  glands, 70
  sphincter, 65, 70, 72
    innervation of, 286, 287
    movements, 72
      control of, 88
  part of stomach, 65, 70
Pylorus, 70, 72
Pyramidal tract
  see Cortico-spinal tract
Pyramids, 259
  decussation of, 259
Pyridoxine (vitamin B_6), 36
Pyrimidine, 38
Pyruvic acid, 39, 41, 42

QRS complex (ECG), 98
Quadriceps femoris, 279

Radiation
  heat loss from skin, 43–45
  optic, 249
Radioactive iodine, 169
Rage, 225
Random thermal motion, 52
Ranvier, node of, 17, 18
  in conduction of nerve impulse, 59
Rathke's pouch, 178
Reabsorption, bile salts, 76
Reabsorption, kidney, 149, 151, 153
  regulation, 159, 160
  see also under Renal tubule
Receptive relaxation of stomach, 72
Receptors, 6, 19, 229–235
  adrenergic, 289
  atrial, 109, 160

cholinergic, 289
cutaneous, 260
hearing, 252, 253
heart, 100, 101
hydrogen ion (respiration), 144
in autonomic reflexes, 288
in control of motor activity, 273
in reflex action, 229–234
membrane, 60, 61
nuclear, 60
muscle spindle, 259
of saccule, 256
of semicircular canals, 256
of utricle, 256
olfactory, 236
pharyngeal for water intake, 159
stretch receptors in the lungs, 143, 145
  see also under Stretch receptors
taste, 237, 238
urinary bladder, 162
V_2 and vasopression, 186
vision, 239, 246–248
Reciprocal innervation, 282
Rectum, 64, 65, 85, 86
  defaecation, 86
  innervation, 86, 88
Red blood corpuscles (cells), 13, 116
  blood group factors, 120–122
  carriage and transfer of O_2 and CO_2, 139–142
  destruction in spleen, 124
  enzymes in, 141, 142
  fetal, 122
  formation of, 119
  nutritional requirements for, 24, 34–36, 118
  haemoglobin, 141, 142
  life span, 119
  volume and numbers, 116
Red nucleus, 224, 232, 272, 273, 276
Red pulp, spleen, 124
Reflex
  accommodation for near vision, 243, 244
  action, 229–233
  arc, 229–233
  autonomic, 288
  cardiac, 100, 101
  centre, 229
  conditioned, 67, 71, 72, 75, 77, 234
  control of gastric motility, 72
  control of muscle movement, 277
  defaecation, 86
  effects on anterior pituitary, 179
  effects on secretion of antidiuretic hormone, 186
  emptying of small intestine, 81
  expulsion of bile from gall bladder, 77
  eye movements, 242
  factors in regulation of respiration, 145
  gastro-colic, 86
  gastro-ileo-colic, 81
  Hering-Breuer, 145
  inhibition and facilitation of, 282

light, 251
micturition, 162
muscles of middle ear, action of, 252
pathways, 234
postural, 258, 272, 273
reciprocal innervation in, 282
response, 233
salivation, 67
secretion of gastric juice, 71
  intestinal juice, 80
  pancreatic juice, 75
spinal, 231
stretch, 230
suckling, 211, 212, 214
swallowing, 69
unconditioned, 67
vasomotor, 108, 109
vomiting, 73
Regulator proteins
  G proteins, 61
  tropomyosin, 284
  troponin, 284
Relaxation of muscle, 284
Relaxin, 199, 208, 210
Release of energy, 30, 39–43, 266, 277
  role of vitamins in, 35, 36
  thyroid control of, 167–169
Releasing hormones, 167, 173, 179, 213–215, 217
Renal artery, 148, 149
Renal blood flow, 156
Renal circulation, 91
Renal corpuscle, 149–156
Renal function test, 155, 156
Renal plasma flow, 156
Renal tubules, 149
  conservation of base, 151, 157, 158
  hormones, action of ADH, 184, 186–188
    corticoids, 173–175, 182, 188
    insulin, 189
    parathormone, 170–172
  reabsorption in regulation of water balance, 152, 160, 188
  reabsorption of urea, 155
  secretion of ammonia, 158
  secretion of creatinine, 154
  secretion of diodine, 154
  secretion of hydrogen ions, 157, 158
  secretion of PAH, 156
Renal vein, 149
Renin, 160, 166, 175, 188
Renin-angiotensin system, 160, 173, 182, 188
Rennin, 70, 74
Repair (endometrium), 197, 217
Reproduction, 3, 22, 192–217
  organs
    development of, 47
    female, 196
    male, 192
  role of hormones in control of, 217
Residual volume (lungs), 136
Resistance, peripheral, 104, 108

Respiration, 3, 10, 22, 130–145
  action of adrenaline on, 177
  artificial, 135
  carbon cycle, 30
  control
    nervous, 143
    chemical, 144
  inhibition
    in swallowing, 69
    in vomiting, 73
  lungs, capillary bed, 113
  muscles of, 130, 143, 278
  rate, 136, 145
    in thyroid deficiency, 168
    in thyroid overactivity, 169
  regulation of, 143, 144, 145
    reflex factors, 145
    voluntary factors, 145
  respiratory
    bronchiole, 132
    centres, 143, 144
      in vasomotor reflexes, 108
    depth, 145
    epithelium, 12
    gases, 116
      composition of, 137, 138
      interchange, 132
      movement of, 138
    minute volume, 136
    movements, 130, 132, 133
      control of, 143–145
      role in movements of lymph, 123
      passages, 12, 14, 15, 131, 236
    pump (venous return), 111
  rhythm, 145
  surfaces, 132, 137
  system, 22
    heat loss from, 43
    water loss from, 115
  units, 131, 132, 136
  uptake of oxygen in blood in lungs, 102
Resting membrane potential, 55, 56, 57
Resting metabolic rate, 48
Retching, 73
Rete testis, 193
Reticular activating system, 223, 225, 255, 258
Reticular formation, 225, 262, 275, 276, 277
Reticular nuclei (of brain stem), 271, 272
Reticular tissue, 13
  in haemopoiesis, 119
  lymph node, 123
  spleen, 124
Reticulocyte, 118, 119
Reticulo-endothelial system, 13
  in bone marrow, 119
  lymph nodes, 123
  spleen, 124
Reticulo-spinal tract, 271, 272
Retina, 239, 243–246
  blood vessels, 239, 245

cell layers of, 246
fundus oculi, 245
light reflex, 251
mechanism of vision, 247, 248
nerve pathways from, 249
visual purple (rhodopsin), 35, 247
Retinene, 247, 248
Retropulsion (stomach), 72
Rhesus (Rh) factor, 122
pregnancy and, 122
Rhodopsin, 247
Rhombencephalon, 221
Riboflavine (vitamin B_2), 36
Ribonucleic acid (RNA), 6, 7, 28
messenger (mRNA), 6, 7, 167
ribosomal (rRNA), 7, 167
Ribose, 28
Ribosomal subunit, 7
Ribosomes, 5, 7, 8, 28
Ribs, 130, 133, 134
Rickets, 35
Rigidity, 272
Rods of Corti, 253
Rods of retina, 245–247
Rotation, and semicircular canals, 257
Round window, ear, 254
Rubro-spinal tract, 224, 271, 272, 276
Ruffini endings, in skin, 261
Rugae (stomach), 70
Rupture of amniotic membranes, 210

Saccule, 256, 258, 275
Sacral
nerves, 227
parasympathetic outflow, 88, 162, 286
Saliva, 66, 70, 72
control of secretion, 67
functions of, 66
glands forming, 11, 64, 65–67
control of blood supply, 107, 109
nerves to, 67, 226, 286
role in taste, 237
role in water balance, 115, 159
substances secreted in, 66
Salivary centre, 67
Saltatory conduction, 59
Sarcolemma, 16
Sarcomere, 283, 284
Sarcoplasm, 16
Sarcoplasmic reticulum, 61, 284
Scala, media, tympani, vestibuli, 253, 254
Schwann cell, 17, 18
Sclerotic coat of eye, 239, 241
Scotopic vision, 247
Scotopsin, 247, 248
Scrotum, 192, 194, 195
Scurvy, 36
Secondary sex characteristics
at menopause, 216
development at puberty, 47, 192–196, 200

role of adrenal cortex, 173–175
role of pituitary, 180, 183
Second heart sound ('dup'), 96
Second messengers, 61
Secretin, 71, 72, 75, 77, 80, 166
Secretion (kidney), 149
Secretomotor nerves, 67
Segmentation in small intestine, 81
in large intestine, 86
Sella turcica, 178
Semicircular canals, 252, 256, 258
links with cerebellum, 275
mechanism of action, 257
nerve pathways, 226
pathways to brain, 258
Semilunar valves, 93, 95, 96
Seminal fluid (semen), 194
Seminal vesicles, 192, 194
Seminiferous tubules, 193, 195
Senility, 183
Sense organs, 235, 273
Sensory cortex, 222, 232, 260, 263, 265, 266, 273, 289
Sensory decussation, 224, 260, 263, 265
Sensory input, controlling motor activity, 273
Sensory neurons, 17, 19, 220, 229, 231
visceral, 288
see also under Afferent nerves and Receptors)
Serosa, in digestive tract, 79
Serotonin (5-HT), 87, 117
Serous acini, 11, 66, 74, 131
Serous coat
duodenum, 78
intestine, 79
Sertoli, cells of, 193, 195
Serum albumin, 37, 116
Serum globulin, 116
Sex glands (see Gonads)
Sex hormones
at puberty, 47
of adrenal cortex, 173–175
(see Oestrogen, Progesterone, Testosterone)
Sexual behaviour, 216, 225
Shivering, 45
Sighing and respiration, 145
Sigmoid flexure
colon, 85
Simmond's disease, 183
Singing and respiration, 145
Sinuses in skull, 131, 236
Sino-atrial node, 97, 98, 101
Skeletal growth and thyroid deficiency, 168
Skeletal muscle, 16, 279–284
actions of adrenaline, 177
action of insulin, 189
and adrenal cortex, 173–175
contraction, 283, 284
control of blood supply, 107, 108, 287
control of movements, 277
control of postural tone, 276
heat production and tone in, 43–45

in Addison's disease, 174
in giantism and acromegaly, 181
initiation of movement, 267
innervation, 270, 271
metabolism, 39, 43
motor pathways to, 269
muscular movements, 281–284
nerve endings in, 17, 18
neuromuscular transmission, 270, 289
oesophagus, 68
proprioceptors in, 259
  pathways from, 260
reciprocal innervation, 282
reflex contractions, 229–231
respiratory, 133, 134
role in movement of lymph, 123
role in venous return to heart, 111
role of cerebellum in control of postural tone, 276
role of multineuronal motor pathways, 272
sphincters
  anal, 85, 86, 88
  bladder, 161, 162
sympathetic innervation of blood vessels to, 287
Skeleton, 15, 278
Skin, 44, 45, 261
adipose tissue in, 14
afferents in autonomic reflexes, 288
control of blood supply, 107–109
effects of adrenaline on, 177
fluid loss from, 115
heat loss from, 43–45
in Addison's disease, 174
in overactivity of the anterior pituitary, 181
in panhypopituitarism, 183
in thyroid deficiency, 168
nerve pathways from, 226, 263–265
sensation, 261
sense organs in, 44, 45, 235, 261
stratified squamous epithelium, 12
sweat glands, 43–45, 115
vitamins A and D and, 35
Sleep
and reticular activating system, 225
heat production during, 45
urine volume in, 156
Sliding filaments (muscle), 284
Small intestine, 11, 64, 37–40, 78–84
absorption, 82, 84
innervation, 87, 88, 286, 287
movements of, 81
secretion, 78, 80
Smell, 225, 235, 236
centres for, 225
olfactory nerve (cranial), 226
Smooth muscle (visceral, involuntary), 16
and adrenaline, 177
  oxytocin (on uterus), 185
  vasopressin, 184
chemical transmission, 289
innervation, 18, 286–289

in blood vessels, 103, 104, 107
  bronchioles, 131
  digestive tract, 68–87
  oesophagus, 68
  urinary passages, 161, 162, 194
  uterine tubes, 197
  uterus, 197, 202, 206, 208
  vas deferens, 194
Sodium, 24, 50, 52–58
and adrenal cortex, 152, 159, 173–175, 188
and antidiuretic hormone, 186, 188
and body fluid volume, 160
and kidney, 151–153, 157, 158
in body fluids, 152, 153, 159
in blood, 116
in urine, 163
membrane permeability, 289
reabsorption, 153, 157, 158
role in passage of nerve impulse, 55–59, 289
Soma of nerve, 226
Somatostatin, 87, 179, 189
Somatotroph cells, 179, 180, 181
Somatotrophin see Growth hormone
Sound waves, 235, 252, 254, 255
Special proprioceptors, 256–258
Specific dynamic action of food, 48
Specific gravity
blood, 116
urine, 163
Speech, 131
interruptions in respiration, 145
motor area of cerebral cortex (centre for), 222, 267
role of saliva in, 66
tongue in, 237
Spermatids, 193
Spermatocytes, 193
Spermatogenesis, 192, 193, 195
Spermatogonia, 193
Spermatozoa, 192–194, 204
development of, 193
maturation of, 20
motility of, 194
Sphenopalatine ganglion, 286
Sphincters
digestive tract, 65, 72, 78, 81, 85, 86, 88
Oddi, 65, 74, 77
pupillae, 243, 251
urinary bladder, 161, 162
Sphygmomanometer, 105
Spinal cord, 220, 225, 227
anterior horn cell, 17, 18, 227, 232, 269–271
as reflex centre, 229–233
cerebrospinal fluid, 127
coverings, 127
development of, 221
in autonomic reflexes, 288
in control of muscle movement, 273, 277
in reciprocal innervation, 282
links with cerebellum, 275
links with multineuronal motor pathways, 272

lower motor neuron, 270
nerve supply to respiratory muscles, 143
pathways
    final common, 270, 271
    motor, 268, 269
    proprioceptor, 260
    sensory, 263–265
sacral parasympathetic outflow, 286
sympathetic outflow, 287
vestibular links, 258
Spinal nerves, 227, 269
    posterior root ganglion, 19, 229
    sympathetic outflow, 287
Spinal reflexes, 231
    final common pathway, 271
    stretch, 230, 282
Spiral ganglion (hearing), 253, 255
Spirometer, 102
Spleen, 124
    capsule, 124
    circulation, 84, 79, 124
    haemopoiesis, 118, 119, 125
    reticular framework, 13
    sinusoids, 124
    trabeculae, 124
Splenic flexure (colon), 85
Standing gradient mechanism, 153
Stapedius muscle (ear), 252
Stapes (hearing ossicle), 252, 254
Starch, 25
    dietary, 33
    digestion of, 37, 66, 74, 78
    metabolism (see also under Carbohydrate), 39, 42
Stercobilin, 76
Stercobilinogen, 76
Stereoscopic vision, 250
Sterility, 35, 195
Sternum, 130, 133, 134
    in haemopoiesis, 118
Steroids
    bile acids, 26, 76
    cholesterol, 26, 173
    in blood, 116
    of adrenal cortex, 173
Stimulus, sensory, 233
Stomach, 11, 64, 65, 68–73, 78
    absorption from, 84
    emptying, 72
    filling, 72
    functions of, 70
    hormones of, 71, 166
    innervation, 88, 286, 287
    intrinsic factor formation, 118
    movements, 72
    muscle coats, 70
    secretion, 37–40, 70, 71
Straight arteries (kidney), 148
Stratified squamous epithelium, 12
    in oesophagus, 68
    skin, 44, 45, 261

vagina, 197, 201
Streptokinase, 117
Stress
    and ADH, 186
    and adrenaline, 176, 177
    and ACTH, 173, 182
    and lactation, 214
    effect on anterior pituitary, 179
    effect on cardiovascular system, 99, 108, 109
    effect on respiration, 145
Stretch receptors
    in arch of aorta, 101, 108, 109, 145
        bronchioles, 145
        carotid sinus, 101, 108, 109, 145
        great veins, 100
        intestine, 81
        skeletal muscle and tendons, 230, 259
Stretch reflexes, 230, 282
Stria vascularis, 253
Striated border epithelium, 11
    in kidney tubules, 149, 150
        large intestine, 85
        small intestine, 78, 82
Striated muscle (see Skeletal muscle), 16, 283, 284
Stroke volume (heart), 102
Subarachnoid space, 127
Sublingual salivary glands, 66
Submandibular ganglion, 286
Submandibular salivary glands, 66
Submucosa (Gastro-intestinal tract), 78, 79, 85, 87
Substance P, 109
Substantia gelatinosa, 262
Substantia nigra, 223, 224, 272
Subthalamic nucleus, 223
Suckling, 185, 211, 212, 214
Sucrase, intestinal, 37, 39, 78
Sucrose, 25, 83
Sugars, 25
    absorption, 82–84
    dietary, 33
    metabolism, 39, 41, 42
        (see also under Carbohydrate)
Sulphate, 29
Sulphur, 24, 32
Sulci of brain, 222
Superior cervical ganglion, 243
Superior mesenteric ganglion, 88, 287
Superior oblique muscle of eye, 226, 241
Suppressor areas, in cerebral cortex, 272
Supraoptic nucleus, 159, 184, 186
Supraopticohypophyseal tract, 184–187
Suspensory ligament of lens, 239, 243
Swallowing, 68, 69
    respiration in, 69, 145
    role of saliva, 66
    stages, 69
    tongue in, 237
Sweat
    fluid loss, 115
    glands, 43–45

chemical transmission, 289
innervation, 287, 288
in body temperature control, 43–45
Sympathetic chain, 18
Sympathetic ganglion cells, 18, 107
Sympathetic nervous system, 177, 286, 287
chemical transmission in, 289
effect of stimulation, 287
in autonomic reflex, 288
organs supplied, 287
outflow, 227, 286, 287
adrenal medulla, 176
blood vessels, 107–109
digestive tract, 80, 81, 87, 88
heart, 99–101
iris, 243
rectum, 86
urinary bladder, 162
veins, 103, 111
vasoconstrictor tone in body temperature regulation, 44, 45
Sympathetic trunk, 287, 288
Sympathomimetic amines, 289
Synapse, 228, 229
chemical transmission at, 60, 228, 289
Synaptic
buttons, 228
cleft, 60
transmitters, 289
Synergists, 281
Synovial membrane, 281
Synthesis, kidney tubule, 149
Systemic circulation, 91
blood flow, 112
Systems, 22
Systole, 95
atrial and ventricular, 95
blood pressure in, 104, 105
elastic erteries in, 103, 106
heart sounds in, 96
Systolic blood pressure, 104, 105

Taenia-coli, 85
Target cells, 60
Tarsal glands, 240
Taurine, 76
Taurocholic acid, 76
Taste, 71, 235, 237
buds, 66, 237
pathways and centres, 226, 237, 238
receptors, 237, 238
T-cells (transmission), 262
T cytotoxic cells, 125
Tears, 240
Tectoral membrane of cochlea, 253, 255
Tecto-spinal tract, 271, 272
Tectum (superior colliculus), 272
Teeth, 24, 34, 35, 65, 66, 131
Telencephalon, 221
Telophase, 9, 20
Temperature

body, control of, 43–45
cutaneous sensation, 262, 264
effect on uptake and dissociation of O_2 from haemoglobin, 139, 141
extremes on blood vessels, 108
Temporal lobes, 222
centres for hearing, 255
Tendon, 14, 281
proprioceptors in, 235, 259
Tensor tympani (muscle in ear), 252
Testis, 166, 192–195
action of anterior pituitary on, 179
influence of adrenals, 173
temperature of, 194
Testosterone, 192, 193, 195
and anterior pituitary, 179
Tetany, 171
Tetraiodothyronine (T_4), 167
Thalamus, 223–225
centres
autonomic, 288
pain and temperature, 262, 264
chemical transmission in, 289
development of, 221
links with other centres, 232, 272, 273, 276
pathways
proprioceptor, 260
sensory, 263, 265
taste, 238
T-helper (T_4) cells, 125
Thermodilution (cardiac output), 102
Thiamine (vitamin B_1), 36
Third heart sound, 96
Third ventricle (brain), 127, 223–225
Thirst, 159, 160, 172, 187, 189
Thoracic
cord, 143, 227, 287
duct, 82, 84, 123
nerves, 227
Thorax, 130, 133, 134
Three-dimensional (stereoscopic) vision, 250
Threshold
potential, 57, 59
stimulus, 57
Thrombin, 117
Thrombocytopenia, 119
Thromboplastin, 117
Thymine, 28
Thymosine, 125
Thymus, 125, 126
Thyroglobulin, 167, 168
Thyroid, 34, 166–169
and anterior pituitary, 179, 182
development of, 21
hormones, 167, 169
in haemopoiesis, 118
overactivity, 169
regulation, 167
underactivity, 168
Thyroid stimulating hormone (TSH, thyrotrophin),

INDEX

167–169, 179, 182
Thyroid stimulating immunoglobulins, 169
Thyrotroph cells, 178
Thyrotrophin (*see* Thyroid stimulating hormone)
Thyrotrophin releasing hormone (TRH), 167, 169, 179
Thyroxine, 167, 168, 179
Tidal volume, 136
Tight junctions in kidney, 153
Timbre (of sound), 254
Tissue fluids, 114, 115, 123
   exchange of water and electrolytes, 110, 152
   heat dispersal, 43
   (*see also under* Body fluids)
Tissue oxidation, 39–42
   role of thyroid, 167–169
Tissue phospholipids, 117
Tissues, 11–19
   connective, 11, 13
   constituents of, 24, 28
   development of, 21
   differential growth of, 46, 47
   epithelia, 11
   exchange of O_2 and CO_2, 138–142
   heat production, 43–45
   metabolism of, 38–42
   muscular, 11, 16
     cardiac, 16
     skeletal, 16
     smooth, 18
   nervous, 11, 17–19, 220
   supporting, 11
   water of metabolism of, 115
T-lymphocytes, 123–126
   T-cytotoxic cells, 125, 126
   T-helper (T_4) cells, 125, 126
   T-memory cells, 126
   T-suppressor cells, 125, 126
Tone
   blood vessels
     arterioles, 104, 108, 109
     veins, 111
   large intestine, 86
   rings in stomach, 72
   skeletal muscle, 44, 45, 259, 272
   stomach, 72
   waves in colon, 86
   waves in stomach, 72
Tonicity of body fluids, 159
Tongue, 65, 66, 69, 131, 226, 237, 238
Total lung capacity, 136
Touch
   afferents and pain, 262
   Meissner's corpuscle, 19, 260
   pathways, 263, 265
Trabeculae of spleen, 124
Trace elements, 24
Trachea, 130–132, 167, 170
Tracts, nerve
   cortico-nuclear, 268
   cortico-pontocerebellar, 275

cortico-rubrospinal, 273
cortico-spinal, 232, 269, 271, 273, 276
cuneo-cerebellar, 275
medial longitudinal bundles, 242
olivo-spinal, 271
pyramidal (*see* Cortico-spinal)
reticulo-spinal, 271, 272
rubro-spinal, 232, 271, 272, 276
spino-cerebellar, 232, 275
spino-thalamic, 262, 264, 265
tecto-cerebellar, 275
tecto-spinal, 271, 272
vestibulo-cerebellar, 275
vestibulo-spinal, 232, 258, 271, 272, 276
Transcellular pathway (kidney), 153
Transcription, 6
Transfusion
   blood, 120
   incompatible, 120
Transitional epithelium, 12, 161
Translation, 6
Transmission of nerve impulse, 54, 289
Transplanted tissue, 126
Transport, 22
   absorbed foods, 84
   through membranes, 52
Transthoracic impedance and cardiac output, 102
Trapezoid body, 255
Tremor, 272
Triacylglycerol, 26
Triceps, 280, 281
Trichromatic vision, 248
Tricuspid valve, 93, 95, 96
Trigeminal nerve (cranial V), 224, 226, 260, 263, 268
Triglyceride, 26, 37
Triiodothyronine (T_3), 167
Tripeptides, 83
Trochlear nerve (cranial IV), 226, 241, 242, 268
Trophic hormones, 179, 180
   lack of, 183
Trophoblast, 205, 207
Tropomyosin, 284
Troponin, 61, 284
Trousseau's sign, 171
Trypsin, 37, 38, 74
Trypsin inhibitor, 74
Trypsinogen, pancreatic, 74
T-suppressor cells, 125
T-tubule, 284
Tunica adventitia, intima media, 103
Turbinate bones, 236
Tympanic membrane, 252, 254
Tyrosine, 27, 167

Ultra-violet rays, 35, 235
Umbilical blood vessels and cord, 207–209
Uncoordinated movements, 272
Uncus (centre for smell), 236
Unicellular organisms, 2–4
Unipolar nerve cells, 19

(*see also* Posterior root ganglion)
Unstirred water layer, 76, 83
Unstriped muscle, 16
Upper motor neurons, 232, 268, 269, 277
  chemical transmission in, 289
Up regulation, 60
Uracil, 28
Uraemia, 174
Urea, 38, 41
  blood, 116
  clearance, 154, 155
  excretion, 150, 151
  role in water reabsorption, 152
  urinary, 163
Ureter, 148, 149, 161, 194
Urethra, 148, 161, 162, 194
Uric acid
  blood, 116
  excretion, 150, 151
  urinary, 163
Urinary bladder, 148, 161, 194
  actions of adrenaline on, 177
  innervation, 162, 286, 287
  lining, 12, 161
  storage and expulsion of urine, 162
Urinary tract, 12, 148–163
  heat loss from, 43
  water loss from, 115
Urine, 163
  clearance of inulin, 154
  composition, 163
  control of Ca^{2+} and P loss, 170–172
  control of output, 115, 152, 159, 160, 186–188
  control of salt loss, 173–175, 182
  creatinine clearance, 154
  diodone clearance, 156
  excretion of urea, 155
  expulsion, 148, 161, 162
  formation, 148–152
  heat loss in, 43
  hypertonic, 152
  in diabetes mellitus, 189
  in maintenance of acid-base balance, 157, 158
  inulin clearance, 154
  in water balance, 115, 152, 159, 160
  ketone bodies in, 40
  PAH clearance, 156
  pH, 163
  specific gravity, 163
  storage, 148, 161, 162
  urea and ammonia in, 38, 157, 163
  volume, 163
Urobilin, 76
Urobilinogen, 76
Urochrome, 163
Urogenital system (*see* Reproduction and Excretion)
  development of, 21
Urokinase, 117
Uterine tube (*see also* Fallopian tube), 194, 196, 197,
    201–204

hormonal control of changes in menstrual cycle, 200
Uterus, 196, 197, 200–212
  action of oxytocin, 184, 185
  at and after menopause, 212, 216
  at and after parturition, 210, 211
  childhood, 202
  hormonal control of changes in menstrual cycle, 201, 217
  implantation, 21, 206
  involution of, 185, 212
  maturity, 203
  menstrual cycle, 203
  motility, 194
  placentation, 206
  puberty, 200, 202
  puerperium, 211
Utricle, 256
  links with cerebellum, 275

Vacuole
  contractile, 4
  food, 2, 3, 4
Vagal, tone, 99
Vagina, 194, 196, 197, 200, 201, 210, 216
  effects of semen, 194
Vagus nerve (cranial X), 69, 224, 226, 238, 263, 268, 286
  contraction of gall bladder, 77
  gastric motility, 72
  gastric secretion, 71
  in regulation of heart action, 99–101
  in regulation of respiration, 143–145
  in swallowing, 69
  in vomiting, 73
  outflow to
    blood vessels, 107–109
    gastro-intestinal tract, 88
    heart, 100
  parasympathetic fibres, 286
  relaxation of sphincter of Oddi, 77
  secretion of pancreatic juice, 75
Valves (*see* Heart)
  lymph, 123
  vein, 103, 111
Vasa recta (kidney), 152
Vas deferens, 192–194
Vasoactive intestinal polypeptide (VIP), 67, 77, 87, 107
Vasoconstriction, 107, 108, 286, 287
  role of adrenaline, 108, 177
  role of noradrenaline, 176
  role of posterior pituitary, 184
Vasodilatation, 107, 109, 286, 287
Vasomotor centre, 100, 101, 107–109
Vasopressin (antidiuretic hormone), 159, 160, 184, 186, 188
Veins, 90, 91, 103
  area of cross-section, 112
  baroreceptors in vasomotor reflexes, 108
  great, 93, 103, 112
  innervation, 107
  mesenteric, 82, 84
  muscular, 103
  portal, 84

pressure in, 104, 111
pulmonary, 113, 132
renal, 149
splenic, 124
stretch receptors in, 100
subclavian, 123
subdural sinuses, 127
umbilical, 207, 209
velocity of blood flow, 112
vena cava
  inferior, 94
  superior, 94
venous return to heart, 100, 111
Venoconstriction, 100, 107
Venous return, 100, 111
Ventral spino-thalamic tract, 265
Ventral vestibulo-spinal tract, 271
Ventricles of brain, 127, 223–225
development of, 221
Ventricles of heart, 92–96, 106
conduction of heart beat to, 97, 98
in cardiac reflexes, 101
output, 102, 113
Ventricular
diastole, 96
systole, 96
Venules, 103
blood pressure in, 104
Vermiform appendix, 85
Vermis (zone of cerebellum), 274, 276
Vertebrae, 130, 133, 134, 278
Vertebral canal, 227
Vesicles
secretory, 8
Vestibular
branch of VIII cranial nerve, 226, 256–258
membrane, 253, 254
nuclei, 258, 272, 276
pathways to brain, 258
Vestibular nerve ganglion, 258
Vestibulocochlear nerve (cranial VIII), 224, 226, 252, 255
Vestibulo-spinal tract, 272
links with cerebellum, 276
Vibration sense, 235
Villi
chorionic, 206, 207
intestinal, 78, 82
Villus pump mechanism, 82
Virilism, 175, 182
Viruses, 123
Vis a fronte, 111
Vis a tergo, 111
Viscera
abdominal, 64, 65
innervation of, 88, 286, 287
control of blood supply, 107–109
in giantism and acromegaly, 181
interoceptors, 235
pelvic (*see* External genitalia, Urinary bladder, Rectum
  and Large intestine)

innervation of, 286, 287
sensations, 235
smooth muscle of, 16
visceral sensory afferents, 288
Visceral pericardium, 92
Visceral pleura, 133
Viscosity of blood, 104, 116
Vision, 239–251
centres for, 67, 222, 239, 249
colour, 247, 248
fields of, 249
mechanism of, 247, 248
monochromatic, 247
nature of stimulus, 235
optic nerve, 226, 239
pathways for, 226, 249
photochemical mechanism, 247, 248
pigments in, 35, 247, 248
receptors for, 246
spectrum, 235, 247
stereoscopic, 250
Visual purple, 35, 247
Vital capacity, 136
Vitamins, 32, 35, 36, 50
A, 35
absorption and transport, 82, 84
B, 12, 36, 50, 70, 118
C, 36, 50, 118, 151
D, 35, 50, 170–172
dietary sources, 50
E, 35
in haemopoiesis, 118
K, 35, 117
metabolites, 171, 172
reabsorption in kidney, 151
rhodopsin regeneration, 247
sources, 35, 36
Vitreous humour, 239
Vocal cords, 69, 131
Voice production, 131
Volume of body fluids and sodium, 159
Voluntary movement
control, 277
initiation, 267, 272, 277
Voluntary muscle, 16, 279–284
(*see also* Skeletal muscle)
Vomiting, 73, 188
Von Ebner's glands, 237
Vulva, 185

Waking state and reticular
  activating system, 225
Warmth
skin sensation, 261, 264
body temperature, 43–45
Water
absorption and transport, 84–86
balance, 115
  internal secretions in, 115
  regulation of, 160

role of parathyroid, 172
role of posterior pituitary, 152, 186–188
role of adrenal cortex, 173–175, 188
daily requirements, 50
distribution in body, 114
excretion of, 148, 151, 153, 156, 159, 160
filtration in kidney, 150
proximal tubule reabsorption, 153
reabsorption in kidney tubules, 151–153
vapour, 137
Water-soluble vitamins, 36, 84
White blood cells (corpuscles), 13, 116, 118, 119, 123, 125, 126
White fibro-cartilage, 15
White matter
brain, 223
spinal cord, 227
White pulp, spleen (*see* Periarterial lymphoid sheaths)
Womb (*see* Uterus)
Work, 30, 32
energy requirements for, 48, 49
release of energy for, 41, 42

Xerophthalmia, 35

Yolk sac, 21

Zona glomerulosa, 159, 173, 188
Zymogen granules, 74
Zygote, 21, 204